Degrading Enzymes for Marine Polysaccharides and Their Applications

Xiangzhao Mao
Editor

Degrading Enzymes for Marine Polysaccharides and Their Applications

Springer

Editor
Xiangzhao Mao
State Key Laboratory of Marine Food Processing
and Safety Control, College of Food Science and Engineering
Ocean University of China
Qingdao, China

ISBN 978-981-96-9516-4 ISBN 978-981-96-9517-1 (eBook)
https://doi.org/10.1007/978-981-96-9517-1

This Springer imprint is published by the registered company Springer Nature Singapore Pte Ltd.
The registered company address is: 152 Beach Road, #21-01/04 Gateway East, Singapore 189721, Singapore

Preface

Polysaccharides are a class of carbohydrates that consist of repeated units of monosaccharides that are linked by glycosidic bonds. Degrading enzymes for marine polysaccharides have recently been studied extensively. These enzymes catalyze the cleavage of glycosidic bonds in polysaccharides and produce oligosaccharides. Numerous findings indicate that marine polysaccharides and their biotransformed products possess beneficial properties including antiviral, antibacterial, and anti-inflammatory activities, and they have wide applications in healthcare, food industry, and agriculture. Exploitation of degrading enzymes for marine polysaccharides is important for use of marine biomass resources. Here, we describe the classification, biochemical properties, and catalytic mechanisms of the main types of degrading enzymes for marine-polysaccharides and corresponding oligosaccharides. In Chap. 1, we introduced structures, sources, physicochemical properties, and applications of marine polysaccharides. In Chap. 2, marine oligosaccharides including their source, structure, preparation methods, physiological activity, and applications are described. In Chap. 3, tools for mining, expression, and modification of marine polysaccharide degrading enzymes are described. From Chap. 4 to Chap. 12 marine polysaccharide degrading enzymes including chitinase, chitosanase, alginate lyase, agarase, chondroitinase, carrageenase, fucanase, porphyranase, and ulvan lyase are introduced.

Xiangzhao Mao
State Key Laboratory of Marine Food Processing
and Safety Control, College of Food Science and Engineering
Ocean University of China
Qingdao, China

Contents

Contributors

Yaoguang Chang State Key Laboratory of Marine Food Processing and Safety Control, College of Food Science and Engineering, Ocean University of China, Qingdao, China

Laboratory for Marine Drugs and Bioproducts, Pilot National Laboratory for Marine Science and Technology (Qingdao), Qingdao, China

Yongzhou Chi State Key Laboratory of Marine Food Processing and Safety Control, College of Food Science and Engineering, Ocean University of China, Qingdao, China

Chunying Du State Key Laboratory of Marine Food Processing and Safety Control, College of Food Science and Engineering, Ocean University of China, Qingdao, China

Yan-jun Gong Weifang Key Laboratory of Marine Functional Oligosaccharides Green Biomanufacturing, Weifang Macroalga Biotechnology Co., Ltd, Weifang, China

Si-Miao Di State Key Laboratory of Marine Food Processing and Safety Control, College of Food Science and Engineering, Ocean University of China, Qingdao, China

Wen-Can Huang State Key Laboratory of Marine Food Processing and Safety Control, College of Food Science and Engineering, Ocean University of China, Qingdao, China

Chengcheng Jiang State Key Laboratory of Marine Food Processing and Safety Control, College of Food Science and Engineering, Ocean University of China, Qingdao, China

Hong Jiang State Key Laboratory of Marine Food Processing and Safety Control, College of Food Science and Engineering, Ocean University of China, Qingdao, China

Ling Jiang College of Food Science and Light Industry, State Key Laboratory of Materials-Oriented Chemical Engineering, Nanjing Tech University, Nanjing, China

Fu-Li Li State Key Laboratory of Photoelectric Conversion and Utilization of Solar Energy, Shandong Provincial Key Laboratory of Synthetic Biology, Shandong C1 Refinery Engineering Research Center, Qingdao New Energy Shandong Laboratory, Qingdao Institute of Bioenergy and Bioprocess Technology, Chinese Academy of Sciences, Qingdao, China

Huining Li State Key Laboratory of Marine Food Processing and Safety Control, College of Food Science and Engineering, Ocean University of China, Qingdao, China

Chunhui Liu State Key Laboratory of Marine Food Processing and Safety Control, College of Food Science and Engineering, Ocean University of China, Qingdao, China

Ming Lu State Key Laboratory of Photoelectric Conversion and Utilization of Solar Energy, Shandong Provincial Key Laboratory of Synthetic Biology, Shandong C1 Refinery Engineering Research Center, Qingdao New Energy Shandong Laboratory, Qingdao Institute of Bioenergy and Bioprocess Technology, Chinese Academy of Sciences, Qingdao, China

Xiangzhao Mao State Key Laboratory of Marine Food Processing and Safety Control, College of Food Science and Engineering, Ocean University of China, Qingdao, China

Xiao-Qing Ma State Key Laboratory of Photoelectric Conversion and Utilization of Solar Energy, Shandong Provincial Key Laboratory of Synthetic Biology, Shandong C1 Refinery Engineering Research Center, Qingdao New Energy Shandong Laboratory, Qingdao Institute of Bioenergy and Bioprocess Technology, Chinese Academy of Sciences, Qingdao, China

Haijin Mou State Key Laboratory of Marine Food Processing and Safety Control, College of Food Science and Engineering, Ocean University of China, Qingdao, China

Hui Ni College of Ocean Food and Biology Engineering, Jimei University, Xiamen, China

Limin Ning College of Medicine, Nanjing University of Chinese Medicine, Nanjing, China

Nghi-Van Phung State Key Laboratory of Photoelectric Conversion and Utilization of Solar Energy, Shandong Provincial Key Laboratory of Synthetic Biology, Shandong C1 Refinery Engineering Research Center, Qingdao New Energy Shandong Laboratory, Qingdao Institute of Bioenergy and Bioprocess Technology, Chinese Academy of Sciences, Qingdao, China

Hang Su State Key Laboratory of Photoelectric Conversion and Utilization of Solar Energy, Shandong Provincial Key Laboratory of Synthetic Biology, Shandong C1 Refinery Engineering Research Center, Qingdao New Energy Shandong Laboratory, Qingdao Institute of Bioenergy and Bioprocess Technology, Chinese Academy of Sciences, Qingdao, China

Huihui Sun Yellow Sea Fisheries Research Institute, Qingdao, China

Peng Wang State Key Laboratory of Marine Food Processing and Safety Control, College of Food Science and Engineering, Ocean University of China, Qingdao, China

Yanxia Wang College of Food Science and Light Industry, State Key Laboratory of Materials-Oriented Chemical Engineering, Nanjing Tech University, Nanjing, China

Shaoqing Yang College of Food Science and Nutritional Engineering, China Agricultural University, Beijing, China

Xinyu Yuan College of Food Science and Light Industry, Nanjing Tech University, Nanjing, China

Yi Zhao College of Food Science and Light Industry, Nanjing Tech University, Nanjing, China

Wenting Zhou State Key Laboratory of Marine Food Processing and Safety Control, College of Food Science and Engineering, Ocean University of China, Qingdao, China

Benwei Zhu College of Food Science and Light Industry, Nanjing Tech University, Nanjing, China

Changliang Zhu State Key Laboratory of Marine Food Processing and Safety Control, College of Food Science and Engineering, Ocean University of China, Qingdao, China

Liying Zhu College of Chemical and Molecular Engineering, Nanjing Tech University, Nanjing, China

Zhengming Zhu College of Food Science and Light Industry, State Key Laboratory of Materials-Oriented Chemical Engineering, Nanjing Tech University, Nanjing, China

Structures, Sources, Physicochemical Properties, and Applications of Marine Polysaccharides

Si-Miao Di, Xiangzhao Mao, and Wen-Can Huang

1 Chitin

1.1 Structure of Chitin

Chitin is a naturally occurring polysaccharide found in the exoskeletons of arthropods, such as crustaceans and insects, as well as in the cell walls of fungi and certain algae [1]. It is the second most abundant biopolymer after cellulose [2].

Chitin is a linear homopolymer, meaning it consists of identical repeating units. The N-acetyl-D-glucosamine units are linked together by β-(1 → 4) glycosidic bonds, similar to the bonds in cellulose [3]. This linkage forms a long, unbranched chain. The structure of chitin is stabilized by extensive hydrogen bonding between the hydroxyl groups and the acetamido groups of adjacent chains. These hydrogen bonds contribute to the high crystallinity and insolubility of chitin in water and most organic solvents.

1.2 Source of Chitin

Chitin is a naturally occurring polysaccharide found in various organisms. It is primarily sourced from marine animals, terrestrial arthropods, and fungi. Each source provides unique characteristics to the chitin, making it suitable for different applications [4].

S.-M. Di · X. Mao · W.-C. Huang (✉)
State Key Laboratory of Marine Food Processing and Safety Control, College of Food Science and Engineering, Ocean University of China, Qingdao, China
e-mail: hwc@ouc.edu.cn

X. Mao (ed.), *Degrading Enzymes for Marine Polysaccharides and Their Applications*, https://doi.org/10.1007/978-981-96-9517-1_1

1.2.1 Marine Sources

Shrimp shells are one of the most abundant and economically significant sources of chitin. After processing, shrimp shells can yield high amounts of chitin. Crab shells are another major source of chitin. Similar to shrimp, crabs have a high chitin content in their exoskeletons. Lobster shells are also rich in chitin. The exoskeletons of lobsters are often utilized in chitin production. Squid pens are composed primarily of β-chitin, which is more flexible and has different solubility characteristics compared to the α-chitin found in crustaceans. Squid chitin is particularly valued for its unique properties [4].

1.2.2 Arthropods

Beetles have exoskeletons that contain significant amounts of chitin. While less commonly used in commercial production, insect chitin is gaining interest, especially in sustainable and alternative sources. Grasshoppers and other orthopterans have chitin-rich exoskeletons that can be processed for chitin extraction [4].

1.2.3 Fungal Sources

Certain mushrooms have cell walls rich in chitin, making them a viable nonanimal source of this biopolymer. This is particularly important for applications where nonanimal-derived materials are required. Yeasts and other filamentous fungi also contain chitin in their cell walls [4].

1.3 Preparation of Chitin

1.3.1 Chemical Method

The chemical method for chitin preparation involves the sequential treatment of raw materials such as crustacean shells or fungal biomass with acid and alkali solutions to remove inorganic materials and proteins. Initially, the raw material is demineralized by treating it with a dilute hydrochloric acid (HCl) solution to dissolve calcium carbonate and other minerals. The demineralized material is then deproteinized using a dilute sodium hydroxide (NaOH) solution to solubilize and remove proteins. This process typically involves heating the material in the alkaline solution under controlled conditions. Following deproteinization, the chitin is thoroughly washed with water to remove residual chemicals and impurities. The resultant chitin is then dried and can be further processed into various forms depending on the intended application [4].

The chemical method for chitin preparation is widely used due to its efficiency in producing high-purity chitin. This method is relatively fast and can be easily scaled up for industrial production, making it suitable for large-scale operations. Despite its effectiveness, the chemical method has several disadvantages. The use of strong acids and bases poses significant environmental and safety concerns, as the process generates harmful waste that requires proper disposal and treatment. The harsh conditions employed can also degrade the chitin structure, potentially reducing its molecular weight and affecting its functional properties.

1.3.2 Fermentation Method

The fermentation method for chitin preparation involves the use of microorganisms to selectively degrade non-chitin components such as proteins and minerals from raw materials like crustacean shells, insect exoskeletons, and fungal biomass. This biotechnological process begins with the collection of raw materials, such as shrimp, crab, or lobster shells, insect exoskeletons, or fungal biomass, followed by thorough cleaning to remove dirt, debris, and soluble organic matter, and subsequent drying to reduce moisture content through methods like air drying, oven drying, or freeze-drying. The dried and prepared raw materials are then inoculated with specific microorganisms known for their chitin extraction capabilities, such as lactic acid bacteria (e.g., *Lactobacillus* species), *Bacillus* species, and certain fungi (e.g., *Aspergillus* species). Fermentation takes place in a controlled environment, typically at temperatures between 25 and 37 °C and pH levels maintained between 5 and 7, where microorganisms produce organic acids (like lactic acid) and enzymes (proteases and chitinases) that demineralize calcium carbonate (CaCO3) and other minerals while deproteinizing proteins into soluble peptides and amino acids. Post-fermentation processing involves filtering the mixture to separate solid chitin from the liquid containing dissolved proteins and minerals, followed by thorough washing of the solid chitin to remove microbial biomass and fermentation by-products. The purified chitin is then dried using various methods like air drying, oven drying, or freeze-drying and subsequently ground into a fine powder for further processing, enhancing its surface area for various applications [5].

The fermentation method for chitin preparation offers several advantages, including environmental friendliness through the use of biological agents instead of harsh chemicals, enhancing safety for workers by avoiding corrosive acids and bases, and operating under mild conditions such as ambient temperature and neutral pH, reducing the risk of chitin degradation. Additionally, the resulting chitin tends to be more biocompatible, making it suitable for sensitive applications in fields like medicine and pharmaceuticals. However, this method also has its drawbacks, such as longer processing times compared to chemical methods, potentially lower yields due to incomplete degradation of non-chitin components, complexity requiring precise control of fermentation conditions and microbial strain selection, and a risk of contamination by unwanted microorganisms, which can impact the efficiency and purity of the final product.

1.3.3 Enzymatic Method

The enzymatic method for chitin preparation involves the use of specific enzymes to selectively degrade proteins and other organic materials, leaving behind high-purity chitin. Commonly used enzymes in this process include proteases, such as papain, pepsin, and trypsin, which are effective in breaking down the protein components in the raw materials, such as crustacean shells or fungal biomass. The raw material is first pretreated to remove inorganic materials, often through mild acidic treatment to dissolve minerals. The demineralized material is then incubated with the protease enzymes under optimal conditions, such as appropriate pH, temperature, and enzyme concentration, to facilitate the enzymatic degradation of proteins. The enzymatic treatment is followed by thorough washing with water to remove the solubilized proteins and any residual enzymes. The resulting chitin is then dried and can be further processed as needed [5].

The enzymatic method for chitin preparation offers several advantages, including high specificity where enzymes target only non-chitin components, preserving the chitin structure intact. Operating under mild conditions reduces the risk of chitin degradation and ensures high purity, meeting stringent quality standards for various applications. Additionally, this method is environmentally friendly, avoiding harsh chemicals and generating less waste, and results in chitin that is often more biocompatible, ideal for medical and pharmaceutical uses. However, there are drawbacks such as costliness due to expensive enzymes, potential enzyme inactivation requiring careful handling and storage, lower yields if enzymatic activity is suboptimal or raw materials are inadequately pretreated, complexity in controlling reaction conditions and enzyme handling, and challenges in scaling up for industrial production, necessitating significant investment in bioreactor technology and enzyme production.

1.4 Physicochemical Properties of Chitin

Chitin exists in several crystalline forms, with the most common being α-chitin, β-chitin, and γ-chitin [6]. α-chitin, found in crustacean shells, exhibits strong inter-chain hydrogen bonding, making it highly crystalline and stable. β-chitin, found in squid pens, has weaker hydrogen bonding, resulting in a more flexible and less crystalline structure. γ-chitin, a less common form, exhibits properties intermediate between α- and β-chitin. The crystalline nature of chitin can be analyzed using X-ray diffraction, revealing its highly ordered structure. Chitin is insoluble in water due to extensive hydrogen bonding between chains and is also insoluble in most organic solvents [7].

1.5 Applications of Chitin

Chitin and its derivatives, such as chitosan, have natural antimicrobial properties that help in extending the shelf life of food products by inhibiting the growth of spoilage and pathogenic microorganisms. Chitin can be used as food additives to enhance the texture, stability, and shelf life of various food products, including sauces, dairy products, and beverages. Chitin is used as biostimulants to enhance plant growth, increase yield, and improve resistance to diseases. They stimulate the natural defense mechanisms of plants and promote root development. Chitin-based products can be used to improve soil structure, water retention, and nutrient availability. They enhance soil fertility and support the growth of beneficial soil microorganisms. Chitin is used to coat seeds, protecting them from fungal infections and enhancing germination rates. Seed coatings improve seedling vigor and resistance to environmental stresses. Chitin is incorporated into fish and shrimp feed to enhance growth rates, improve feed conversion efficiency, and boost the immune system of aquatic animals [8].

2 Chitosan

2.1 Structure of Chitosan

Chitosan is a linear polysaccharide derived from chitin, composed of β-(1 → 4)-linked D-glucosamine and N-acetyl-D-glucosamine units. D-Glucosamine (GlcN) is the primary monomer unit in chitosan, where the amino group (-NH_2) replaces the acetamido group (-$NHCOCH_3$) found in chitin. N-acetyl-D-glucosamine (GlcNAc) is the monomer unit retained from chitin, where the amino group is acetylated [9].

The degree of deacetylation (DDA) refers to the proportion of D-glucosamine units to the total number of units (D-glucosamine and N-acetyl-D-glucosamine) in the chitosan molecule. It is usually expressed as a percentage. The DDA significantly influences the solubility, chemical reactivity, and biological properties of chitosan. Higher DDA indicates more free amino groups, enhancing solubility in acidic solutions and increasing reactivity [9, 10].

Molecular weight refers to the total mass of the chitosan polymer. It is determined by the number of monomer units in the polymer chain. Molecular weight affects the viscosity, mechanical strength, and film-forming ability of chitosan. High-molecular-weight chitosan exhibits higher viscosity and better film-forming properties. Different applications require chitosan of specific molecular weights. Low-molecular-weight chitosan is often used in drug delivery systems for its lower viscosity and easier absorption, while high-molecular-weight chitosan is used in applications requiring strong mechanical properties, such as wound dressings and biodegradable films.

The free amino groups present in the D-glucosamine units are responsible for chitosan's unique properties. These groups can form hydrogen bonds, ionic bonds, and covalent bonds, contributing to chitosan's biocompatibility, bioactivity, and chemical reactivity. Hydroxyl groups are present in both the D-glucosamine and N-acetyl-D-glucosamine units. They participate in hydrogen bonding, affecting the solubility and interaction of chitosan with other molecules. The acetamido groups present in the N-acetyl-D-glucosamine units influence the overall hydrophobicity and crystallinity of chitosan. The presence of these groups decreases solubility in acidic solutions compared to fully deacetylated chitosan [9, 10].

2.2 *Preparation of Chitosan*

2.2.1 Chemical Method

Chitosan is chemically prepared by alkaline deacetylation of chitin. Purified chitin is suspended in concentrated sodium hydroxide solution (40–50% w/w) at a solid-to-liquid ratio of about 1:10–1:20. The mixture is heated at 80–120 °C for 0.5–6 h under constant stirring. During this treatment, hydroxide ions cleave the acetamide bonds of N-acetyl-D-glucosamine residues, releasing acetate ions and converting them into D-glucosamine units, thereby increasing the DDA.

2.2.2 Enzymatic Production of Chitosan

The enzymatic deacetylation process for chitosan production involves several key steps. The key enzyme used in this process is chitin deacetylase, which specifically catalyzes the removal of acetyl groups from the N-acetylglucosamine units of chitin. The enzymatic deacetylation process is carried out under mild conditions. These conditions are optimized to maintain the activity and stability of the enzyme while ensuring efficient deacetylation. After the enzymatic reaction, the chitosan is separated from the reaction mixture, washed to remove any residual enzyme and by-products, and then dried.

Enzymatic production of chitosan offers several advantages, including environmental friendliness, as it avoids the use of strong acids and bases, thereby reducing harmful waste disposal and minimizing environmental impact. Additionally, the mild conditions help maintain the molecular weight of the chitosan, preserving its intrinsic properties and functionality. However, the enzymatic production method also has its disadvantages. The high cost of enzymes can be a significant barrier to its widespread adoption, as enzyme production, purification, and stabilization add to the overall cost. Enzymatic reactions can be slower compared to chemical methods, leading to longer production times and reduced throughput in industrial settings. Enzymes can be sensitive to process conditions and may lose activity over time, requiring careful optimization and potentially frequent replenishment. The process requires precise control and monitoring of various parameters, such as pH, temperature, and enzyme concentration, which can complicate production. Additionally, the limited availability of specific enzymes, such as chitin deacetylase, poses a challenge, as their production and purification can be technically demanding.

2.3 *Physicochemical Properties of Chitosan*

Chitosan is soluble in acidic aqueous solutions due to the protonation of its amino groups ($-NH_2$) to form $-NH_3^+$ groups. This solubility in acidic environments allows chitosan to be easily processed into various forms such as films, fibers, and hydrogels.

The molecular weight of chitosan can vary significantly, typically ranging from 50,000 to 1,000,000 Da. The molecular weight influences many of its properties, including viscosity, mechanical strength, and biodegradability. Higher molecular weight chitosan generally exhibits higher viscosity and stronger mechanical properties.

The degree of deacetylation (DDA) refers to the percentage of acetyl groups removed from chitin during the conversion to chitosan. This value usually ranges from 50% to 95%. A higher DDA results in more free amino groups, enhancing chitosan's solubility, reactivity, and ability to form gels. The DDA is a crucial parameter for tailoring chitosan's properties for specific applications [10].

Chitosan solutions exhibit high viscosity, which is influenced by its molecular weight and concentration. The viscosity of chitosan solutions increases with higher molecular weight and concentration. This property is important for applications requiring thickening, gelling, or film-forming capabilities [10].

Chitosan exhibits semicrystalline properties due to its molecular structure. The degree of crystallinity is affected by the DDA and the molecular weight. The crystalline regions contribute to the mechanical strength and thermal stability of chitosan.

Chitosan is biodegradable, breaking down into nontoxic, natural by-products through the action of enzymes such as lysozymes. This biodegradability makes chitosan an attractive material for biomedical and environmental applications, such as drug delivery systems, wound dressings, and biodegradable packaging.

Chitosan is biocompatible, meaning it is nontoxic and does not elicit an immune response when used in biomedical applications. Its biocompatibility, combined with its biodegradability, makes it suitable for use in tissue engineering, drug delivery, and wound healing.

Chitosan possesses intrinsic antimicrobial properties against a wide range of bacteria, fungi, and viruses. This activity is attributed to the interaction between the positively charged amino groups of chitosan and the negatively charged microbial cell membranes, leading to disruption of the cell membrane and leakage of intracellular contents.

Chitosan can form strong, flexible films when cast from its acidic solutions. These films are transparent, with good mechanical strength and barrier properties against oxygen and moisture. This makes chitosan films suitable for applications in food packaging and biomedical dressings.

The free amino groups in chitosan provide sites for chemical modifications, allowing the introduction of various functional groups. This reactivity enables the customization of chitosan for specific applications by grafting or cross-linking with other molecules.

2.4 Applications of Chitosan

Chitosan has natural antimicrobial properties against a wide range of bacteria, fungi, and yeasts. It is used as a preservative to extend the shelf life of perishable foods such as fruits, vegetables, meats, and seafood. Chitosan-based edible coatings and films are applied to food surfaces to create a protective barrier against moisture, oxygen, and microbial contamination, thus enhancing food quality and shelf life. Chitosan is used to produce biodegradable food packaging films that are environmentally friendly alternatives to synthetic plastics. These films offer good mechanical properties and barrier functions. Chitosan is used as a dietary fiber in functional foods and supplements. It is known to aid in cholesterol reduction and weight management. Chitosan is used as a thickening and stabilizing agent in various food products such as sauces, dressings, and dairy products. Chitosan is used as a clarifying agent in the production of beverages like wine and fruit juices. It helps to remove suspended particles, improve clarity, and stabilize the final product. Chitosan is used as a seed coating material to enhance seed germination, protect against soil-borne pathogens, and promote healthy plant growth. Chitosan exhibits antifungal and antibacterial properties, making it an effective natural biopesticide. It helps to control plant diseases and reduce the reliance on chemical pesticides. Chitosan-based products improve soil health by promoting beneficial microbial activity, enhancing soil structure, and increasing water retention. This leads to improved plant growth and productivity. Chitosan is used in the formulation of controlled-release fertilizers that provide a steady supply of nutrients to plants over time, reducing nutrient leaching and improving fertilizer efficiency. Chitosan is used in aquaculture to improve water quality by removing suspended solids, heavy metals, and other pollutants. It helps maintain a healthy environment for fish and other aquatic organisms. As a feed additive, chitosan enhances the growth and immunity of fish and shellfish, improving overall health and reducing the need for antibiotics. Chitosan is used in the development of controlled-release drug delivery systems. Its biocompatibility and ability to form hydrogels make it suitable for delivering drugs at a sustained rate. Chitosan's mucoadhesive properties enhance the bioavailability of drugs administered through mucosal routes, such as nasal, ocular, and oral delivery. Chitosan is used as a hemostatic agent to control bleeding and promote clot formation in wounds and surgical sites. Chitosan-based wound dressings provide a moist healing environment, inhibit microbial growth, and accelerate tissue regeneration. Chitosan is used to fabricate scaffolds for tissue engineering applications. These scaffolds support cell attachment, proliferation, and differentiation, aiding in the regeneration of damaged tissues and organs [11–13].

3 Agarose

3.1 Structure of Agarose

Agarose is a linear polysaccharide composed of repeating units of agarobiose, which consists of alternating residues of D-galactose and 3,6-anhydro-L-galactopyranose. These sugar units are linked by glycosidic bonds, specifically β-(1 → 4) and α-(1 → 3) linkages [14]. The three-dimensional structure of agarose in solution forms a double helix, which aggregates to create a gel network upon cooling. This gelation property is due to hydrogen bonding and hydrophobic interactions between the agarose chains [15]. The resulting gel matrix has a porous structure, which makes agarose particularly useful in gel electrophoresis and chromatography applications, where it serves as a medium for the separation of biomolecules such as nucleic acids and proteins [16].

3.2 Source and Preparation of Agarose

Agarose is derived from agar, a gelatinous substance extracted from the cell walls of certain species of red algae, primarily from the genera *Gelidium* and *Gracilaria* [17]. The preparation of agarose involves several detailed steps to isolate it from agar, starting with the extraction from red algae such as *Gelidium* and *Gracilaria*. After harvesting, the algae are washed to remove sand and salts. The cleaned algae are then boiled in water at temperatures ranging from 90 to 100 °C to dissolve the agar. This hot extraction process typically lasts for several hours to ensure complete dissolution of the agar. Once dissolved, the hot agar solution is filtered to remove insoluble residues and impurities. The filtered solution is allowed to cool to form a gel. This crude agar gel contains both agarose and agaropectin, along with other minor components. To purify agarose, the gel is subjected to multiple cycles of dissolution and precipitation.

In the first cycle, the crude agar gel is dissolved again in hot water or a mild alkaline solution, often using a base such as sodium hydroxide (NaOH) at a concentration of about 0.1–0.5%. The solution is heated to around 60–80 °C to ensure complete dissolution. Following dissolution, the solution is cooled to precipitate the agarose while keeping the agaropectin in solution. The precipitated agarose is collected by filtration or centrifugation. This process is repeated several times to increase the purity of agarose. In some preparations, additional steps may include washing the agarose with ethanol or acetone to remove any remaining impurities and pigments. Finally, the purified agarose is dissolved one last time in hot water, cooled, and allowed to gel. The gel is then dehydrated, typically through lyophilization, to produce a dry, powdered form of agarose that is easy to handle and store.

3.3 Physicochemical Properties of Agarose

Agarose is renowned for its ability to form gels. When dissolved in hot water (typically around 85–95 °C) and then cooled below 35–40 °C, agarose solutions form firm gels [18]. This reversible gelation is due to the formation of a double helix structure, stabilized by hydrogen bonds and hydrophobic interactions. The gel strength and melting temperature can be adjusted by altering the concentration of agarose in the solution.

Agarose gels are characterized by their porous structure, with pore sizes that can be controlled by the concentration of agarose. Higher concentrations produce gels with smaller pores, while lower concentrations result in larger pores. This property is particularly useful in electrophoresis and chromatography for the separation of molecules based on size.

Agarose is insoluble in cold water but dissolves readily in boiling water. It is also soluble in certain organic solvents such as dimethyl sulfoxide (DMSO) and formamide. The solubility in hot water is a key factor in its use for gel formation.

Agarose is nontoxic and biocompatible, making it suitable for use in biomedical applications, including cell culture and drug delivery systems. Its inert nature ensures that it does not interact adversely with biological systems [19].

3.4 Applications of Agarose

Agarose gel electrophoresis is a widely used technique for the separation and analysis of nucleic acids (DNA and RNA) [20]. The porous structure of agarose gels allows for the effective separation of molecules based on size. By adjusting the concentration of agarose, researchers can tailor the pore size to optimize the resolution of different-sized fragments. Agarose gels provide clear, sharp bands that facilitate the identification and quantification of biomolecules, making it a fundamental tool in molecular biology, genetics, and biochemistry.

Agarose beads are used in various types of chromatography, including gel filtration, affinity chromatography, and ion-exchange chromatography [21]. In gel filtration chromatography, agarose beads separate molecules based on size, allowing for the purification of proteins, nucleic acids, and other macromolecules. In affinity chromatography, agarose beads are functionalized with ligands that specifically bind to target molecules, enabling highly selective purification. The chemical stability and modifiability of agarose make it suitable for these diverse chromatographic applications.

In microbiology, agarose is a key component of solid culture media, particularly in the form of agar plates [22]. These plates provide a stable and nutrient-rich environment for the growth of microorganisms, enabling the isolation, identification, and enumeration of bacterial and fungal species. The transparency and firmness of agarose gels facilitate the observation of microbial colonies and the testing of antimicrobial agents.

Agarose is explored as a material for drug delivery systems due to its biocompatibility, biodegradability, and ability to form hydrogels [23]. It can be used to encapsulate drugs, providing controlled and sustained release. Agarose-based hydrogels can be engineered to respond to environmental stimuli, such as pH or temperature, enhancing the targeted delivery of therapeutic agents. This application is particularly relevant in the development of localized drug delivery systems for cancer treatment and wound healing.

In molecular cloning, agarose gels are used to analyze and purify DNA fragments. After electrophoretic separation, desired DNA bands can be excised from the gel and purified for subsequent cloning steps. This application is essential for genetic engineering, recombinant DNA technology, and synthetic biology. Agarose's inert nature ensures that it does not interfere with downstream enzymatic reactions, maintaining the integrity of DNA samples [24].

4 Alginate

4.1 Structure of Alginate

Alginate is a natural polysaccharide composed mainly of linear chains of 1–4 linked β-D-mannuronic acid (M) and α-L-guluronic acid (G) residues [25]. These monomeric units are arranged in a blockwise pattern, forming homopolymeric sequences of consecutive M or G residues (M-blocks and G-blocks) and alternating MG blocks. The relative proportion and sequential arrangement of these blocks influence the physical properties of alginate, such as its gelling behavior and mechanical strength. Alginate is extracted primarily from brown seaweeds and is known for its biocompatibility, making it suitable for various biomedical and industrial applications [26].

4.2 Source of Alginate

Alginate is primarily sourced from brown seaweeds (*Phaeophyceae*), such as *Laminaria hyperborea*, *Laminaria digitata*, *Ascophyllum nodosum*, and *Macrocystis pyrifera* [27]. These seaweeds are abundant in coastal regions and are harvested for their high alginate content. The alginate is present in the cell walls of these seaweeds, where it plays a crucial role in maintaining the structural integrity and flexibility of the seaweed tissue.

4.3 Preparation of Alginate

The preparation of alginate involves several key steps to extract and purify the polysaccharide from the brown seaweed. Initially, the dried and milled seaweed is treated with a dilute acid solution, such as hydrochloric acid, to remove minerals and other impurities. This step is followed by washing the seaweed residue with water to neutralize the pH. The alginate is then extracted using a dilute alkali solution, typically sodium carbonate or sodium hydroxide. The alkaline solution solubilizes the alginate, forming a viscous extract. This extract is filtered to remove insoluble residues, and the filtrate is treated with calcium chloride or another divalent cation to precipitate the alginate as calcium alginate. The calcium alginate is collected and washed to remove impurities and then converted back to sodium alginate by treating it with a sodium salt solution. The sodium alginate is further purified through filtration, dialysis, or other purification methods to remove any remaining contaminants. Finally, the purified sodium alginate is dried and milled into a fine powder, ready for use in various applications. This detailed preparation process ensures that the alginate retains its desirable physicochemical properties, such as gelation capability and biocompatibility.

4.4 Physicochemical Properties of Alginate

Alginate is hydrophilic, which allows it to readily absorb water and form viscous solutions. This property makes alginate an excellent thickening, gelling, and stabilizing agent. One of the most remarkable features of alginate is its ability to form gels in the presence of divalent cations such as calcium (Ca^{2+}). This gelation process is driven by ionic cross-linking between the guluronic acid (G) blocks in the alginate chains and the calcium ions, resulting in a robust three-dimensional network [28]. The mechanical strength and rigidity of the alginate gel can be fine-tuned by adjusting the M/G ratio and the concentration of the cross-linking ions.

Additionally, alginate is a pH-responsive polysaccharide. At low pH, protonation of the carboxyl groups reduces electrostatic repulsion between polymer chains, leading to decreased swelling, shrinkage of hydrogels. At high pH, deprotonation of carboxyl groups increases chain repulsion and enhances swelling [29].

Moreover, alginate is biocompatible and nontoxic, making it suitable for biomedical applications such as drug delivery systems, wound dressings, and tissue engineering scaffolds [30]. It can encapsulate a wide range of molecules, protecting them from degradation and facilitating their controlled release. Alginate's ability to form hydrogels that mimic the extracellular matrix provides a supportive environment for cell growth and differentiation, which is critical for tissue engineering.

4.5 Applications of Alginate

In the biomedical field, alginate is widely used for drug delivery systems due to its ability to form hydrogels that can encapsulate a variety of drugs, protecting them from degradation and allowing for controlled release [31, 32]. Alginate microspheres and beads are employed to deliver proteins, peptides, and small molecule drugs, particularly for oral and injectable formulations [33]. Additionally, alginate is used in wound dressings due to its ability to maintain a moist environment, promote healing, and provide a barrier against infections. The biocompatibility and mild gelation conditions of alginate make it suitable for tissue engineering applications, where it is used as a scaffold material to support cell growth and differentiation. Alginate hydrogels mimic the extracellular matrix, providing a conducive environment for cells in regenerative medicine and 3D bioprinting [34].

In the food industry, alginate serves as a thickening, gelling, and stabilizing agent [35]. It is commonly used in the production of desserts, such as jellies and puddings, where it forms a gel matrix that improves texture and consistency. Alginate is also used in the production of restructured foods, such as artificial caviar and meat analogs, by forming stable gels that encapsulate flavors and nutrients. In addition, alginate acts as a stabilizer in ice creams and dairy products, preventing ice crystal formation and improving mouthfeel. Its ability to form films and coatings makes it useful for edible coatings on fruits and vegetables, extending their shelf life by reducing moisture loss and providing a barrier against microbial contamination.

In the cosmetics industry, alginate is valued for its moisturizing and skin-conditioning properties [36]. It is used in a variety of skincare products, such as face masks, creams, and lotions, where it helps to hydrate and soothe the skin. Alginate's film-forming ability makes it an excellent ingredient for peel-off masks, providing a gentle and effective way to remove impurities from the skin's surface. Its biocompatibility and nonirritating nature make it suitable for use in products designed for sensitive skin.

5 Carrageenan

5.1 Structure of Carrageenan

Carrageenan is a natural polysaccharide extracted from red seaweeds (Rhodophyceae), primarily from species like *Kappaphycus alvarezii* and *Eucheuma denticulatum*. It consists of repeating galactose units and 3,6-anhydrogalactose, both of which are sulfated to varying degrees [37]. Carrageenan is categorized into three main types: kappa (κ), iota (ι), and lambda (λ), based on their sulfate content and gelling properties [38]. Kappa-carrageenan has one sulfate group per disaccharide unit and forms strong, rigid gels in the presence of potassium ions. Iota-carrageenan contains two sulfate groups per disaccharide and forms soft, elastic gels with calcium ions. Lambda-carrageenan, with three sulfate groups per disaccharide, does not gel but increases the viscosity of solutions. The polysaccharide chains of carrageenan have alternating 1–3 linked β-D-galactose and 1–4 linked α-D-galactose units, which may be in the form of 3,6-anhydro-α-D-galactose in κ and ι types.

5.2 Source of Carrageenan

Carrageenan is extracted from various species of red seaweeds, predominantly from the genera *Kappaphycus*, *Eucheuma*, *Chondrus*, and *Gigartina*. After collection, the seaweeds are cleaned to remove sand, salt, and other impurities. They are then dried to reduce moisture content and either used directly for carrageenan extraction or exported in dried form to processing facilities.

5.3 Preparation of Carrageenan

The preparation of carrageenan begins with the collection and drying of red seaweeds such as *Kappaphycus alvarezii* and *Eucheuma denticulatum*. The dried seaweed is milled into a coarse powder to increase the surface area for extraction. The powdered seaweed is then subjected to an alkaline treatment, typically using potassium hydroxide or sodium hydroxide, at elevated temperatures. This process solubilizes the carrageenan, extracting it from the seaweed cell walls. The alkaline solution containing the solubilized carrageenan is filtered to remove insoluble residues. The filtrate is then cooled, and carrageenan is precipitated by adding alcohol, such as ethanol or isopropanol. The precipitated carrageenan is collected by centrifugation or filtration, washed with alcohol to remove impurities, and dried. The dried carrageenan is milled into a fine powder and packaged for use in food, pharmaceutical, and cosmetic applications. This process ensures the production of high-quality carrageenan with desirable gelling and thickening properties.

5.4 Physicochemical Properties of Carrageenan

Carrageenan's solubility depends on its type (kappa, iota, lambda) and the presence of cations. Kappa-carrageenan is soluble in hot water but only partially soluble in cold water. Iota-carrageenan is more soluble in cold water compared to kappa, but its solubility improves significantly in hot water. Lambda-carrageenan is fully soluble in cold water, forming viscous solutions without the need for heating. The solubility is influenced by the degree of sulfation, higher sulfation increases solubility in water [39].

One of the most notable properties of carrageenan is its ability to form gels. Kappa-carrageenan forms strong, brittle gels in the presence of potassium ions. These gels exhibit syneresis, where they expel water over time. Iota-carrageenan forms soft, elastic gels in the presence of calcium ions, with minimal syneresis. Lambda-carrageenan does not form gels but increases the viscosity of solutions. The gelation process involves the formation of double helices and subsequent aggregation into a three-dimensional network, influenced by temperature and the presence of specific cations [18].

5.5 Applications of Carrageenan

Carrageenan is used as a stabilizer and thickener in dairy products such as milk, yogurt, cheese, and ice cream. In processed meats like sausages, ham, and poultry, carrageenan improves water retention, texture, and sliceability. It helps in binding water, thus reducing cooking losses and enhancing juiciness. It acts as a stabilizer in various beverages, including fruit juices and alcoholic drinks, to prevent the settling of particles and improve consistency [40].

Carrageenan is used as an excipient in tablets and capsules, aiding in the controlled release of active ingredients. Its gel-forming ability ensures the sustained release of medications. Carrageenan-based hydrogels are employed in wound dressings due to their biocompatibility and ability to maintain a moist environment, which promotes healing [39].

Carrageenan hydrogels are used in tissue engineering and regenerative medicine. Their biocompatibility and ability to form stable gels make them suitable for cell encapsulation and tissue scaffolds.

6 Chondroitin Sulfate

6.1 *Structure of Chondroitin Sulfate*

Chondroitin sulfate is a sulfated glycosaminoglycan composed of repeating disaccharide units. Each disaccharide unit consists of N-acetylgalactosamine and glucuronic acid, linked by β-1,3 glycosidic bonds [41]. The N-acetylgalactosamine residues are variably sulfated at the four or six positions, leading to different forms of chondroitin sulfate, such as chondroitin-4-sulfate and chondroitin-6-sulfate. The glucuronic acid residues are linked to the N-acetylgalactosamine residues by β-1,4 glycosidic bonds, forming the linear polymer chain [41]. This structure imparts high negative charge density to chondroitin sulfate, allowing it to interact with a variety of proteins and contributing to its role in providing structural support and lubrication in cartilage, tendons, and other connective tissues.

6.2 *Physicochemical Properties of Chondroitin Sulfate*

Chondroitin sulfate is soluble in water due to the presence of multiple hydrophilic sulfate groups and carboxyl groups. It forms clear, viscous solutions when dissolved in water. The solubility can vary with pH, temperature, and ionic strength of the solution. In general, chondroitin sulfate is more soluble in neutral and slightly alkaline conditions [42]. It is insoluble in organic solvents such as ethanol, acetone, and chloroform.

The molecular weight of chondroitin sulfate affects the viscosity of solutions, the biological activity, and the pharmacokinetic properties of chondroitin sulfate. Higher molecular weight chondroitin sulfate tends to form more viscous solutions and may have prolonged biological effects [43].

The degree of sulfation of chondroitin sulfate varies depending on its biological source. Chondroitin sulfate can be sulfated at the four-position (chondroitin-4-sulfate) or six-position (chondroitin-6-sulfate) of the N-acetylgalactosamine residue [44]. The degree of sulfation affects the molecule's charge density, solubility, viscosity, and interaction with proteins and other biomolecules. Higher sulfation levels generally enhance the biological activity of chondroitin sulfate, including its anti-inflammatory and chondroprotective effects.

6.3 *Applications of Chondroitin Sulfate*

Chondroitin sulfate is widely used in dietary supplements aimed at improving joint health and alleviating symptoms of osteoarthritis. It supports cartilage structure, reduces pain, and enhances joint function. Chondroitin sulfate is used in prescription medications and over-the-counter products for treating osteoarthritis. It helps in slowing cartilage degradation and reducing inflammation [45].

7 Fucoidan

7.1 *Structure of Fucoidan*

Fucoidan is a complex sulfated polysaccharide found mainly in the cell walls of brown seaweeds such as *Fucus vesiculosus* and *Ascophyllum nodosum*. Its structure is characterized by a backbone composed primarily of α-(1 → 3)-linked L-fucose residues, with some regions containing α-(1 → 4) linkages [46]. These fucose units are extensively sulfated at various positions, which contribute to the biological activity of fucoidan. In addition to fucose, fucoidan may contain other monosaccharides, such as galactose, mannose, xylose, and glucuronic acid, in smaller amounts. The degree of sulfation, as well as the presence of other monosaccharides, can vary depending on the seaweed species and extraction methods, resulting in heterogeneity in fucoidan's structure and properties.

7.2 *Source and Preparation of Fucoidan*

Fucoidan is prepared from brown seaweeds such as *Fucus vesiculosus* and *Ascophyllum nodosum*. The seaweed is first harvested, washed to remove sand and debris, and then dried. The dried seaweed is milled into a coarse powder to facilitate extraction. The powdered seaweed is extracted using hot water or dilute acid at controlled temperatures to solubilize the fucoidan. The extraction solution is filtered to remove insoluble residues. The fucoidan-rich filtrate is then precipitated by adding ethanol or methanol. The precipitated fucoidan is collected by centrifugation or filtration, washed with alcohol to remove impurities, and dried. The dried fucoidan is further purified using techniques such as dialysis, ultrafiltration, or ion-exchange chromatography to achieve high purity. The purified fucoidan is dried again and milled into a fine powder, suitable for use in dietary supplements, pharmaceuticals, and cosmetic formulations for its bioactive properties.

7.3 *Physicochemical Properties of Fucoidan*

Fucoidan is highly soluble in water, forming viscous solutions. Its solubility is influenced by the degree of sulfation and the presence of other monosaccharides. It is generally insoluble in organic solvents like ethanol, acetone, and chloroform [47].

Fucoidan is characterized by a high degree of sulfation, which is essential for its biological activity. The sulfate groups are usually attached at the C-2 and C-4 positions of the fucose residues [48]. The degree of sulfation influences its solubility, charge density, and interaction with other molecules. Higher sulfation typically enhances its anticoagulant, antiviral, and immunomodulatory properties. Due to its high content of sulfate groups, fucoidan carries a significant negative charge. This high charge density allows fucoidan to interact strongly with positively charged molecules.

7.4 Applications of Fucoidan

Fucoidan is included in dietary supplements for its overall health benefits, including boosting the immune system, improving gut health, and providing antioxidant protection. In addition, fucoidan is used as a biostimulant in agriculture to enhance plant growth, improve resistance to stress, and increase crop yield.

8 Porphyran

8.1 Structure of Porphyran

Porphyran is characterized by a unique repeating disaccharide unit composed of alternating α-(1 → 3)-linked L-galactose-6-sulfate and β-(1 → 4)-linked D-galactose [49]. This structure is similar to that of agarose but includes significant sulfation. The L-galactose units in porphyran are often six-sulfated, while the D-galactose units can also be methylated or carry other substituents. The degree of sulfation can vary, which influences its solubility and gelling properties. The presence of 3,6-anhydro-α-L-galactose in some of the L-galactose units further modifies the structure. This complex arrangement allows porphyran to form strong gels and provides it with unique bioactive properties, such as antioxidant, anti-inflammatory, and antiviral activities. The molecular structure of porphyran, with its specific sulfation patterns and potential for various modifications, is key to its functional properties in both food and pharmaceutical applications [50].

8.2 Source and Preparation of Porphyran

Porphyran is predominantly sourced from red algae, specifically from species within the genus *Porphyra*. The preparation of porphyran from *Porphyra* species involves several steps to ensure the efficient extraction and purification of the polysaccharide. First, the harvested algae are thoroughly washed to remove salts, sand,

and other impurities. The cleaned algae are then dried and milled into a fine powder. This powdered algae is subjected to an extraction process, typically involving hot water extraction, where the algae are boiled in water to solubilize the porphyran. The extraction conditions, such as temperature, pH, and duration, are optimized to maximize yield and preserve the structural integrity of porphyran. After extraction, the solution is filtered to remove solid residues. The filtrate, containing dissolved porphyran, is then precipitated using alcohol, commonly ethanol or isopropanol, which induces the formation of a gel-like substance. This precipitate is collected and washed several times with alcohol to remove any remaining impurities.

8.3 Physicochemical Properties of Porphyran

Porphyran is highly soluble in water, forming clear, viscous solutions. Its solubility is due to the presence of multiple sulfate groups and hydrophilic hydroxyl groups. It is insoluble in most organic solvents, including ethanol, acetone, and chloroform [50]. Porphyran is characterized by significant sulfation; higher sulfation enhances its solubility, gelling properties, and biological activity. The presence of sulfate groups contributes to its antioxidant, anti-inflammatory, and antiviral effects. Porphyran has excellent gelling properties, forming strong and thermally reversible gels. The gel strength and melting temperature are influenced by the degree of sulfation and molecular weight.

8.4 Applications of Porphyran

Porphyran acts as a thickener and stabilizer in various food applications, including sauces, dressings, and dairy products, improving consistency and shelf life.

Porphyran's biocompatibility and gel-forming properties make it suitable for use in drug delivery systems, where it can control the release of active ingredients. It is explored for its potential in wound healing applications, where porphyran-based hydrogels can provide a moist environment and support tissue regeneration.

9 Ulvan

9.1 Structure of Ulvan

Structure of ulvan is complex and heterogeneous, comprising repeating disaccharide units that include various types of uronic acids and sulfated rhamnose. The primary building blocks of ulvan are α-(1 $\rightarrow$ 4)- and β-(1 $\rightarrow$ 4)-linked uronic acids (such as glucuronic and iduronic acids) and α-(1 $\rightarrow$ 2)-linked rhamnose-3-sulfate

[51]. These disaccharide units can be arranged in different sequences, leading to the formation of ulvanobiuronic acids (e.g., A3s and B3s) and ulvanobiose. The structure is further complicated by the presence of sulfate groups at various positions, which contribute to its high negative charge density. Additionally, ulvan can contain minor amounts of other monosaccharides, such as xylose and glucose.

9.2 Source and Preparation of Ulvan

Ulvan is sourced from green algae, particularly from species belonging to the genus Ulva. Common species include *Ulva lactuca* and *Ulva rigida*. The preparation of ulvan from *Ulva* species involves a series of steps to extract and purify the polysaccharide [51]. Initially, the collected algae are thoroughly washed with fresh water to remove salts, epiphytes, and other contaminants. The cleaned algae are then dried, either under the sun or using drying equipment, and subsequently ground into a fine powder. The powdered algae undergo an extraction process to solubilize ulvan, typically through hot water extraction or alkaline extraction. For hot water extraction, the algae are heated in water at temperatures ranging from 80 to 100 °C for several hours. Alkaline extraction involves treating the algae with a dilute alkaline solution, such as sodium hydroxide, to enhance the solubilization of ulvan.

The extraction solution is then filtered to separate the liquid containing dissolved ulvan from the solid residues. The filtrate is subjected to a series of purification steps, including precipitation with alcohol (ethanol or isopropanol), to isolate ulvan. The precipitate is collected and washed with alcohol to remove impurities. Further purification can be achieved through dialysis or ultrafiltration to eliminate low-molecular-weight contaminants and salts [52].

9.3 Physicochemical Properties of Ulvan

Ulvan is soluble in water, forming viscous solutions. Its solubility is due to the presence of sulfate groups and uronic acids, which contribute to its hydrophilicity. Ulvan is insoluble in most organic solvents, including ethanol, acetone, and chloroform. The water solubility of ulvan makes it suitable for various aqueous applications.

9.4 Applications of Ulvan

Ulvan's biocompatibility and gel-forming properties make it suitable for use in drug delivery systems, where it can control the release of active ingredients and enhance bioavailability. Ulvan is explored for its potential in wound healing applications, where ulvan-based hydrogels can provide a moist environment and support tissue regeneration.

References

1. Huang W-C, Zhao D, Guo N, Xue C, Mao X (2018) J Agric Food Chem 66:11897. https://doi.org/10.1021/acs.jafc.8b03847
2. He Y, Liu Y, Luo J, Mao X, Huang W-C (2024) ACS Sustain Chem Eng 12:751
3. Wang W, Liu J, Xu H, Zhang Y, Mao X, Huang W-C (2024) Int J Biol Macromol 256:128322
4. Joseph SM, Krishnamoorthy S, Paranthaman R, Moses J, Anandharamakrishnan C (2021) Carbohydr Polym Technol Appl 2:100036
5. Mao X, Guo N, Sun J, Xue C (2017) J Clean Prod 143:814
6. Lv J, Lv X, Ma M, Oh D-H, Jiang Z, Fu X (2023) Carbohydr Polym 299:120142
7. Rinaudo M (2006) Prog Polym Sci 31:603
8. Piekarska K, Sikora M, Owczarek M, Jóźwik-Pruska J, Wiśniewska-Wrona M (2023) Polymers 15:793
9. Kou SG, Peters LM, Mucalo MR (2021) Int J Biol Macromol 169:85
10. Mohan K, Ganesan AR, Ezhilarasi P et al (2022) Carbohydr Polym 287:119349
11. Wang W, Xue C, Mao X (2020) Int J Biol Macromol 164:4532
12. Wang W, Meng Q, Li Q et al (2020) Int J Mol Sci 21:487
13. Li Q, Dunn E, Grandmaison E, Goosen MF (2020) Applications of Chitan and Chitosan. CRC Press
14. Araki C (1966) Proceedings of the fifth international seaweed symposium, Halifax, August 25–28, 1965. Elsevier
15. Arnott S, Fulmer A, Scott W, Dea I, Moorhouse R, Rees D (1974) J Mol Biol 90:269
16. Greenspan P, F-w Mao B-H, Ryu RLG (1995) J Chromatogr A 698:333
17. Armisen R, Gaiatas F (2009) Handbook of hydrocolloids. Elsevier
18. Zhao J, Sun C, Li H, Dong X, Zhang X (2020) Int J Biol Macromol 154:878
19. Manivasagan P, Oh J (2016) Int J Biol Macromol 82:315
20. Renn DW (1984) Ind Eng Chem Prod Res Dev 23:17
21. Gustavsson P-E, Axelsson A, Larsson P-O (1999) J Chromatogr A 830:275
22. Zhu Y, Liang M, Li H et al (2020) Food Chem 320:126652
23. Yazdi MK, Taghizadeh A, Taghizadeh M et al (2020) J Control Release 326:523
24. Vogelstein B, Gillespie D (1979) Proc Natl Acad Sci 76:615
25. Ertesvåg H, Valla S (1998) Polym Degrad Stab 59:85
26. Lee KY, Mooney DJ (2012) Prog Polym Sci 37:106
27. Smidsrød O, Skja G (1990) Trends Biotechnol 8:71
28. Braccini I, Pérez S (2001) Biomacromolecules 2:1089
29. Pawar SN, Edgar KJ (2012) Biomaterials 33:3279
30. George M, Abraham TE (2006) J Control Release 114:1
31. Sanchez-Ballester NM, Soulairol I, Bataille B, Sharkawi T (2019) Carbohydr Polym 207:224
32. Wei X, Xiong H, Zhou D, Jing X, Huang Y (2018) Carbohydr Polym 186:45
33. Zhou M, Hu Q, Wang T, Xue J, Luo Y (2018) Int J Biol Macromol 120:859
34. Rowley JA, Madlambayan G, Mooney DJ (1999) Biomaterials 20:45
35. Senturk Parreidt T, Müller K, Schmid M (2018) Foods 7:170
36. Bae SB, Nam HC, Park WH (2019) Int J Biol Macromol 133:278
37. Jiao G, Yu G, Zhang J, Ewart HS (2011) Mar Drugs 9:196
38. Stone AK, Nickerson MT (2012) Food Hydrocoll 27:271
39. Liu J, Zhan X, Wan J, Wang Y, Wang C (2015) Carbohydr Polym 121:27
40. Gu YS, Decker EA, McClements DJ (2005) Food Hydrocoll 19:83
41. Zhang W, Xu R, Chen J et al (2023) Int J Biol Macromol:126551
42. Sharma R, Kuche K, Thakor P et al (2022) Carbohydr Polym 286:119305
43. Baeurle S, Kiselev M, Makarova E, Nogovitsin E (2009) Polymer 50:1805
44. Restaino OF, Schiraldi C (2022) Carbohydr Polym 292:119690

45. Bali J-P, Cousse H, Neuzil E (2001) Seminars in arthritis and rheumatism. Elsevier
46. Wu L, Sun J, Su X, Yu Q, Yu Q, Zhang P (2016) Carbohydr Polym 154:96
47. Zayed A, El-Aasr M, Ibrahim A-RS, Ulber R (2020) Mar Drugs 18:571
48. Mak W, Hamid N, Liu T, Lu J, White W (2013) Carbohydr Polym 95:606
49. Qiu Y, Jiang H, Fu L, Ci F, Mao X (2021) Food Chem 349:129209
50. Wang H, Zhu B, Yao Z (2023) Algal Res 73:103167
51. Kidgell JT, Magnusson M, de Nys R, Glasson CR (2019) Algal Res 39:101422
52. Lahaye M, Robic A (2007) Biomacromolecules 8:1765

Marine Oligosaccharides: The Source, Structure, Preparation Methods, and Physiological Activity and Applications

Benwei Zhu, Yi Zhao, Xinyu Yuan, and Limin Ning

1 Introduction

Marine polysaccharide is a biological macromolecule with various physiological activities originating from the ocean. Due to its antitumor, anti-oxidation, antivirus, and immunomodulatory functions, marine polysaccharide has been widely used in many fields such as medicine, food, chemical industry, agriculture, and environment [1]. For example, polysaccharides extracted from the red algae *Porphyra yezoensis* and *Gracilaria verrucosa* have macrophage-stimulating activity both in vitro and in vivo. Isolated from brown algae *Undaria pinnatifida sporophyll* of fucoidan can stimulate macrophages to produce cytokines such as IL-6 and TNF-α [2]. Carrageenan polysaccharides isolated from red algae can promote the production of TNF-α in mouse white blood cells [3]. A sulfated water-soluble polysaccharide isolated from green algae *Capsosiphon fulvescens* can induce macrophages to release cytokines such as TNF-α and IL-6 and promote the expression of inducible nitric oxide synthase (iNOS) and cyclooxygenase-2 (COX-2) [4]. However, due to the poor solubility of polysaccharides, the bioavailability of polysaccharides is also low, which greatly limits the further application of polysaccharides. Marine oligosaccharides are degradation products of polysaccharides after physical, chemical, or enzymatic hydrolysis methods. While retaining various activities of polysaccharides, the solubility and bioavailability of oligosaccharides have been greatly improved [5]. Therefore, marine oligosaccharides have become a research hotspot in the field of marine biological resources exploitation [6]. In this chapter, the

B. Zhu (✉) · Y. Zhao · X. Yuan
College of Food Science and Light Industry, Nanjing Tech University, Nanjing, China
e-mail: zhubenwei@njtech.edu.cn

L. Ning (✉)
College of Medicine, Nanjing University of Chinese Medicine, Nanjing, China
e-mail: ninglimin@njucm.edu.cn

X. Mao (ed.), *Degrading Enzymes for Marine Polysaccharides and Their Applications*, https://doi.org/10.1007/978-981-96-9517-1_2

sources, preparation, purification, and application of several seaweed functional oligosaccharides were systematically summarized, which provided some references for promoting the utilization of marine polysaccharide resources and the development of marine biological resources.

2 Alginate Oligosaccharides

Alginate oligosaccharides are the products of alginate polysaccharide by acid or enzymatic degradation, and their degree of polymerization (DPs) is mostly between 2 and 20. Structurally, alginate oligosaccharides contain two monosaccharide units: α-L-mannuronic acid and its differential isomer of C5, β-D-guluronic acid [7]. These two monosaccharide units can be randomly arranged to form oligosaccharide fragments of various structures, such as mannuronic oligosaccharides containing only α-L-mannuronic acid, guluronic oligosaccharides containing only β-D-guluronic acid, and heterozygous block oligosaccharides containing both monosaccharide units (Fig. 1).

Chemical method is the most common method for preparing alginate oligosaccharides, such as hydrolyzing alginate with hydrochloric acid, oxalic acid, formic acid, and sulfuric acid to prepare alginate oligosaccharides [8]. Larsen et al. used 1 mol/L of oxalic acid to degrade 30% of the substrate into alginate oligosaccharides with polymerization degree of 10–30 [9]. Chandia et al. hydrolyzed alginate with three different acids and found that alginate could be completely hydrolyzed with 90% formic acid at 100°C for 6 hand 1.5 mol/L formic acid at 100 °C for 2 h [10]. In addition to acid degradation, alkaline hydrolysis can also be used to prepare alginate oligosaccharides. For example, Niemela et al. hydrolyzed alginate at 95 °C and 135 °C with strong alkali, and a series of products, such as dehydrated isoglycolic acid, were obtained, indicating that the structure of alginate was completely destroyed during the hydrolysis process, so it was not suitable for the preparation of alginate oligosaccharides [11]. In addition, H_2O_2 and other oxidants can also be used to prepare alginate oligosaccharides. Zhang et al. hydrolyzed alginate with 5% H_2O_2 at 90 °C to obtain a series of oligosaccharide products [12]. Compared with acid-base method, the oligosaccharides prepared by oxidation method have higher purity and fewer by-products and impurities. However, compared with acid-base method, oxidation method has higher cost and poor product specificity, which also limits its application in industrialization. In addition, the microwave method can

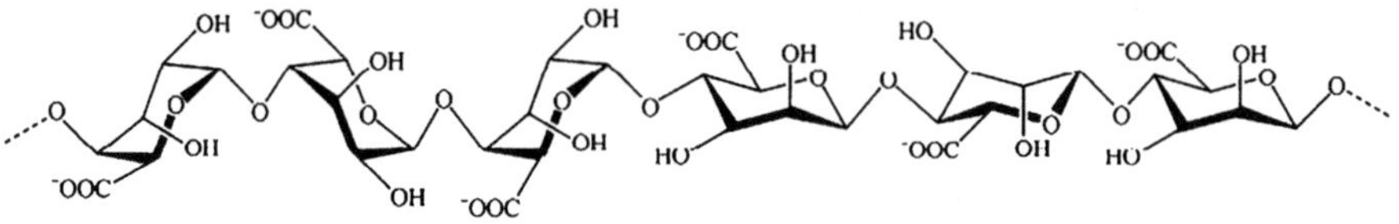

Fig. 1 The structure and composition of alginate oligosaccharides (AOS)

also be used to degrade alginate to produce alginate oligosaccharides. For example, Hu Ting et al. used the microwave method to degrade alginate and obtain saturated oligosaccharides with polymerization degree of 1 ~ 7. This method can realize the controllable preparation of alginate oligosaccharides [13], so it is the development direction of the preparation technology of alginate oligosaccharides with specific polymerization degree in the future. In addition to the above physical and chemical methods, alginate oligosaccharides can also be prepared by alginate lyase, which is a polysaccharide lyase that can produce a series of unsaturated alginate oligosaccharides through β-elimination reaction [14]. According to the mode of action, alginate lyase can be divided into endolytic and exolytic enzymes. Endolytic enzyme can randomly degrade alginate into oligosaccharides with different polymerization degrees, while exolytic enzyme can degrade alginate one by one from one end of the polysaccharide chain of substrate to produce uniform monosaccharide products. Therefore, endogenous alginate lyase is mainly used to prepare alginate oligosaccharides with different polymerization degrees [15]. Our research team has done a lot of work in this field and obtained a series of alginate lyases with different product distributions [16–21]. For example, we obtained a tool enzyme from Marine *Vibrio* sp. W13 that can specifically prepare alginate triose, and the yield of triose in its degradation products is about 83%. Therefore, it can be used for large-scale preparation of trisaccharides [22]. In addition, Zhang et al. used alginate lyase from Marine *Vibrio* sp.510 to prepare alginate oligosaccharides with polymerization degree 2–5 [23]. Li et al. used alginate lyase derived from *Pseudomonas* sp. HZJ216 to prepare oligosaccharides with polymerization degree 2–7 [24]. With the development of enzyme engineering, alginate oligosaccharides with different structures and polymerization degrees can be realized by various methods such as rational design and directed evolution. Alginate oligosaccharides have various biological activities, such as anti-oxidation, anti-inflammatory, antitumor and immunomodulatory [8]. Sun et al. found that alginate oligosaccharides can remove different types of free radicals, so they can be used as antioxidants [25]. Xu et al. found that unsaturated guluronic acid oligosaccharides could inhibit inflammation by inducing the production of NO and the expression of NO synthase and had a significant dose-dependent relationship [26]. Hu et al. found that the clearance rate of alginate oligosaccharides on Sarcoma cells 180 could reach 70.4% [27]. In addition, alginate oligosaccharides can also regulate the production of cytokines by immune cells, so as to have the role of immune regulation. Yamamoto et al. found that alginate oligosaccharides with different polymerization degrees could induce the production and secretion of various cytokines (such as TNF-α, GCSF, etc.) [28]. Iwamoto et al. found that mannuronic oligosaccharides with polymerization degree 3–7 had significant cytokine inducing activity [29]. Recent studies have shown that alginate oligosaccharides also play a prebiotic role, which can promote the proliferation of beneficial microorganisms such as *Bifidobacterium*, while inhibiting the proliferation of harmful bacteria [30]. GV-971, a recently reported derivative of alginate oligosaccharides, can slow down and inhibit Alzheimer's disease by regulating gut microbiota [31]. Therefore, alginate oligosaccharides have great potential in the development of novel functional foods and drugs. In agriculture, alginate oligosaccharides can promote seed germination and root elongation of plants such as tobacco, soybean, and potato. For example, Natsume et al. found that alginate oligosaccharides can promote the root elongation of cereal [32]. Ma et al. found that a certain concentration of alginate oligosaccharides can promote the germination of pea seeds [33]. Alginate oligosaccharides can be used as seed leaching agents in agriculture and can be used for root regeneration and elongation of plants after transplantation. In addition, alginate oligosaccharides can also improve the drought resistance of crops and induce the disease resistance of plants as signal molecules [32, 34]. For example, Liu et al. found that 0.20% alginate oligosaccharides can enhance the drought resistance of tomatoes [35]. Zhang et al. found that leaf spraying alginate oligosaccharides could enhance the drought resistance of wheat seedlings [36]. Therefore, alginate oligosaccharides can be used as plant nutrients or developed as new seaweed fertilizers.

3 Agar Oligosaccharides

Agar is a kind of linear polysaccharide extracted from the cell wall of rhodocarpus such as *Gracilaria* and litholaria. Agarose is a mixture of two polysaccharides, agarose and thioagarose (Fig. 2), where the agarose molecule is made up of a repeating disaccharide unit of β-D-galactose and 3, 6-α-l-galactose inner ether, in addition to some sulfuric acid groups in the agarose structure [8]. The structure of agarose is very similar to that of agarose and is also composed of repeating disaccharide units, except that some of the 3, 6-α-L-galactolactone in the structure is replaced by sulfuric acid groups or methoxyl and acetone residues. The molecular weight of agarose is above 100 000 Da, and the content of sulfuric acid group is below 0.15%, while the molecular weight of sulfur agar gel is generally below 200 000 Da, and the content of sulfuric acid group is 5–8% [37].

Similar to alginate oligosaccharides, the preparation methods of agaric oligosaccharides also include chemical and enzymatic methods. Xu Qiang et al. used 0.05 M hydrochloric acid to hydrolyze agarose to prepare low-molecular-weight agarose oligosaccharides, but the conditions of acid hydrolysis were not easy to control, and the structure of the product was easily damaged [38]. Therefore, enzymatic preparation of agarose oligosaccharides has attracted the attention of researchers. Agarase can be divided into α-agarase and β-agarase according to the way it breaks its

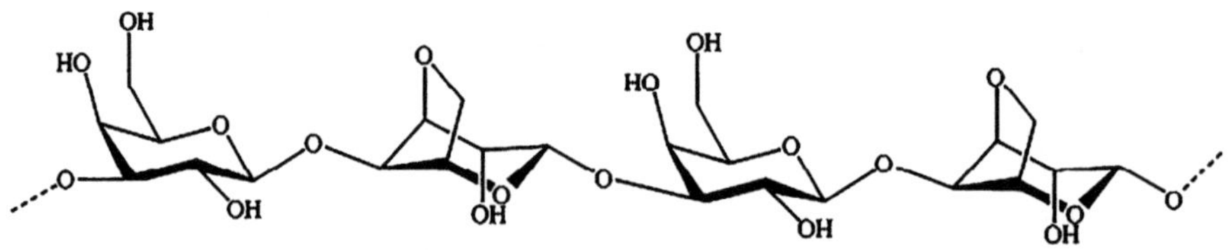

Fig. 2 The structure and composition of agar

substrate. The β-agarase can recognize the disaccharide repeating unit, cut the β-1, 4-glucoside bond, and release a series of new oligosaccharides with δ-galactose as the reducing end. However, α-agarase can hydrolyze the α-1, 3-glucoside bond in the repeating unit of disaccharides to produce the reducing agar glucose oligosaccharides with inner ether galactose [39]. Sun et al. used β-agarase derived from *Streptomyces coelicolor* A3 to prepare agar oligosaccharides and evaluated the in vivo and in vitro activity [40]. Xu et al. compared the distribution of polymerization degree of agarose oligosaccharides prepared by enzymatic method and acid method, and the results showed that the polymerization degree of agarose oligosaccharides prepared by enzymatic method was 4, 6, and 8, which was more specific than that of agarose oligosaccharides prepared by acid method [41]. As a hot spot in the development of marine functional food, the study on the activity of agaric oligosaccharides has attracted the attention of many researchers. Chiber et al. found that agarose oligosaccharides could be used for food preservation [42]. Li et al. found that agarose oligosaccharides prepared by enzymatic method can inhibit the growth of bacteria and promote the growth and reproduction of *Bifidobacterium* and *Lactobacillus*, so it has the potential to be developed as a new prebiotic [43]. Sutapa et al. found that agarose oligosaccharides can prevent the adhesion of influenza virus B to cells, thus significantly inhibiting the proliferation of influenza virus B [44]. In addition, agar oligosaccharides can also inhibit the growth of tumor cells by inhibiting the production of inflammatory factors prostaglandin PEG2 and TNF-α. Wang et al. found that the inhibition rate of agar oligosaccharides on sarcoma cells S180 was as high as 48.7%. Studies have shown that agar oligosaccharides can also regulate the immune function of human venous endothelial cells by promoting apoptosis, inhibiting angiogenesis, and inducing the expression of cytokines (such as TNF-X, IL-1β, IL-6, etc.) [45]. In addition, agar oligosaccharides have various biological activities such as antioxidant, anti-inflammatory, and promoting plant root growth. It is widely used in food, agriculture, environment, and other fields [46, 47].

4 Carrageenan Oligosaccharides

Carrageenan is the general name of a class of sulfureted linear polysaccharides extracted from the cell walls of red algae (*Gracilaria*, etc.). Its structure contains a disaccharide unit consisting of 3-linked β-D-galactopyranose and 4-linked α-D-galactopyranose, or a 4-linked 3, 6-galactopyranose ether (Fig. 3). According to the connection mode of monosaccharides and the number and location of sulfate groups, carrageenan can be divided into six different structures, namely, Kappa(κ), Iota(ι), Lambda(λ), Mu(μ), Nu(ν), and Theta(θ) carrageenan [48].

At present, the main methods for preparing carrageenan oligosaccharides are chemical, physical, and enzymatic methods. Li et al. found that H_2O_2 could degrade carrageenan into oligosaccharides with molecular weight of 1000 ~ 10,000 Da [49],

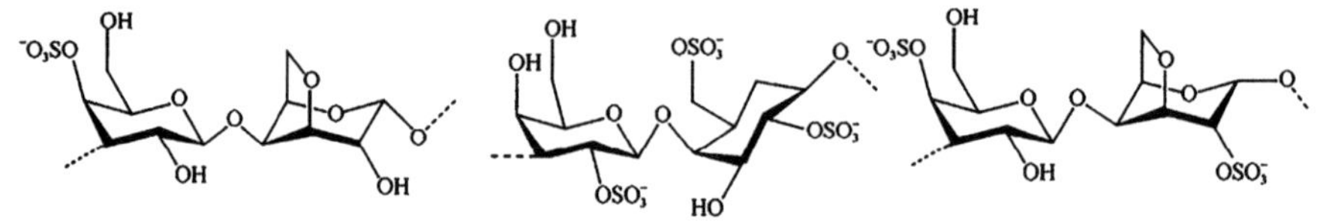

Fig. 3 The structure and composition of different kinds of carrageenan

methanol, hydrochloric acid, and sulfuric acid could be used to prepare carrageenan disaccharides, and Tony et al. obtained carrageenan oligosaccharides with different polymerization degrees by gently degrading carrageenan with acid method [50]. At the same time, Li et al. obtained higher molecular weight carrageenan oligosaccharides by microwave treatment of carrageenan [51]. However, because the conditions of chemical and microwave methods are difficult to control, the product distribution is complicated, and it is not easy to separate, so its application is greatly limited. Carrageenase is a glycoside hydrolase that degrades carrageenase with specific structure. According to the different substrates, carrageenase can be divided into carrageenase (EC3.2.1.83), carrageenase (EC3.2.1.157), and λ-type carrageenase. Among them, the carrageenases belong to the 16th and 82nd families of the glycoside hydrolase family [52]. Carrageenan is an endo-glycoside hydrolase, which can degrade carrageenan into a series of even number of carrageenan oligosaccharides. For example, Potin et al. prepared a series of carrageenan oligosaccharides with low-molecular-weight and different polymerization degrees by using carrageenase derived from *Cytophaga* sp.MCA-2 [53]. Duan et al. established an efficient preparation process of carrageenan oligosaccharides by using recombinant carrageenase and cellulase [54]. Yao et al. made use of *Cellulophaga lytica* strain N5-2 carrageenase to prepare disaccharides, hexasaccharides and octadecaccharides [55]. Due to the advantages of mild reaction conditions and high specificity of product distribution, enzymatic preparation of carrageenan oligosaccharides has attracted more and more attention. Due to the special structure of carrageenan oligosaccharides, their biological activity has also been widely concerned and studied. For example, Mou et al. found that carrageenan oligosaccharides can inhibit the growth of sarcoma cell S180 [56]. Zhou et al. found that carrageenan oligosaccharides with different molecular weights showed multiple inhibitory effects on sarcoma cell S180 [57]. Yuan et al. observed in S180-infected mice that carrageenan oligosaccharides could significantly inhibit the growth of transplantable sarcoma cells S180, promote phagocytosis of macrophages, promote the proliferation of splenic lymphocytes and the secretion of antibodies by splenic cells, promote the activation of NK cells, and increase the levels of IL-2 and TNF-α in serum [58]. In addition, carrageenan oligosaccharides also have antiviral activity and can significantly inhibit the proliferation of influenza virus (H_1N_1), which may be due to the fact that carrageenan oligosaccharides inhibit the adsorption process of the virus, thus blocking the proliferation of the virus in cells [59].

5 Fucoidan Oligosaccharides

Fucoidan, also known as fucoidan sulfate, is mainly derived from algae such as kelp, fucoidan and sea cucumber, sea urchin, and other marine animals. There is a large amount of sulfated fucose and other monosaccharides in the polysaccharide of rock algae, and its structure is very complex (Fig. 4). The oligosaccharide product obtained from the degradation of fucoidan is called fucoidan oligosaccharides. Fucoidan oligosaccharides from different sources differ greatly in structure [8]. For example, Qu et al. found that fucoidan oligosaccharides from *Fucus vesiculosus* mainly contain α-1, 3-L-fucose, while fucoidan oligosaccharides from other marine animals have a skeleton composed of fucose. It also contains other kinds of monosaccharides such as mannose, galactose, and uronic acid [60].

The preparation of fucoidan oligosaccharides mainly includes three methods: physical degradation, chemical hydrolysis, and enzymatic preparation. Chen et al. conducted ultrasonic degradation of fucoidan polysaccharide and found that the ultrasonic degradation rate of fucoidan solution increased with the increase of ultrasonic power and reaction temperature and the decrease of initial pH value of solution. In addition, under the test conditions, ultrasonic degradation did not change the main structure of polysaccharide [61]. Choi et al. prepared fucoidan oligosaccharides by gamma ray irradiation and found that when the sample was irradiated at an intensity of 10 kGy, the molecular weight of fucoidan oligosaccharides rapidly dropped to 38 kDa [62]. Natalie et al. used the purified new fucoidan polysaccharide enzyme to hydrolyze fucoidan polysaccharide. It was found that the polysaccharide-degrading enzyme could hydrolyze the polysaccharide without desulfurization, and no sulfate release was detected in the first 4 h of the polysaccharide-degrading enzyme action, and fucose was not detected until the later stage of hydrolysis [63]. Wang et al. used fucoidan polysaccharide-degrading enzyme to degrade fucoidan sulfate to obtain oligosaccharides and determined its antioxidant activity [64]. Fucoidan-degrading enzymes are called fucoidan-degrading enzymes, because they can break the glucoside bond between fucoidan sulfated in fucoidan molecules. However, the structure of fucoidan from different sources is too different, so it is impossible to classify fucoidan-degrading enzymes according to substrate types. However, according to the action characteristics and hydrolysis products, fucoidan-degrading enzymes can be divided into three categories: fucoidase (EC3.3.1.44), α-L-fucoidase (EC3.2.1.51), and sulfatase (EC3.1.6.-), in which fucoidase can act

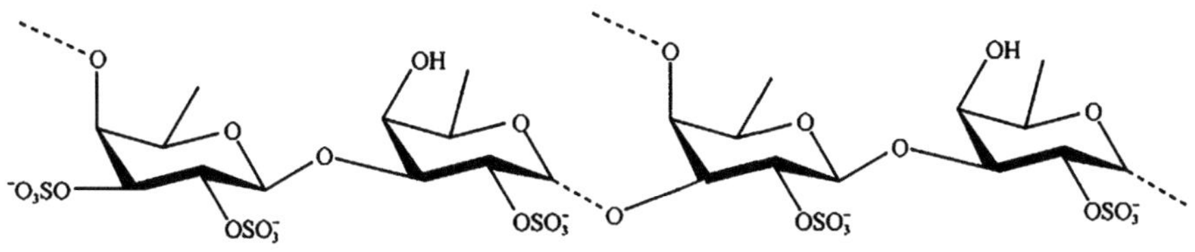

Fig. 4 The structure and composition of fucoidan

on the α-1, 3-, or β-1, 4-glucoside bonds within or at one end of fucoidase molecules and degrade them to oligosaccharides. Fucoidan oligosaccharides have a variety of biological activities and are becoming the new darling of the development of a new generation of natural functional health food, drugs, and cosmetics. In terms of medical and healthcare products, among the products developed and marketed with fucoidan as raw material at home and abroad, Japan's "Power Fucoidan Gold" has better antitumor effect, and "Best Fucoidan" produced by Doctor Best have strong antioxidant/immune-enhancing functions. In the field of cosmetics, the Ocean Legend series of moisturizing mask and moisturizing water developed by Beijing Leili and Marine Biological Company are mainly composed of fucoidan oligosaccharide functional active factors, and studies have shown that the biological activity of fucoidan oligosaccharide is related to its structure, sulfate content, and location. However, the structural differences, high molecular weight, and viscosity of the polysaccharide limit its application, especially as a therapeutic agent. The bioactive oligosaccharides with lower molecular weight will help to overcome these problems, and the polysaccharide-degrading enzyme of fucoidan can specifically degrade the polysaccharide of fucoidan and has the advantages of controllable enzymatic hydrolysis, mild reaction conditions, uniform product and no side reaction, etc. Therefore, the polysaccharide-degrading enzyme of fucoidan is the most ideal tool for the preparation of specific fucoidan oligosaccharides [8].

6 Ulvan Oligosaccharides

The chemical composition of Ulva polysaccharide is affected by the species of Ulva as extraction source, harvest season, extraction and purification methods of Ulva polysaccharide, etc., so its composition is relatively complex [65, 66]. Kidgella et al. [67] investigated the foliated Ulva (*U. australis*, *U. rigida*, and Ulva sp.) and *Ulva filamentum* (*U. flexuosa*, *U. compressa*, *U. prolifera*, and *U. ralfsii*) were analyzed. The results showed that there were significant differences in the composition of monosaccharides in the polysaccharides derived from leaf ulva and *Ulva filamentum* ulva. The polysaccharide from ulva leaf was mainly composed of rhamnose (about 49 mol%), followed by glucuronic acid (about 23.83 mol%), and the proportion of xylose to iduronic acid varied with the different species of ulva leaf. The polysaccharide derived from *Ulva filamucta* was significantly different. The content of rhamnose in the polysaccharide derived from *U. pellifer* and *U. flexuosa* could reach 56 mol% and 60 mol%, respectively. In contrast, polysaccharides derived from *U. compress* contain about 43 mol% rhamnose. The proportion of iduronic acid in the polysaccharide of *Ulva filamentum* (about 7 mol%) was generally lower than that of *Ulva foliata* (about 14 mol%). In addition, polysaccharides from *U. regid* alpha had the highest content of iduronic acid, reaching 18 mol%. The proportion of galactose in the polysaccharide derived from cultured *U. mutabilis* reached 16 mol%, which was much higher than that of other *Ulva* species. Samarasinghea et al. [68]

analyzed the monosaccharide composition of ulva polysaccharide harvested at different times and found that there was no difference in the main monosaccharides of ulva polysaccharide, but the content of monosaccharides of different kinds was significantly different. For example, the dry matter contents of rhamnose, xylose, galactose, glucose, and uronic acid in ulva polysaccharide harvested in June were 3.65, 0.43, 0.41, 0.32, and 0.62 g/100 g, respectively. The dry matter contents of rhamnose, xylose, galactose, glucose, and uronic acid in ulva polysaccharide collected in August were 0.84, 0.33, 0.22, 0.75, and 1.92 g/100 g, respectively. It was obvious that the content of glucose and uronic acid of ulva polysaccharide harvested in August was higher than that harvested in June. In addition, Olsson et al. [69] studied the effects of cultivation conditions (such as temperature, irradiance, pCO_2, nitrogen, and phosphate) on the monosaccharide composition of ulva and found that lower sulfate concentration and high temperature could promote the increase of monosaccharide content, while increasing irradiance and temperature could increase the concentrations of rhamnose and iduronic acid. Guidara et al. [70] extracted ulva polysaccharide from by acid extraction (CA) and enzyme chemical method (EE), respectively, and analyzed their monosaccharide components. The results showed that the contents of rhamnose and xylose in the ulva polysaccharide CA1 obtained by acid extraction were higher than those obtained by enzyme chemical method. The content of glucuronic acid and glucose in EE is higher than that in CA1. The above results indicate that the chemical composition of ulva polysaccharide varies with the species, growth environment, harvest time, and extraction methods, which is also a common phenomenon in the composition analysis of other algal polysaccharides [71].

Due to the complex composition of Ulva polysaccharide monosaccharides, different ways of linking between monosaccharides, complex and diverse group modifications, existence of branching structures, etc., the structure complexity of Ulva polysaccharide is far more complex than that of other algal polysaccharides such as alginate, carrageenan and agar [72, 73] (Fig. 5). Lahaye et al. [74] studied the structure of ulva polysaccharide derived from *U. regid*. We have a better understanding of the structure of ulvan. The water-soluble sulfated polysaccharide consists of a linear backbone with l-rhamnose-3-sulfate (Rha3S), d-glucuronic acid (GlcA), l-iduronic acid (IdoA), and d-xylose (Xyl), and the sulfate group is linked to the rhamnose [26]. The two major repeating disaccharide units of ulvan are →4)-β-d-glucuronic acid (1 → 4)-α-l-rhamnose-3-sulfate (1 → (A3S) and →4) α-l-iduronic acid (1 → 4)- α-l-rhamnose-3-sulfate(1 → (B3S). In addition, there are cases where a small amount of glucuronic acid or α-l-iduronic acid is replaced by xylose unit, which will form two units of →4)-β-d-xylose(1 → 4)-α-l-rhamnose-3-sulfate(1 → (U3S) and →4)-β-d-xylose-2-sulfate(1 → 4) -α-l-rhamnose-3-sulfate(1 → (U2′S,3S). Chi et al. [75] obtained three degradation products UO-1, UO-2, and UO-3 with different molecular weights through the cleavage of ulva polysaccharide derived from *Ulva linnaeus* via β-elimination reaction with ulva polysaccharide lyase and conducted structural analysis on them, respectively. And UO-1-2 of small molecular weight, respectively, is a disaccharide D-Δ GlcA- (1-4)- alpha/beta-L-Rha3S and 4D-sugar Δ GlcA-(1-4)-

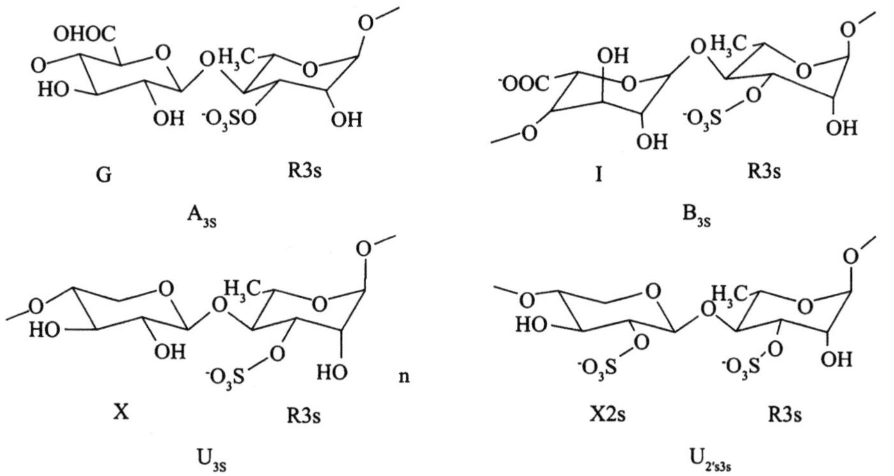

Fig. 5 The structure and composition of ulvan

alpha -L-Rha3S-(1-4)-beta-D-Xyl-(1-4)-alpha/beta-L-Rha3S. The study of UO-3 structure with high molecular weight found that it is mainly composed of A3s and U3s disaccharide repeating units, and there are U2′s and 3s disaccharide repeating units. It can be seen that ulva polysaccharide is a complex polysaccharide mainly composed of type A3s or B3s disaccharide repeating units and a small number of type U3s, U2′s, and 3s disaccharide repeating units. The contents of different disaccharide repeating units are affected by different sources of ulva. Therefore, the study on the structure of ulva polysaccharide is helpful to strengthen the high-value development and utilization of ulva polysaccharide resources.

According to the preparation method of ulva oligosaccharides, it can be classified into three categories: chemical degradation method, physical degradation method, and enzymatic method. The chemical degradation method of ulva oligosaccharides is mainly to prepare ulva oligosaccharides by destroying the glucoside bond in ulva polysaccharide with strong acidic or oxidizing chemical reagents. Feng et al. [76] extracted ulva polysaccharide in Beibei gulf of Guangxi, added 2 mol/L trifluoroacetic acid (TFA) 2 mL according to the ratio of solid to liquid 1:15, and hydrolyzed it at 110 °C for 2 h. Sun et al. [77] oxidized ulva polysaccharide with H_2O_2 and reacted at 60 °C for 2 h; the yield of oligosaccharides was 65.1%. In addition, Xu et al. [78] used sulfuric acid to degrade ulva polysaccharide and obtained the optimal technological conditions for the degradation of ulva polysaccharide: 0.3 mol/L sulfuric acid, degradation temperature 95 °C, polysaccharide concentration 7 mg/mL, degradation time 1 h, and solid-liquid ratio 1:30. The yield of ulva oligosaccharides was 50.94%. Relatively few studies have been conducted on the preparation of ulva oligosaccharides by physical degradation method. Only Li et al. [79] adopted a quadratic orthogonal rotation combination design to optimize ultrasonic-assisted enzyme extraction of ulva oligosaccharides, in which cellulase was only used to destroy the cell wall structure of ulva, and the optimal combination was as follows: The content of cellulase was 430 U/g, the ratio of solid to liquid was 1:30, the extraction temperature was 50°C, and the extraction time was 1.5 h. The yield of oligosaccharides was 2.38%. Yu et al. [80] prepared ulva oligosaccharides by microwave thermal degradation of ulva polysaccharide. In addition, Simona et al. [81] showed that ulva polysaccharide would self-hydrolyze in hot water solution at high temperature, and 78.7% of the rare sugar rhamnose, glucuronic acid, and other degradation products with no significant proportion could be recovered by controlling the optimal temperature for self-hydrolysis. The physical degradation method needs to provide a lot of energy to break the glucoside bond in ulva polysaccharide to prepare ulva oligosaccharides, which has the same problems as the chemical method, such as harsh reaction conditions and long processing time. Compared with the degradation of ulva polysaccharide by chemical and physical methods, enzymatic degradation of ulva polysaccharide has the advantages of mild reaction conditions and good product specificity, which has attracted wide attention from researchers. Currently, the main enzymes used for degrading ulva polysaccharide are the ulva polysaccharide lyases of PL24, PL25, PL28, and PL40 families, which are specific enzymes for degrading ulva polysaccharide. For instance, LOR_107 [82] derived from *Alteromonas* sp. LOR, AsPL [83] derived from *Alteromonas* sp. (AsPL), and *Pseudomonas* sp. strain PLSV [84], Uly1from *Catenovulum maritimum* [85] LOR_29 derived from *Alteromonas* sp. LOR [86], and FaUL from *Formosa agariphila* (KMM 3901T) KMM 3901 [87] and FaPL28derived from *Formosa agariphila* KMM 3901T [88] had the highest activity of ulva polysaccharide lyase between 30 and 45 °C. PLSV_3875 from *Pseudoalteromonas* sp. PLSV [82], ALT3695 from *Alteromonas* sp. A321.ALT3695 [89], and NLR42 [90] from *Persicivirga ulvanivorans* NLR42 exhibit the best activity at 50 °C. The optimal temperature of TsUly25B [91] derived from *Thalassomonas* sp. LD5 reaches 60 °C. In addition, the optimal pH value of ulva polysaccharide lyase is between 7.5 and 9, and it shows high catalytic activity in weakly alkaline environment, which may be the adaptive evolution of marine bacteria to weakly alkaline seawater environment [88]. Ulaganathan studied the structure and degradation mechanism of ulva polysaccharide from PL24, PL25, and PL28 families, respectively. The results showed that ulva polysaccharide lyase cleaved the β-(1 → 4)-glucoside bond between sulfated rhamnose (Rha3S) and mannuronate acid (GluA) or IduA (IduA) through β-elimination mechanism and produced oligosaccharides with 2 and 4 degrees of polymerization (Dp) [70–72]. This is the reason why almost all the degradation products of ulva polysaccharide by Ulva lyase are even oligosaccharides of DP2 and DP4. In addition, the β-elimination mechanism could preserve the characteristics of rare sugar structure in ulva polysaccharide to degrade ulva polysaccharide, which laid a foundation for the high-value development of ulva polysaccharide. In addition to ulva polysaccharide lyase, which can specifically degrade Ulva polysaccharide, there is Cdf79930. A broad-spectrum lyase [92] and the dewatering enzyme P29_PDnc [93] were involved in the degradation of ulva polysaccharide. It provides different ideas for enzymatic degradation of ulva polysaccharide.

At present, the relevant studies and reports on the activity of ulva oligosaccharides are increasing, but the literature reports are scattered, and due to the complexity of the structure of ulva oligosaccharides, the mechanism of the relevant activity and the structure-activity relationship of the oligosaccharides has not been determined. Chi et al. [94] evaluated the antiviral activity of ulva oligosaccharides with different molecular weights prepared by PL25 family ulva polysaccharide lyase against vesicular stomatitis virus (VSV), and the results showed that the degraded Ulva oligosaccharides with lower molecular weight of 100 μL/mL had similar antiviral activity to the undegraded ulva polysaccharide. In the case of similar activity to natural polysaccharides, low-molecular-weight polysaccharides are more conducive to the development of supplements and drugs. Li et al. [95] studied the anti-inflammatory bowel disease (IBD) effect of ulva oligosaccharides prepared by enzymatic method, and the results showed that 50 mg/kg ulva oligosaccharides began to show protective effect on IBD, and the concentration of 100 ~ 120 mg/kg ulva oligosaccharides had the best protective effect. In addition, ulva oligosaccharides can reduce the damage of sodium dextran sulfate (DSS) on colon epithelial tissue. Tabarsa et al. [96] found that ulva oligosaccharides with low-molecular-weight and high sulfate group composition could effectively prolifize RAW264.7 macrophages, indicating that oligosaccharides with low molecular weight had better bioavailability. Moreover, RAW264.7 cells were stimulated to release large amounts of nitric oxide, IL-1β, TNF-α, IL-6, IL-10, and IL-12 cytokines, which had mild immunomodulatory activities. Carvalho et al. [97] studied the cytotoxicity of ulva oligosaccharides on human cervical cancer cells and found that the existence of degraded and prepared ulva oligosaccharides could improve the anticancer activity of ulva polysaccharide. Berria et al. [98] IPEC-1 cell lines purified with *U. armoricana* and treated with low-molecular-weight (4.4 ku) ulva oligosaccharides (5–500 μg/mL) increased mRNA and cytokine protein expression (such as CCL20, IL-8, and TNF-α). Qi et al. [99] prepared ulva oligosaccharides with different molecular weights by H_2O_2 oxidation and found that the degraded Ulva oligosaccharides had higher antioxidant activity than the ulva polysaccharide with high molecular weight. Sun et al. [77] investigated the moisturizing activity of ulva oligosaccharides prepared by H_2O_2 oxidation and found that the degraded ulva oligosaccharides had good moisturizing activity as well. Yu et al. [80] studied the anti-high lipid activity of ulva oligosaccharides in male Wistar rat models and found that the increase of HDL cholesterol and decrease of TG in ulva oligosaccharide fed rats reached 2.0 times and 46.4%, respectively. This suggests that degraded ulva oligosaccharides are more beneficial to diabetes-related hyperlipidemia than undegraded ulva oligosaccharides. Paulert et al. [100] investigated the oxidative burst activity of dicotyledonous plants induced by ulva oligosaccharides prepared by acid hydrolysis. The results showed that ulva oligosaccharides with low molecular weight could affect plant growth like ulva polysaccharide, and its induced activity did not require acid sulfation. This discovery compensates for the difficulty of applying high-molecular-weight polysaccharides in conventional agricultural settings due to their solubility or high viscosity. Abouraichaa et al. [101] evaluated the effects of ulva polysaccharide and ulva oligosaccharides prepared by enzymatic degradation of ulva polysaccharide on defense-related reactions and decay of apples, and the results showed that ulva oligosaccharides could completely inhibit the growth of blue and cyan molds on fruits. In addition, compared with ulva polysaccharide, ulva oligosaccharides could better stimulate the immune regulatory system of apples, such as the activation of anti-oxidation-related enzymes, the enhancement of the activities of phenylalanine aminolyase (PAL), peroxidase (POD), and polyphenol oxidase (PPO). At present, studies on the activity of ulva oligosaccharides are relatively shallow, which is also due to the complexity of the structure of ulva polysaccharide, and there is no appropriate method to obtain the oligosaccharides determined by fine structure to study the structure-activity relationship of the oligosaccharides.

7 Conclusion and Perspective

With the development of marine biological resources, our research and understanding of marine oligosaccharides have become more and more in-depth. With the development of separation techniques and structural characterization methods, researchers have also built a marine characteristic oligosaccharide bank and conducted systematic and detailed studies on the structure, activity, and related structure-activity relationships of marine oligosaccharides, especially seaweed oligosaccharides. In addition, marine functional oligosaccharides have been widely used in food, agriculture, medicine, environmental protection, and other fields due to their significant biological activities such as antioxidant, antibacterial, anti-inflammatory, antitumor, antiviral, and anticoagulant. In particular, marine food and drug development based on marine algal oligosaccharides has gradually become a key development direction of marine biological resources utilization. Taking alginate oligosaccharides as an example, Taisho Pharmaceutical Co., LTD., an internationally renowned functional food manufacturer in Japan, added alginate oligosaccharides to beverages and developed a series of alginate oligosaccharides functional drinks named "marine angle," which can effectively inhibit the absorption of cholesterol. Qingdao Bozhihuili Biotechnology Co., Ltd. developed a functional food "Ocean Angel" alginate oligosaccharide tablet using alginate oligosaccharide. In addition, the development of new drugs is also an important development direction for the application of seaweed oligosaccharides; among which Ocean University of China, Shanghai Pharmaceutical Institute, and Shanghai Green Valley Pharmaceutical Co., Ltd. have developed mannuronic acid disaccharides from brown algae as a precursor, which can effectively inhibit mild to moderate Alzheimer's disease. China's original and the world's first new drug targeting brain-gut axis for the treatment of Alzheimer's disease improves patients' cognitive function. Moreover, carrageenan oligosaccharides derived from red algae have been proved to have unique antiviral activity and can effectively inhibit H1N1 influenza virus, which has great potential for drug development. However, there are still some technical bottlenecks in the development of seaweed functional oligosaccharides, which greatly limit the in-depth development and fine application of marine oligosaccharides, such as the specific mechanism of marine oligosaccharides which is still unclear, and the lack of efficient separation and preparation methods. Therefore, the future development direction is mainly focused on elucidating the mechanism of marine functional oligosaccharides at the molecular level and developing continuous and efficient preparation methods, so as to promote the development of marine biological resources and the utilization of marine oligosaccharides.

Acknowledgments Dr. Zhu Benwei gratefully acknowledges the support of the Excellent Young Backbone Teachers of "Blue Project" in Jiangsu Province.

Declaration of Competing Interest The authors declare that they have no known competing financial interests in this paper.

Data Availability Statement The data used to support the findings of this study are included within the article.

Funding This research was funded by the National Natural Science Foundation of China (Grant No. 32372268) and the China Postdoctoral Science Foundation (2023M743532).

References

1. Jiao G et al (2011) Chemical structures and bioactivities of sulfated polysaccharides from marine algae. Mar Drugs 9(2):196–223
2. Mensi F et al (2012) A statistical approach for optimization of R-phycoerythrin extraction from the red algae Gracilaria verrucosa by enzymatic hydrolysis using central composite design and desirability function. J Appl Phycol 24:915–926
3. de Araújo IWF et al (2011) Effects of a sulfated polysaccharide isolated from the red seaweed Solieria filiformis on models of nociception and inflammation. Carbohydr Polym 86(3):1207–1215
4. Hamid SS et al (2020) Effect of blanching on the concentration of metabolites in two parts of Undaria pinnatifida, Wakame (leaf) and Mekabu (sporophyll). Algal Res 47:101829
5. Nordgård CT, Rao SV, Draget KI (2019) The potential of marine oligosaccharides in pharmacy. Bioact Carbohydr Diet Fibre 18:100178
6. de Borba Gurpilhares D et al (2019) Marine prebiotics: polysaccharides and oligosaccharides obtained by using microbial enzymes. Food Chem 280:175–186
7. Jutur PP, Nesamma AA, Shaikh KM (2016) Algae-derived marine oligosaccharides and their biological applications. Front Mar Sci 3:83
8. Zhu B et al (2021) Marine oligosaccharides originated from seaweeds: source, preparation, structure, physiological activity and applications. Crit Rev Food Sci Nutr 61(1):60–74
9. Larsen B, Haug A (1958) Chemical composition of the brown alga Ascophyllum nodosum (L.) Le Jol.: presence of reducing compounds in ascophyllum nodosum. Nature 181(4617):1224–1224
10. Chandıa N, Matsuhiro B, Vásquez A (2001) Alginic acids in Lessonia trabeculata: characterization by formic acid hydrolysis and FT-IR spectroscopy. Carbohydr Polym 46(1):81–87
11. Niemelä K, Sjöström E (1985) Alkaline degradation of alginates to carboxylic acids. Carbohydr Res 144(2):241–249
12. Zhang H et al (1994) A method for preparation of saturated oligosaccharides from alginate. Chin J Mar Drugs

13. Hu Ting XM, Xiaolin Z et al (2014) Preparation of mannuronic acid oligosaccharides by microwave-assistant method and study of their in vitro antioxidant activities. Chin J Mar Drugs 33(03):27–33
14. Zhu B et al (2016) Enzymatic hydrolysis of alginate to produce oligosaccharides by a new purified endo-type alginate lyase. Mar Drugs 14(6):108
15. Zhu B, Yin H (2015) Alginate lyase: review of major sources and classification, properties, structure-function analysis and applications. Bioengineered 6:125–131
16. Zhu B et al (2019) Elucidation of degrading pattern and substrate recognition of a novel bifunctional alginate lyase from Flammeovirga sp. NJ-04 and its use for preparation alginate oligosaccharides. Biotechnol Biofuels 12:1–13
17. Zhu B et al (2017) Expression and characterization of a new heat-stable endo-type alginate lyase from deep-sea bacterium Flammeovirga sp. NJ-04. Extremophiles 21:1027–1036
18. Hu F et al (2019) Effects of module truncation on biochemical characteristics and products distribution of a new alginate lyase with two catalytic modules. Glycobiology 29(12):876–884
19. Zhu B et al (2018) Biochemical characterization and degradation pattern of a unique pH-stable polyM-specific alginate lyase from newly isolated Serratia marcescens NJ-07. Mar Drugs 16(4):129
20. Li Q et al (2019) Biochemical characterization and elucidation of action pattern of a novel polysaccharide lyase 6 family alginate lyase from marine bacterium Flammeovirga sp. NJ-04. Mar Drugs 17(6):323
21. Zhu B et al (2018) Characterization of a new endo-type alginate lyase from Vibrio sp. NJU-03. Int J Biol Macromol 108:1140–1147
22. Zhu B et al (2019) Preparation of trisaccharides from alginate by a novel alginate lyase Alg7A from marine bacterium Vibrio sp. W13. Int J Biol Macromol 139:879–885
23. Zhang Z et al (2004) Preparation and structure elucidation of alginate oligosaccharides degraded by alginate lyase from Vibro sp. 510. Carbohydr Res 339(8):1475–1481
24. Li L et al (2011) Preparation, purification and characterization of alginate oligosaccharides degraded by alginate lyase from Pseudomonas sp. HZJ 216. Carbohydr Res 346(6):794–800
25. Sun L et al (2005) A study of the antioxidant abilities of alginate oligosaccharides. Period Ocean Univ China (Nat Sci) 35(5):811–814
26. Xu X et al (2015) Morphological and proteomic analyses reveal that unsaturated guluronate oligosaccharide modulates multiple functional pathways in murine macrophage RAW264. 7 cells. Mar Drugs 13(4):1798–1818
27. Hu X et al (2004) Antitumour activities of alginate-derived oligosaccharides and their sulphated substitution derivatives. Eur J Phycol 39(1):67–71
28. Yamamoto Y et al (2007) Stimulation of multiple cytokine production in mice by alginate oligosaccharides following intraperitoneal administration. Carbohydr Res 342(8):1133–1137
29. Iwamoto M et al (2005) Structure–activity relationship of alginate oligosaccharides in the induction of cytokine production from RAW264. 7 cells. FEBS Lett 579(20):4423–4429
30. Miao L et al (2008) Effects of alginate derived oligosaccharide on the growth of bifidobacterium in vitro. Food Res Dev 29(11):16–19
31. Wang X et al (2019) Sodium oligomannate therapeutically remodels gut microbiota and suppresses gut bacterial amino acids-shaped neuroinflammation to inhibit Alzheimer's disease progression. Cell Res 29(10):787–803
32. Natsume M et al (1994) Isolation and characterization of alginate-derived oligosaccharides with root growth-promoting activities. Carbohydr Res 258:187–197
33. Ma L, Bu N, Ma C (2007) Effects of alginate-derived oligosaccharide on seed germination and some physiological characters of seedlings in garden pea (Pisum sativum L.). Plant Physiol Commun 43(6):1097
34. Liu R, Jiang X, Guan H (2009) Effects of alginate-derived oligosaccharide on the low temperature resistance of tobacco leaves. Period Ocean Univ China 39(2):243–248
35. Liu R et al (2009) Promotive effects of alginate-derived oligosaccharides on the inducing drought resistance of tomato. J Ocean Univ China 8:303–311

36. Zhang Y-H et al (2019) Effects of irrigating root with alginate oligosaccharides on rhizosphere soil properties and nutrient absorption and utilization in wheat
37. Hamer GK, Bhattacharjee SS, Yaphe W (1977) Analysis of the enzymic hydrolysis products of agarose by13C-nmr spectroscopy. Carbohydr Res 54(1):C7–C10
38. Xu Q, Xue C-h, Zhao X (2002) Preparation of agar oligosaccharides by acid hydrolysis and determination of their antioxidative effect. Chin J Mar Drugs 21(1):19–22
39. Fu XT, Kim SM (2010) Agarase: review of major sources, categories, purification method, enzyme characteristics and applications. Mar Drugs 8(1):200–218
40. Hong SJ et al (2017) In vitro and in vivo investigation for biological activities of neoagarooligosaccharides prepared by hydrolyzing agar with β-agarase. Biotechnol Bioprocess Eng 22:489–496
41. Xu X-Q et al (2018) Preparation of bioactive neoagaroligosaccharides through hydrolysis of Gracilaria lemaneiformis agar: a comparative study. Food Chem 240:330–337
42. Bo Q et al (2012) Effect of agar-oligosaccharide on quality of frozen shrimp and tilapia fillets. South China Fish Sci 8(6):73–79
43. Li J et al (2007) A simple method of preparing diverse neoagaro-oligosaccharides with β-agarase. Carbohydr Res 342(8):1030–1033
44. Mazumder S et al (2002) Isolation, chemical investigation and antiviral activity of polysaccharides from Gracilaria corticata (Gracilariaceae, Rhodophyta). Int J Biol Macromol 31(1–3):87–95
45. Wang J, Jiang X, Mu H (2007) Optimization of cultivation conditions of marine agarase-producing strain Vibrio sp. QJH-12. Marine Sciences-Qingdao-Chinese edition 31(7):14
46. Higashimura Y et al (2013) Oligosaccharides from agar inhibit murine intestinal inflammation through the induction of heme oxygenase-1 expression. J Gastroenterol 48:897–909
47. WANG X-j, CHEN H-m, YAN X-j (2011) Research on the Mechanism of Agaro-oligosaccharides Induced Reactive Oxygen Related Defense Response against Kidney Bean Phaseolus vulgaris L. Chin J Biol Control 27(2):254
48. Falshaw R, Furneaux RH (1994) Carrageenan from the tetrasporic stage of Gigartina decipiens (Gigartinaceae, Rhodophyta). Carbohydr Res 252:171–182
49. Li Guicun ZZ, Meiyu G (2002) Oxidative degradation of κ-Carrageenan. J Qingdao Inst Chem Technol 04:48–51
50. Malfait T, Van Cauwelaert F (1990) Preparative and analytical separation of oligosaccharides from κ-carrageenan. J Chromatogr A 504:369–380
51. Lii C-y et al (1999) Preliminary study on the degradation kinetics of agarose and carrageenans by ultrasound. Food Hydrocoll 13(6):477–481
52. Zhu B et al (2018) Insight into carrageenases: major review of sources, category, property, purification method, structure, and applications. Crit Rev Biotechnol 38(8):1261–1276
53. Potin P et al (1991) Purification and characterization of a new k-carrageenase from a marine Cytophaga-like bacterium. Eur J Biochem 201(1):241–247
54. Duan F et al (2016) An effective method for the preparation of carrageenan oligosaccharides directly from Eucheuma cottonii using cellulase and recombinant κ-carrageenase. Algal Res 15:93–99
55. Yao Z et al (2013) Characterization of a κ-carrageenase from marine Cellulophaga lytica strain N5-2 and analysis of its degradation products. Int J Mol Sci 14(12):24592–24602
56. Haijin M, Xiaolu J, Huashi G (2003) A κ-carrageenan derived oligosaccharide prepared by enzymatic degradation containing anti-tumor activity. J Appl Phycol 15:297–303
57. Zhou G et al (2004) In vivo antitumor and immunomodulation activities of different molecular weight lambda-carrageenans from Chondrus ocellatus. Pharmacol Res 50(1):47–53
58. Yuan H et al (2006) Immunomodulation and antitumor activity of κ-carrageenan oligosaccharides. Cancer Lett 243(2):228–234
59. Wang W et al (2011) In vitro inhibitory effect of carrageenan oligosaccharide on influenza A H1N1 virus. Antivir Res 92(2):237–246

60. Qu G et al (2014) Isolation and characterization of fucoidans from five brown algae and evaluation of their antioxidant activity. J Ocean Univ China 13:851–856
61. Chen Yajing WW, Suhua K et al (2012) Study on the degradation conditions of fucoidan from Sargassum henslowianum. Sci Technol Food Ind 33(20):284–287
62. Choi J-i, Kim H-J (2013) Preparation of low molecular weight fucoidan by gamma-irradiation and its anticancer activity. Carbohydr Polym 97(2):358–362
63. Thanassi NM, Nakada HI (1967) Enzymic degradation of fucoidan by enzymes from the hepatopancreas of abalone, Haliotus species. Arch Biochem Biophys 118(1):172–177
64. Wang Fengwu QF, Linyun K et al (2018) Enzymatic preparation and activity of fucosylated oligosaccharides. Food Sci 39(4):112–117
65. Shanmugam M et al (2001) Distribution of heparinoid-active sulphated polysaccharides in some Indian marine green algae. Indian J Geo-Mar Sci 30(4):222–227
66. Robic A et al (2009) Seasonal variability of physicochemical and rheological properties of ulvan in two Ulva species (Chlorophyta) from the Brittany coast 1. J Phycol 45(4):962–973
67. Kidgell JT et al (2021) Are all ulvans equal? A comparative assessment of the chemical and gelling properties of ulvan from blade and filamentous Ulva. Carbohydr Polym 264:118010
68. Samarasinghe MB et al (2021) A descriptive chemical analysis of seaweeds, Ulva sp., Saccharina latissima and Ascophyllum nodosum harvested from Danish and Icelandic waters. Anim Feed Sci Technol 278:115005
69. Olsson J et al (2020) Cultivation conditions affect the monosaccharide composition in Ulva fenestrata. J Appl Phycol 32(5):3255–3263
70. Guidara M et al (2020) Smart ulvan films responsive to stimuli of plasticizer and extraction condition in physico-chemical, optical, barrier and mechanical properties. Int J Biol Macromol 150:714–726
71. Benslima A et al (2021) The brown seaweed Cystoseira schiffneri as a source of sodium alginate: chemical and structural characterization, and antioxidant activities. Food Biosci 40:100873
72. Stender EG et al (2019) Structural and functional aspects of mannuronic acid–specific PL6 alginate lyase from the human gut microbe Bacteroides cellulosilyticus. J Biol Chem 294(47):17915–17930
73. Yang B et al (2011) Structural characterisation and bioactivities of hybrid carrageenan-like sulphated galactan from red alga Furcellaria lumbricalis. Food Chem 124(1):50–57
74. Lahaye M, Robic A (2007) Structure and functional properties of ulvan, a polysaccharide from green seaweeds. Biomacromolecules 8(6):1765–1774
75. Chi Y et al (2020) Structural characterization of ulvan extracted from Ulva clathrata assisted by an ulvan lyase. Carbohydr Polym 229:115497
76. Shan-guang FXCYW (2014) Analysis of monosaccharide compositions in Ulva lactuca Lpolysaccharides. Sci Technol Food Ind 35(7):91–94
77. Xifeng S, Xiaolei L (2014) Studies on the moisturising activity of Ulva porea polysaccharides with different molecular weights. World Biotechnol 12:155–156
78. Yannan X, Feng L, Yanmei L (2019) Optimisation of the partial degradation process of Ulva polysaccharides. J Fish Res 41(1):42
79. Yanmei L et al (2018) Optimisation of the key process of ultrasound-assisted enzymatic extraction of functional oligosaccharides from Ulva sp. J Fish Res 40(6):458
80. Pengzhan Y et al (2003) Antihyperlipidemic effects of different molecular weight sulfated polysaccharides from Ulva pertusa (Chlorophyta). Pharmacol Res 48(6):543–549
81. Podolean I et al (2022) Catalytic transformation of the marine polysaccharide ulvan into rare sugars, tartaric and succinic acids. Catal Today 383:345–357
82. Kopel M et al (2016) New family of ulvan lyases identified in three isolates from the Alteromonadales order. J Biol Chem 291(11):5871–5878
83. Qin H-M et al (2020) Biochemical characterization and structural analysis of ulvan lyase from marine Alteromonas sp. reveals the basis for its salt tolerance. Int J Biol Macromol 147:1309–1317

84. Qin H-M et al (2018) Biochemical characterization of a novel ulvan lyase from Pseudoalteromonas sp. strain PLSV. RSC Adv 8(5):2610–2615
85. Xu F et al (2021) Mechanistic insights into substrate recognition and catalysis of a new ulvan lyase of polysaccharide lyase family 24. Appl Environ Microbiol 87(12):e00412–e00421
86. Foran E et al (2017) Functional characterization of a novel "ulvan utilization loci" found in Alteromonas sp. LOR genome. Algal Res 25:39–46
87. Konasani VR et al (2018) Ulvan lyase from Formosa agariphila and its applicability in depolymerisation of ulvan extracted from three different Ulva species. Algal Res 36:106–114
88. Reisky L et al (2018) Biochemical characterization of an ulvan lyase from the marine flavobacterium Formosa agariphila KMM 3901 T. Appl Microbiol Biotechnol 102:6987–6996
89. Gao J et al (2019) Cloning, expression, and characterization of a new PL25 family ulvan lyase from marine bacterium Alteromonas sp. A321. Mar Drugs 17(10):568
90. Collén PN et al (2011) Ulvan lyases isolated from the flavobacteria Persicivirga ulvanivorans are the first members of a new polysaccharide lyase family. J Biol Chem 286(49):42063–42071
91. Wang D et al (2022) Biochemical properties of a new polysaccharide lyase family 25 ulvan lyase TsUly25B from marine bacterium Thalassomonas sp. LD5. Mar Drugs 20(3):168
92. Konasani VR et al (2018) A novel ulvan lyase family with broad-spectrum activity from the ulvan utilisation loci of Formosa agariphila KMM 3901. Sci Rep 8(1):14713
93. Bäumgen M et al (2021) A new carbohydrate-active oligosaccharide dehydratase is involved in the degradation of ulvan. J Biol Chem 297(4):101210
94. Chi Y et al (2020) Ulvan lyase assisted structural characterization of ulvan from Ulva pertusa and its antiviral activity against vesicular stomatitis virus. Int J Biol Macromol 157:75–82
95. Li Y et al (2020) Characterization of low molecular weight sulfate Ulva polysaccharide and its protective effect against IBD in mice. Mar Drugs 18:499
96. Tabarsa M et al (2018) Water-soluble polysaccharides from Ulva intestinalis: molecular properties, structural elucidation and immunomodulatory activities. J Food Drug Anal 26(2):599–608
97. de Carvalho MM et al (2020) Conformational analysis of ulvans from Ulva fasciata and their anticoagulant polycarboxylic derivatives. Int J Biol Macromol 162:599–608
98. Berri M et al (2017) Ulvan from Ulva armoricana (Chlorophyta) activates the PI3K/Akt signalling pathway via TLR4 to induce intestinal cytokine production. Algal Res 28:39–47
99. Qi H et al (2005) Antioxidant activity of different molecular weight sulfated polysaccharides from Ulva pertusa Kjellm (Chlorophyta). J Appl Phycol 17:527–534
100. Paulert R et al (2021) The non-sulfated ulvanobiuronic acid of ulvans is the smallest active unit able to induce an oxidative burst in dicot cells. Carbohydr Polym 270:118338
101. Abouraïcha E et al (2015) Induction of natural defense and protection against Penicillium expansum and Botrytis cinerea in apple fruit in response to bioelicitors isolated from green algae. Sci Hortic 181:121–128

Tools for Mining, Expression, and Modification of Marine Polysaccharide-Degrading Enzymes

Zhengming Zhu, Yanxia Wang, Liying Zhu, and Ling Jiang

1 Exploring the Gene Resources of Marine Polysaccharide-Degrading Enzymes

A large number of marine microorganisms live in the marine environment, which accounts for about 71% of the surface area of earth. Thriving in extreme environments, these microorganisms develop distinctive genetic profiles, specialized physiological mechanisms, and unique metabolites, rendering them indispensable biological resources. Development and utilization of marine microbial enzymes represent a pivotal technology for realizing the high value of marine biological resources. The marine ecological environment is complex, characterized by high salinity, high pressure, low temperature, and special light, which may cause extremophiles to produce enzyme families with unique physiological activities. Polysaccharide-degrading enzymes derived from marine microorganisms are an important branch of novel active enzymes. Their adaptability to extreme environment and multifunctional catalytic activities endows them with significant potential for application in industrial production. Meanwhile, they can also be used as crucial tools for the degradation of marine bio-polysaccharides to obtain new biologically active oligosaccharide components, which can enhance the development of active marine polysaccharides. However, mining of polysaccharide-degrading enzymes from marine extremophiles is currently confronted with numerous challenges.

Owing to the unique marine habitat, the growth of marine microorganisms is relatively slow, accompanied by a lower metabolic activity, which has significantly

Z. Zhu · Y. Wang · L. Jiang (✉)
College of Food Science and Light Industry, State Key Laboratory of Materials-Oriented Chemical Engineering, Nanjing Tech University, Nanjing, China
e-mail: jiangling@njtech.edu.cn

L. Zhu
College of Chemical and Molecular Engineering, Nanjing Tech University, Nanjing, China

X. Mao (ed.), *Degrading Enzymes for Marine Polysaccharides and Their Applications*, https://doi.org/10.1007/978-981-96-9517-1_3

influenced the development and application of marine microorganisms. The challenges in the development and utilization of marine biological resources are primarily centered around the intricate processes of isolating, cultivating, and screening marine microorganisms. The main difficulty lies in replicating the natural environment of deep-sea regions with precision in terms of pressure, salinity, temperature, and nutrient conditions. In the cultivation of marine microorganisms, the commonly employed medium is formulated based on the growth traits of terrestrial microorganisms, posing a challenge in fully adapting to the nutritional requirements of marine microorganisms. Currently, the available pure culture strains of marine microorganisms are limited to less than 1%. Therefore, obtaining purely cultured strains holds great significance in basic research and the exploration of bioactive compounds.

1.1 Exploring the Gene Resources from Cultured Microorganisms

1.1.1 General Screening Strategies

Microbial plate and microplate culture methods, two most commonly used traditional screening methods, can be widely employed for isolating advantageous strains from deep-sea environmental samples. Agar plate activity screening based on transparent circles and color circles or agar plate growth selection based on resistance can be served as a straightforward preliminary screening method, facilitating the exclusion of numerous inactive or minimally active enzyme-producing strains. Nevertheless, the agar plate screening method is not universally applicable to all screening targets. Owing to the challenges in precise quantification, a microtiter-plate (MTP) screening method that accurately detects the targets based on fluorescence or absorbance has emerged and been widely used in screening of enzyme-producing strains.

The agar plate method serves as a straightforward and effective screening method which has been widely used in the initial assessment of enzyme-producing strains encompassing diverse hydrolytic enzymes (e.g., lipase, esterase, protease) and oxidoreductase (e.g., laccase). The plate screening method is employed for the identification of enzyme-producing strains, assessing the extent of substrate hydrolysis within the solid medium by the enzyme expressed in the cell or the diameter of the halo resulting from variations in enzyme activity. The agar plate method primarily engages in phenotypic activity screening and phenotypic growth selection. Activity screening involves measuring the hydrolysis circle, color circle, or fluorescent product generated around the colony. Simultaneously, growth selection is applied by evaluating the resistance of cells to antibiotics or other harmful substances based on the growth profiles in the selection medium. Presently, the plate screening method finds extensive use in the screening of high-performance enzyme-producing strains. For instance, lipase decomposes triglyceride substrates, forming a transparent circle around the colony. Cellulase decomposes cellulose into glucan, forming a red halo with Congo red. Phospholipase decomposes lecithin, forming a white precipitation ring. In addition, the plate screening method can be employed to enhance the resistance of specific enzymes to antibiotics, facilitating the investigation of bacterial resistance mechanisms. For example, DNA shuffling technology is utilized to construct a cephalosporin mutation library, and LB plates with 34 μg/ml laoxycephalosporin are employed for screening resistance genes [1]. However, a notable issue with the plate screening method is its accuracy. The size of the halo is not solely linked to enzyme activity but is closely associated with the expression level of the enzyme, rendering the agar plate method less effective in visualizing the differences between strains, resulting in the difficulty of accurate quantification of the enzyme activity. Moreover, the method's capability to screen enzyme activity is limited, impeding various applications. Finally, it is suitable for the preliminary screening of enzyme-producing strains. Consequently, for the screened strains, additional detection methods such as microplate detection are still needed for accurate quantification. Interestingly, the application of digital imaging spectrophotometer in the agar plate screening improves the sensitivity of the method, achieving a throughput of 10^5 CFU/d [2]. If the agar plate screening method can be established according to the characteristics of the target enzyme, there is no necessity for resorting to an ultrahigh-throughput screening method that relies on intricate equipment.

Compared to the plate screening method, the microplate screening method demonstrates enhanced accuracy and throughput [3]. Typically, 96, 384, and 1536-well plates are employed for detecting and screening individual enzyme-producing strains in the library. In the microplate, changes in the absorbance or fluorescence resulting from the substrate or target product are quantitatively analyzed to ensure screening accuracy and sensitivity. This method presently stands as the most widely used screening approach. Usually, bacterial species are stored in the original plate, with a single strain cultivated in the first plate, subsequent evaluation of enzyme activity conducted in the second plate. The reaction volume typically ranges from 100 to 200 μL, resulting in a reduced solution requirement. Currently, the 96-well plates are the most prevalent in microplate applications. Additionally, to increase screening throughput and save screening time, the 384, 1536, and 3456-well plates have been used for enzyme screening, offering savings in the consumption of expensive substrates. In summary, microplate screening has significant advantages over the plate screening, enabling accurate quantification of the enzyme activity and facilitating enzyme reactions with greater complexity. Due to its relatively fixed experimental process, mechanized operations such as single colony picking, pipetting, shaking incubation, centrifugal collection, liquid addition reaction, microplate reader detection, etc., are iteratively performed. This lends itself to relatively straightforward automation. A full set of screening workstations can thereby contribute to a reduction in labor costs.

1.1.2 Strategies to Improve the Cultivability of Microorganisms

Due to the changes in living conditions, the conventional culture of marine extremophiles may lead to the failure of some to adapt and result in death, while others may enter a dormant state through the production of spores. They enter a living environment that maintains a certain metabolic activity but does not grow and reproduce, which also named viable but nonculturable state (VBNC). The noncultivability of microorganisms encompasses several dimensions. Firstly, the laboratory setting often falls short of fully simulating the conditions prevalent in extreme marine environments. Secondly, the concentration of artificial nutrient substrate may be excessively high, or the oxidative pressure may impede microbial growth. Thirdly, the intricate interrelationships between microorganisms are frequently overlooked. Lastly, it is challenging to observe the growth performances of slow-growing microorganisms with the existing standards. Therefore, there is a pressing need to enhance traditional culture methods and develop new culture techniques to improve the cultivability of microorganisms [4].

Conventional methods for culturing microorganisms primarily involve optimizing both the culture medium and conditions. However, distinct deep-sea environments exhibit diverse characteristics. For example, the seabed area is characterized by oligotrophic, the hydrothermal area exhibits high concentrations of sulfur and heavy metals, while the cold seep areas reveal elevated levels of methane and sulfate. During separation process, it is necessary to optimize the concentration and composition of the nutrient matrix in accordance with the specific environmental traits. Additionally, the dilution culture method is frequently employed to forestall the dominance of a few microorganisms, thus inhibiting the growth of other species. In this method, diluting the microbial population in seawater to trace amounts prevents oligotrophic microorganisms from succumbing to the competitive pressures exerted by dominant species, significantly enhancing the likelihood of cultivating the primary oligotrophic microorganisms [5]. Moreover, the majority of the deep-sea environments maintain a consistently low temperature (1 ~ 4 °C), with only the hydrothermal vents displaying elevated temperature. Therefore, it becomes essential to simulate the in situ temperature of the sample during the separation and cultivation of deep-sea microorganisms, which is a technically feasible task. Pressure represents another distinctive environmental feature of the deep sea. All microorganisms inhabiting these regions possess pressure-tolerant properties, including a small number of piezotrophic bacteria. Typically, a special pressure simulation device is required for the cultivation of these piezotrophic bacteria [6].

Currently, in addition to the abovementioned microbial culture method aimed at enhancing cultivability, the interaction among microorganisms in the environment is also a very critical factor. As a harmonious community, the metabolites produced by different microorganisms have a non-negligible effect on the growth of other species within the community. In a previous study, an in situ culture technique, also known as a diffusion chamber, was employed to trick cells by simulating the natural environment to make them feel that they were growing in the natural environment. Epstein et al. who designed a diffusion chamber facilitating the passage of substances from the natural environment (intertidal marine sediments) through the membrane successfully isolated some previously uncultivated bacteria. These bacteria were subsequently cultured in a solid medium but demonstrated growth only in the presence of other bacteria [7]. To date, the in situ culture technology has been successful applied to obtain numerous unique uncultivated bacteria. Nevertheless, challenges persist in isolating specific bacterial strains of interest from these populations to achieve pure cultures. The high-throughput separation culture technology developed in recent years provides a viable alternative strategy [8]. Firstly, agarose (or other embedding matrix) microcapsules embedding a single microbial cell were prepared. Then, the microbes embedded in the microcapsules were cultured in slow-flowing seawater. After completing the culture, the flow cytometer was used to sort microcapsules containing microcolony. Finally, the sorted microcolony was expanded and cultured. This strategy strives to closely simulate the natural environment, allowing all microorganisms to be sustained in an open and continuous culture environment. Consequently, metabolites and signal molecules can be exchanged without disrupting the ecological connection between species. Meanwhile, individual species are cultured separately, avoiding nutritional competition during mixed culture, and enabling the acquisition of a substantial number of microorganisms through pure culture.

Recently, a promising strategy known as culturomics has been widely used to obtain different microorganisms from environmental samples. Culturomics, referred to high-throughput bacterial isolation and culture, serves the purpose of isolating and identifying bacteria. This method employs various culture techniques in conjunction with matrix-assisted laser desorption ionization time-of-flight mass spectrometry (MALDI-TOF MS) and 16SrRNA gene sequencing to maximize the diversity of microorganisms from samples, especially those that are challenging to cultivate [9]. The critical aspect of this strategy lies in the design of the culture method. The difficult-to-cultivable microorganisms were acquired depending on the appropriate culture strategy. The selection of nutrients, environment, temperature, and cultivation time is the most basic aspect of cultivation conditions. Additionally, the adoption of the culture strategy often aligns with the results of genomics analysis. And the metabolic characteristics of the strain could be predicted to identify additives that inhibit or support its growth. On the other hand, ensuring the rapid cultivation of collected samples is imperative. Besides selecting appropriate culture strategies, swiftly identifying phenotypes is equally critical. With the development of sequencing technology, 16S rRNA gene sequencing has become the gold standard for bacterial identification, especially when no spectrum is available. However, molecular identification of bacteria is time-consuming and expensive, limiting its application in high-throughput culturomics. Contrastingly, based on the MALDI-TOF MS technology, the specific protein mass spectrum of the bacteria is compared with the reference library to identify the species, emerging as an identification method with high accuracy. Additionally, the bacteria can be identified rapidly, especially the price is 100 times cheaper than 16S rRNA technology, making MALDI-TOF MS technology to be an effective method for bacterial identification in culturomics [10]. Previously, the culturomics technology found extensive application in studying human gut microbes. Over the past 5 years, 531 previously uncultivated microorganisms had been isolated from the intestines through a variety of culture method, including hundreds of new bacteria, resulting in a doubling of the total bacterial count [11]. Additionally, it is believed that this technology can significantly contribute to the mining of marine extremophile resources.

1.2 Exploring the Gene Resources from Uncultured Microorganisms

To date, numerous marine extremophiles have been successfully acquired in the laboratory through the application of cultivable strategies. This significantly contributes to advancing the understanding of microbial life activities and the exploration of microbial resources. However, basing on these cultivable microorganisms to explore new marine polysaccharide-degrading enzyme resources is often time-consuming. This is due to the necessity of identifying and obtaining the nature of relevant enzymes and their DNA sequences. Moreover, in many cases, it is sufficient to obtain the DNA sequences encoded by the enzymes, as these sequences can be overexpressed in industrial microbial cells (*Escherichia coli*, *Bacillus subtilis*, *Yeast*, *Aspergillus*, etc.). Although various cultivable strategies have been developed, challenges persist in cultivating certain microorganisms and obtaining target sequences from these species. Fortunately, the evolution of metagenomics and high-throughput sequencing technologies has provided an alternative, enabling the study of coding genes without cultivating microorganisms. While the direct cultivation of marine extremophiles remains challenging, metagenomics technology allows for the direct extraction of the total genome from their living environment. Metagenomics technology plays a pivotal role in deep-sea microbial research and resource development, including investigations into microbial diversity, population structure, evolutionary relationships, functional activities, and mutual cooperation.

1.2.1 Metagenomics Technology

Generally, the metagenome contains genes from cultivable, non-cultivable, and uncultivated microorganisms, offering the advantage of circumventing the isolation and cultivation of microorganisms, which significantly broadens the scope for the utilization of microbial resources. The typical process for metagenomics technology involves extracting metagenomic DNA from environmental samples, cloning the DNA into an appropriate vector, introducing host bacteria, and screening target transformants (Fig. 1).

The extraction of metagenomic DNA from environmental samples is a crucial initial step, providing a high-quality foundation for subsequent research. Various kits are now available for the effective extraction of metagenomic DNA. It is worth

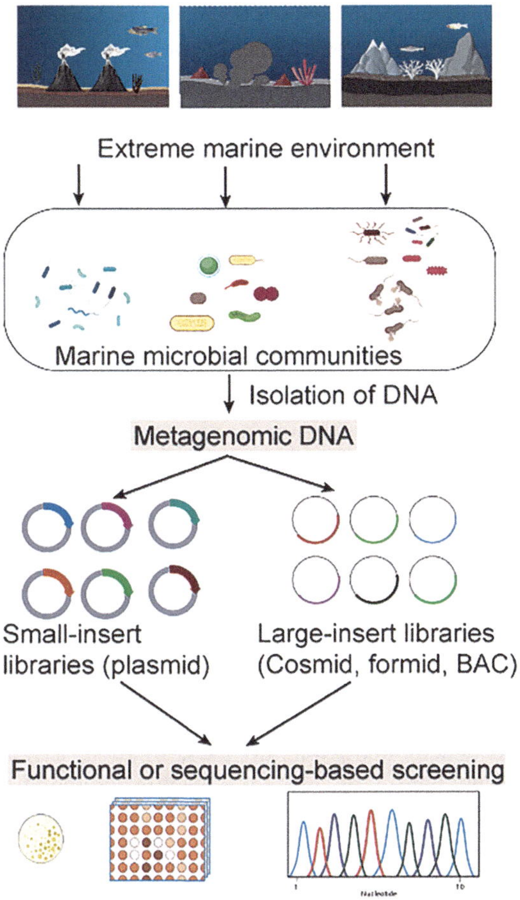

Fig. 1 Flow chart of screening new enzymes based on metagenomics technology

noting that the number of microorganisms in deep-sea sediment samples, for the same weight or volume, is considerably lower than that in land and offshore sediments [12]. Consequently, there remains a need to refine the extraction method based on specific research objectives.

Vector selection plays a pivotal role in metagenomic studies, particularly since the inception of the environmental DNA library for marine microplankton. Subsequently, a plethora of metagenomic libraries has been developed from marine environmental samples [13], including Cosmid, Fosmid, BAC, K2 phage, and various shuttle vectors. Among these, Cosmid, Fosmid, and BAC are prevalent choices, with Cosmid and Fosmid being particularly prominent in the investigation of active substances within deep-sea microorganisms. Notably, the insert fragments of Cosmid and Fosmid vectors are relatively small (35–40 kb), making them suitable for research on active substances controlled by single genes, such as extreme enzymes. Furthermore, the Fosmid vector demonstrates higher cloning efficiency and stability for inserted fragments in *E. coli* [14]. As a result, the Fosmid vector has emerged as the prevailing choice among the vectors in current usage.

Screening of target transformants in the metagenomic library has become a bottleneck during new enzymes exploration due to the substantial number of samples. The current screening methods based on metagenomic library can be categorized into two main types, namely, sequence-driven screening and function-driven screening. High-throughput sequencing technology is predominantly utilized for sequencing the obtained metagenomics and conducting sequence analysis. The potential gene sequence of the target enzyme in the metagenome is then identified through homology alignment of the protein sequence. Subsequently, some new protein sequences with unique properties can be identified by synthesizing these genes, expressing the corresponding protein in host cells, and studying its properties. While this strategy ensures high accuracy and efficiency, it faces challenges in identifying genes with entirely novel sequences and functions based on existing protein sequence information [15]. In contrast, functional screening methods involve the direct construction of metagenomic gene fragments into vectors and expression hosts for protein expression. High-throughput screening techniques are then applied to acquire enzyme resources with new functions, allowing the discovery of resources with brand new sequences and functions without relying on currently known protein sequence information [16]. However, heterologous protein expression usually exhibits low efficiency, resulting in a low positive rate in metagenomics and the potential oversight of unknown genes. Therefore, to effectively explore new enzyme resources in the metagenomic library and address the limitations of conventional sequence prediction methods, there is a necessary to develop efficient, high-throughput, low-cost, and versatile screening methods for functional screening of metagenomics.

1.2.2 Single-Cell Genome Technology for Uncultivated Microorganisms

Metagenomics technology has emerged as a potent tool for researching and developing genetic resources in deep-sea extreme ecosystems. However, the target gene is typically obtained from the mixed DNA of the microbial community, lacking association with a specific individual, and precluding the acquisition of its genome sequence. Consequently, this limits the information available for the exploration and utilization of gene resources. Interestingly, single-cell genome analysis technology introduces a novel strategy for analyzing the background of metagenomic information and the physiological functions of microorganisms. The fundamental process involves single-cell sorting, DNA extraction, multiple displacement amplification, and genome analysis. Among these processes, single-cell sorting and multiple displacement amplification are the most critical steps. Various methods, including serial dilution, flow sorting, micromanipulation, laser-tweezers-Raman-spectroscopy (LTRS), and microfluidic chip technology, can be employed to capture single cells in environmental samples [17]. The amplification of the captured single-cell genome is commonly achieved through multiple displacement amplification

(MDA). Optical trapping and Raman spectroscopy, combined to be LTRS, presents a nondestructive, highly sensitive and automated technique, which has been widely applied in various single-cell studies. The microfluidic chip developed by Quake's group stands out as the most complex but efficient technology [18]. It enables the single-cell sorting and subsequent MDA collection to be completed on one chip, and real-time monitoring to be achieved through microscopic observation and real-time PCR. Currently, these two technologies are gradually being applied in the research of marine extremophiles.

1.2.3 Exploring Proteins Through Synthetic Biology Strategies

At present, the scalable mining of protein resources predominantly relies on physical samples, while synthetic biology strategies provide innovative ideas for scalable protein researches [19]. In terms of the enzyme catalytic system, the "bottom-up" strategy of synthetic biology enables the identification of amino acid sequences in the database. Subsequently, the DNA sequence of the protein and regulatory elements can be designed and synthesized based on the adaptability of host cell. Concurrently, standardized DNA assembly methods facilitate the construction of protein expression modules. Finally, these elements can be transformed, expressed, and functionally characterized in host cells. Interestingly, this approach is not constrained by biological entities, and the development efficiency of novel elements can be enhanced through synthetic biology strategies. Furthermore, DNA synthesis can be modified by sequence engineering to enhance the adaptability of protein elements to the host. Researches centered on standardized hosts, processes, and experimental conditions contribute to the accumulation of high-quality components and data, thereby laying the groundwork for protein component mining guided by artificial intelligence.

Accurately predicting the key properties of proteins, such as catalytic activity, substrate selectivity, and soluble expression based on protein sequence information is challenging. To improve the efficiency of functional screening, the new algorithms need to be developed to prioritize candidate elements. Meanwhile, the functional diversity of similar proteins can be explored based on minimal experimental resources. Firstly, various protein resource databases provide valuable resources for high-throughput mining of protein elements. As of December 2021, approximately 230 million protein sequences were stored in the UniProt database. Unfortunately, at least 50% of those comments were inaccurate or even wrong [20]. Therefore, it is necessary to integrate different levels of functional annotation tools and experimental data to comprehensively analyze candidate protein elements from multiple levels such as sequence, structure, evolution, and protein interaction. Among them, based on visualization methods such as sequence similarity network (SSN) [21] and CATH (Class, Architecture, Topology, Homologous superfamily) [22] analysis tools, the functional annotation of unknown proteins can be conducted from multiple dimensions.

There are still a large number of candidate sequences after preliminary screening of various protein databases through cluster analysis, which is far beyond the ability of experimental verification. Therefore, to improve the efficiency of functional verification experiments, it is necessary to explore prioritization criteria and algorithms to rank candidate enzymes. Among these criteria, soluble protein expression and substrate promiscuity are important for prioritization [23]. Soluble expression is a prerequisite for the functional characterization of enzymes, which can be predicted by some algorithms based on energy calculations, machine learning, and evolutionary analysis, currently. In addition, it is necessary to select enzymes with different substrate heterogeneity according to the research objectives. At present, researchers have developed a variety of algorithms to predict the promiscuity of enzyme substrates, such as BNICE (predicting new chemical structures and reactions based on enzymatic classification) [24], SVM (predicting novel substrates for enzymes based on physicochemical properties, monopeptide/dipeptide distribution, molecular weight, isoelectric point, and amino acid sequence information) [25], and BioGPS (predicting the promiscuity and structure based on protein three-dimensional structure information) [26].

As candidate proteins from different species are mined, the codons, regulatory elements, cell culture conditions, etc. need to be systematically optimized to improve the suitability of proteins with heterologous cells [27]. To this end, various high-throughput technologies including high-throughput molecular cloning, high-throughput protein expression, and high-throughput characterization of enzyme catalytic functions have been developed for the scalable characterization of proteins and the combinatorial optimization of experimental protocols. Mainly based on these technologies, the developed algorithms, databases, and high-throughput experimental technologies have effectively promoted the high-throughput mining of proteins.

1.3 Integration of Gene Resources Databases

Microorganisms serve as the primary source of enzymes that degrade marine polysaccharide. The databases are crucial in modern biological research. For example, the NCBI database is one of the most commonly used biological databases all over the world. With the substantial reduction in the cost of high-throughput sequencing and gene synthesis, and the development of bioinformatics tools, it is no longer necessary to rely on live species to identify new enzymes. Within these databases, ideal protein sequences can be identified from whole genome or metagenomic sequences, enabling the synthesis of target genes and the establishment of recombinant expression systems. Metagenomics and enzyme resource databases, providing comprehensive enzymatic information, have become an indispensable part of new enzymes discovery and application. These databases not only offer a standard reference for enzyme classification but also facilitate the exploration, modification, and de novo design of new catalytic elements.

1.3.1 Metagenomic Databases of Marine Microorganisms

Currently, two major projects, the Global Ocean Sampling (GOS) and Census of Marine Life (CoML), have been implemented for the global marine microbial survey. The GOS focuses on exploring the genomes of marine organisms [28]. It aimed to assess the genetic diversity of marine microbial communities and understand their role in natural processes. Since 2003, the oceans around the world had been sampled by researchers from the J. Craig Venter Institute, who aimed to understand the mystery of the endless ocean through DNA sampling, sequencing, and analysis of marine microorganisms. Then, these data were submitted to NCBI and CAMERA database, online marine metagenomic resource databases. In contrast, CoML project evaluated the diversity, distribution, and abundance of marine life around the world [29]. As the first comprehensive survey of marine life, its findings were published in London in 2010, accompanied by a special collection of articles (http://www.ploscollections.org/static/comlCollections.action). Both GOS and CoML have significantly enriched the field with numerous samples and data, leading to the establishment of corresponding databases. Additionally, other specialized databases incorporate the metagenomics data of marine microorganisms (Table 1).

As a large amount of data generated by metagenomics technology needs to be analyzed, the analysis methods and tools have been rapidly developed. These tools can be classified as two categories: independent downloadable and installable metagenomic analysis software packages, which typically require a robust bioinformatics background and user-friendly online metagenomic data analysis platforms. For instance, the MG-RAST server (the Metagenomics Analysis Server), an automatic metagenomic analysis platform, provides quantitative analysis of microbial communities based on sequence information [30]. Quality control, automatic annotation, and analysis of prokaryotic metagenomic shotgun sequencing data can be performed after data upload. Another analysis platform is EBI Metagenomics, which was built by the European Institute of Bioinformatics (EBI) [31], and enables submission and automatic archiving of the metagenomic data in the ENA (European Nucleotide Archive). CoMet, a convenient online platform, allows for the rapid comparison and analysis of metagenomic profiles [32], without requiring registration. It processes DNA sequences uploaded by users in FASTA format for gene prediction, analyzes the encoded Pfam functional structure domains, and conducts statistical analysis and comparison.

1.3.2 Enzyme Databases from Microorganisms

Currently, three strategies, employed to construct enzyme resource databases, mainly include sorting and extracting annotation information from public sequence/structure databases such as PDB and UniProt, manually extracting literature information and computer mining using bioinformation methods such as sequence homology alignment, phylogenetic analysis, etc. The initial step in building an enzyme resource database involves creating a core library with detailed annotations,

Table 1 List of available marine microbial macrogenomics databases

Abbreviations	Full name	Address	Function introduction
MeganDB	The MEtaGenome Analyzer DataBase	megan-db.org/megan-db/	Storage and analysis of macro-genomic data
EBI Metagenomics	European Bioinformatics Postgraduate Metagenomics Website	ebi.ac.uk/metagenomics	Platform for storing and analyzing macro-genomic data
CoMet	Metagenomics comparative analysis platform	comet2.gobics.de/	Metagenome comparative analysis platform based on protein domain characteristics
MG-RAST	The Metagenomics RAST	metagenomics.anl.gov/	Platform for storing and analyzing metagenomic data
EnvBase	The EnvBase Data Catalogue	envgen.nox.ac.uk/cgi-bin/envbase.cgi	A searchable index of environmental omics data
IMG	Integrated Microbial Genomes	img.jgi.doe.gov/	Storage, annotation, analysis, and publication of microbial genomic and macrogenomic data
GOLD	Genomes Online Database	gold.jgi-psf.org/	Genome and metagenomic sequencing data
MgOI	Metagenomes Online	metagenomesonline.org/	Environmental metagenomic protein database
CAMERA	Advanced Microbial Ecological Analysis System	camera.calit2.net/	Storage of environmental genomic data
HMP	Hydrocarbon Metagenomics Project	hydrocarbonmetagenomics.com/ hmp.ucalgary.ca/HMP/	Analysis of metagenomic characteristics of microbial communities in petroleum resources
GOS	Global Ocean Sampling	ncbi.nlm.nih.gov/sra/ camera.calit2.net/ data.imicrobe.us/	Probing the genome of marine organisms
CoML	Census of Marine Life	www.coml.org	Explore the diversity of marine life, including multiple marine and regional projects

sequences, and properties of enzyme sources. Following the core library's completion, keywords from manually gathered documents are extracted using various databases, and mining programs are then utilized to retrieve pertinent information from diverse documents. Subsequently, the collected sequences and structures are classified and integrated into the database. To facilitate the subsequent maintenance, many databases have been opened the upload permissions, encouraging users to upload experimental structured data. Presently, the construction of enzyme databases heavily relies on manual effort. Thus, automatic construction of the database is a primary development goal for the enzyme resource databases. By now, the enzyme resource databases could be categorized as follows: basic enzyme information and mechanism databases, comprehensive enzyme databases, structure-function family databases, enzymatic application databases, and enzymatic characteristics and modification database (Table 2).

2 The High-Efficiency Expression Systems for Enzymes

2.1 Construction of Host Cells for Heterologous Expression

Extremozymes are synthesized by a multitude of extremophiles, predominantly found within the archaeal and bacterial domains. These organisms thrive in various harsh environments including the high temperatures of hot springs or high-temperate zones, the low temperatures of glaciers or the deep sea, the acidic and alkaline conditions of industrial or mine wastewater, the high salt concentration conditions of salt lakes, the highly radioactive conditions, and the extremely dry desert conditions. Extremophiles have developed sophisticated mechanisms to adapt to these conditions, ensuring the stability and functionality of their cellular components. These adaptations typically include the production of enzymes resilient to extreme conditions. As is demonstrated in recent research, most extremozymes are generally capable of heterologous expression in normothermic microbial host cells, such as *E. coli*. The extremozymes widely used in the food industry are mainly classified as thermophilic, cryophilic, piezophilic, salinophilic, acidophilic, and alkalophilic enzymes [33]. Among them, thermophilic and cryophilic enzymes are more extensively utilized, often targeted for enzyme engineering modification research.

In the last several years, the pursuit of enzyme discovery in extreme environments has significantly expanded, enhancing microbial culture techniques in these settings. However, cultivating natural extremophiles poses challenges, including markedly slow growth rates, low enzyme yields, and the need for specialized equipment. These factors diverge from standard industrial fermentation and downstream purification processes, thereby increasing the cost of producing extremozymes. The classical route of microbial enzyme production involves isolation of pure culture organisms, cloning, and expression of the corresponding genes from the natural host, which has been applied to obtain extremozymes. The most important of all is the transfer of genes from extremophiles into common host bacteria, which allows the production of extremozymes under mild conditions using standard microorganisms [34]. Proliferation of methods and production pathways for preparing extremophiles will facilitate their broad application in industrial manufacturing.

Table 2 Information on microbial enzyme databases

Database	Address	Information
Databases of basic enzyme information and mechanism		
ExplorEnz	enzyme-database.org	Enzyme classification and nomenclature
IntEnz	ebi.ac.uk/intenz	Enzyme classification and nomenclature
EzCatDB	ezcatdb.cbrc.jp	Enzyme reaction database
ENZYME	expasy.org/enzyme	Enzyme classification and nomenclature
M-CSA	ebi.ac.uk/thornton-srv/m-csa	Enzyme reaction mechanisms and active Sites
SFLD	sfld.rbvi.ucsf.edu	Sequence-structure-function relation of Enzymes
BKMS-react	bkm-react.tu-bs.de	Enzymatic reactions, kinetic parameters, experimental conditions and metabolic pathways
TargetTrack	dx.doi.org/10.5281/zenodo.821654	Protein solubility
PRINTS	bioinf.man.ac.uk/dbbrowser/PRINTS	Database on protein motif
iProClass	pir.georgetown.edu/iproclass/	Protein classification database
MEROPS	ebi.ac.uk/merops/	Peptidase database
SDRED	sdred.biocatnet.de	Dehydrogenase/Reductase Engineering Database
IRED	ired.biocatnet.de	imine reductase database
TEMLACED	temlaced.biocatnet.de	TEM lactamase database
TTCED	ttced.biocatnet.de	Triterpene cyclase database
Cofactor database	ebi.ac.uk/thorntonsrv/databases/CoFactor	Organic cofactors in enzyme
Comprehensive enzyme databases		
BRENDA	brenda-enzymes.org	Comprehensive enzyme information database
KEGG	genome.jp/kegg/annotation/enzyme.html	Enzyme and metabolism
EnzyMine	rxnfinder.org/enzymine	Database for enzyme function annotation with enzymatic reaction chemical feature
MetaCyc	metacyc.org	Pathway information of enzymes
Uniprot	uniprot.org	Composed of data from three major databases: Swiss-Prot, TrEMBL and PIR-PSD
3DID	3did.irbbarcelona.org	Interaction information of proteins with known 3D structures
DOMINE	manticore.niehs.nih.gov/cgi-bin/Domine	Protein domain interaction
Bionemo	bionemo.bioinfo.cnio.es	Comprehensive protein information

(continued)

Table 2 (continued)

Database	Address	Information
Databases of structure-function family		
CAZy	cazy.org	Carbohydrate-active enzymes datab
dbCAN-seq	bcb.unl.edu/dbCAN_seq	Database of carbohydrate-active enzyme (CAZyme) sequence and annotation
MERPOS	ebi.ac.uk/merops	Database of proteolytic enzymes
CCD	ncbi.nlm.nih.gov/cdd	Conserved domain database
PDB	rcsb.org/	The 3D shapes of proteins, nucleic acids, and complex assemblies
beta-lactamase	ifr48.timone.univ-mrs.fr/betalactamase/public	Database of β-lactamase enzymes
EzCatDb	ezcatdb.cbrc.jp/EzCatDB/	Enzymatic reactions, cofactors, catalytically active domains and structures
Pfam	pfam.xfam.org	A large collection of protein families
FIGFAMs	blog.theseed.org/servers	Artificially annotated protein sequences
CATH	cathdb.info/	A database organized hierarchically in protein domains
SCOP	scop.mrc-lmb.cam.ac.uk	Protein structure classification
CCDC	ccdc.cam.ac.uk	Cambridge crystal center database
BioMagResBank	bmrb.wisc.edu	NMR database on proteins, amino acids and nucleotides
TIGRFAM	tigr.org/TIGRFAMs/	Protein family database
Binding MOAD	BindingMOAD.org	Protein-ligand crystal structure
STITCH	titch.embl.de	Protein-compound interaction network
SABIO-RK	sabio.h-its.org	Biochemical reaction kinetic parameters
TECRDB	randr.nist.gov/enzyme/Default.aspx	Thermodynamics of enzyme-catalyzed reactions
ESTHER	bioweb.supagro.inra.fr/esther	Database of the α/β-hydrolase fold superfamily of proteins
Enzymatic application databases		
EAWAG-BBD	eawag-bbd.ethz.ch	Biocatalysis/Biodegradation database
MetaBioME	metasystems.riken.jp/metabiome	Commercially useful enzymes in metagenomic data
BioFuelDB	metabiosys.iiserb.ac.in/biofueldb/index.html	Enzymes involved in biofuels production
LED	led.uni-stuttgart.de	Lipase engineering database
HYED	hyed.biocatnet.de	Hydratase engineering database
LCCED	lcced.biocatnet.de	Laccase and multicopper oxidase engineering database
oTAED	otaed.biocatnet.de	ω-Transaminase engineering database
TEED	teed.biocatnet.de	Thiamine diphosphate-dependent enzymes engineering database

(continued)

Table 2 (continued)

Database	Address	Information
Databases of enzymatic characteristics and modification		
SABIO-RK	sabio.h-its.org	Biochemical reaction kinetics
ProtaBank	protabank.org	Enzyme engineering data
BioCatNet	biocatnet.de	Biocatalytic data on protein families to facilitate protein engineering
MuteinDB	muteindb.genome.tugraz.at	Substrates, products and enzymatic reactions linked with variants
FireProt	loschmidt.chemi.muni.cz/fireprotdb/	Data on thermal stability changes caused by sequence mutations
ProTherm	iitm.ac.in/bioinfo/ProTherm/	Data on thermal stability changes caused by sequence mutations
eSOL	tanpaku.org/tp-esol/index.php?lang=en	Estimate protein solubility based on protein translation and centrifugation conditions
SoluProtMut	loschmidt.chemi.muni.cz/soluprotmutdb/	Data of solubility changes caused by mutations
ProtaBank	protabank.org	Protein sequence information after mutation
PMD	proteinmicroarray.cn	Database on protein mutation
Phospho.ELM	phospho.elm.eu.org	Protein phosphorylation site

2.1.1 *Escherichia coli*

E. coli, a genetically engineered bacterium, is the most exhaustively studied prokaryotic bacterium, possessing mature expressing recipient cells for genes cloning, rapid multiplication, and easily controlled culture metabolism. Therefore, *E. coli* has been chosen by most researchers as the most widely used host cell for synthetic biology.

The effects of expression systems on exogenous gene expression Expression vectors, central to these systems, are predominantly categorized into fusion, non-fusion, and secretory types within *E. coli* expression systems [35]. Compared to fusion vectors, non-fusion vectors yield proteins more akin to their natural counterparts in structure, function, and antigenicity, thereby facilitating further research. During expression vector designing, the translation efficiency can be improved through various strategies [36], including optimizing the translation start region, constructing multiple expression vectors with unequal SD sequence and start codon spacing for selection when constructing expression vectors, applying "prokaryotic translation enhancer" sequences to constructing expression vectors, and employing a double cis-trans structure combined with the "prokaryotic translation enhancer." Furthermore, the promoter, crucial for exogenous gene expression, consists of two specific nucleotide sequences, TTGACA and TATAAT in the −35 and −10 regions, respectively, which regulate the transcription start site and influence transcription efficiency [37]. Incorporating a tightly controlled, robust promoter in the vector ensures gene expression even under low inducer concentrations, simultaneously mitigating the expression of other proteins, thereby reducing protease release by bacteria and minimizing bacteriophage-induced autolysis [38].

The interactions between host cells and gene circuit As recent research has extensively demonstrated the inextricable link between the gene circuits and host cells. Gene circuits, being synthetic constructs, are regulated by the host cell's resource allocation mechanisms and, conversely, consume the cell's resources. Host cells assimilate substances necessary for growth from their environment, converting these through physiological and biochemical processes into resources essential for biological processes such as replication, transcription, translation, and enzymatic reactions. Gene circuits, as exogenous genes, compete for various shared resources such as RNA polymerase, ribosomes, and amino acids, as well as enzymes and energy required for various biochemical reactions on the physiological state of the host cells, leading to reduced growth and morphological changes in the cells. Additionally, competition within different circuit modules can result in outputs diverging from expectations. The expression of exogenous genes intensifies physiological heterogeneity in host cells, leading to faster growth in cells with lower gene expression. This can confer a competitive advantage during continuous passaging, potentially leading to the failure of the synthetic circuit. In *E. coli*, the genetic stability of three fluorescent proteins, each regulated by different elements, was analyzed, revealing that circuits with lower expression and fewer repetitive sequences exhibit better genetic stability. Furthermore, interactions between exogenous gene expression products and the host genome can occur, such as with certain TetR family proteins aberrantly regulating host gene expression [39]. The high expression of dCas9 also significantly impacts cell physiology. Histological experiments have shown that dCas9, when overexpressed, binds nonspecifically to the genome [40], altering the transcription levels of certain genes, resulting in a drastic reduction in cell growth rate and alteration in cell morphology.

The applications of the *E. coli* expression system A xylose isomerase from the thermophilic microorganism *Thermotoga neapolitana* 5068 was successfully expressed in *E. coli*. Although the assembly of the expressed protein slightly differed from the original, its stability and catalytic properties remained unaffected [41, 42]. Similarly, an extreme thermophilic aldolase from the thermophilic archaeon *Sulfolobus solfataricus* was expressed in *E. coli*, with the enzyme kinetics was consistent with the original enzyme [42]. Genes encoding enzymes with low-temperature activity from cryophilic bacteria have been also successfully expressed. For instance, the aspartate transcarbamylase (ATCase) gene from *Vibrio psychrophicivibvio* strain 7693 and the α-amylase gene from cryophilic bacterium *Alieromonas haloplanktis* were both cloned and expressed in *E. coli* [43]. Expression of extreme salinophilic enzymes in the common presents more challenges. For example, the malate dehydrogenase gene from the salinophilic archaeon *Haloarcula marismortui* was heterologous expressed in *E. coli*, yielding a soluble but inactive product. This product required denaturation and subsequent refolding to achieve a structure akin to the natural enzyme [44]. The literature on the expression of extremophilic, alkalophilic, and acidophilic enzymes is sparse. However, a gene for an alkalophilic cryoprotease was expressed in *E. coli*, with the optimum pH of the expressed product was 9.0 [45].

2.1.2 *Bacillus subtilis*

Bacillus subtilis, a typical Gram-positive bacterium and a model industrial microorganism, has numerous advantages such as nonpathogenicity, robust extracellular protein secretion ability, and no significant codon preference. Recognized as a GRAS (generally recognized as safe) food-safe host, *Bacillus subtilis* is widely applied in producing functional nutritional products, fine chemicals, and enzyme preparations. However, the development of *B. subtilis* host cells lags significantly behind that of *E. coli*. Thus, designing and constructing high-performance *B. subtilis* host cells hold substantial scientific and practical importance. The subsequent sections outline common methods for modifying *B. subtilis* host cells, as detailed in Table 3.

The hosts and plasmids of *Bacillus subtilis* The *B. subtilis* 168 serves as the model strain, with commonly used research derivatives primarily being protease-protease-deficient strains such as 1A751, WB600, and WB800. Plasmids commonly used in *B. subtilis* transformation can be classified as integrative and replicatable types. Integrative plasmids, incapable of autonomous replication in *B. subtilis*, integrate their homologous fragments into corresponding chromosomal sites under selective pressure, thereby replicating alongside the chromosome. This integration ensures high stability and reduces gene loss, a frequent issue in replicable plasmids. However, the low copy number of the integrated target gene often results in reduced yield of the expressed protein. Replicable plasmids, including pUB110 and pC194 derived from *S. aureus*, shuttle plasmids modified from *S. aureus* to *E. coli*, and more stable plasmids combining *B. subtilis* fragments with *E. coli* plasmids, can replicate and express the target protein within the host. Notably, pUB110, a naturally occurring plasmid with a small base and high copy number, is often used as a foundational plasmid for plasmid-based engineering in experiments.

Genome simplification techniques Recent studies on minimal genomes have significantly enhanced our understanding of *B. subtilis* cell functions and the evolutionary patterns of life. However, the minimal genome of *B. subtilis* host cells typically sustains only basic cellular metabolism and may lack desirable characteristics such as rapid growth and resistance [46]. Therefore, targeted simplification of the *B. subtilis* genome to remove interference from nonessential metabolic pathways is needed to reduce genomic complexity, improve the efficiency of substrate and energy utilization by *B. subtilis*, as well as the controllability of cytogenetic stability and physiological phenotype, making it an ideal host cell for the production of natural metabolites or the expression of heterologous proteins. Over the past

Table 3 The methods of transformation of *B. subtilis* cells

Construction methods	Functional characteristics
Gene editing	
Genome simplification	Construction of multiple genomic simplification strains using traceless gene knockout technology
CRISPR/Cas9 Iterative System	Enabling knockouts, point mutations, and knock-ins to avoid off-target effects
CRISPR/Cas9 Toolkit	Enabling gene knockout, multiple knock-in, and transcriptional repression
CAMERS-B system	Multigene editing and regulation by designing crRNA arrays
Endogenous promoter library	An endogenous promoter library based on cell growth constructs
Modular operating units	Co-regulating gene and protein expression at the level of transcription translation and protein hydrolysis
DNA scaffold	Regulating the spatial scaling and orientation of pathway enzymes through DNA scaffolding
RNA scaffold	Transcriptional repression of target gene expression by RNA scaffold binding to specific proteins
Endogenous space scaffold	Construction of FMMs-multienzyme complexes through FMMs spatial scaffolds
Cell-free protein expression systems	In vitro expression of proteins using exogenous DNA as a template with cell lysates
Endogenous regulatory systems	
Sec secretion pathway	Enhancing the secretion of heterologous proteins by improving the Sec secretion pathway
Quorum sensing system	Dynamically regulating the expression of related genes when cell density returns to a threshold
TypeItoxin-antitoxin system	Downregulating gene expression through complementary pairing of antitoxin antisense RNA with toxin mRNA
Carbon catabolic blocking system	Unblocking carbon catabolism, allowing cells to use multiple carbon sources simultaneously

decade, researchers have used homologous recombination techniques to knock out nonessential genes from *B. subtilis* to obtain a series of top-down *B. subtilis* genome simplifying strains. Westers et al. [47] obtained a series of *B. subtilis* genome simplifying strains by deleting five phage regions on the *B. subtilis* genome (*SPβ*, *PBSX*, *pro1*, *pro6 skin*) and the *pks* manipulator region. A *B. subtilis* Δ6 mutant strain lacking 7.7% of the genome was also obtained, with no significant changes in the cellular phenotype when compared to the wild type.

CRISPR gene editing technologies The genomic deletion fragments produced by homologous recombination such as conventional Cre/loxP are dependent on the insertion of resistance genes, leaving a trace of exogenous DNA on the genome after removal of the target gene, which affects the application of *B. subtilis* in food fermentation to some extent. Gene editing tools based on interspaced regularly interspaced short palindromic repeat sequences (CRISPR) are one of the most popular gene editing techniques, which allow precise editing of multiple genes in microorganisms. In the CRISPR gene editing system, a guide RNA (sgRNA) directs Cas proteins (e.g., Cas9, Cpf1, etc.) to target loci in the genome, which subsequently cleave DNA at specific sites, resulting in DNA double-strand breaks (DSBs). The homology-directed repair (HDR) process introduces a specific modified fragment into the genome through the homology arm, resulting in gene knockout, insertion, or mutation, which overcomes the shortcomings of traditional homologous recombination by producing mutants without exogenous DNA remnants, exemplified in model organisms like *B. subtilis* [48]. Various RNA-modified tools, such as cis-riboswitches, antisense RNAs (asRNAs), and other noncoding RNAs (sRNAs), which have been used to recruit structural domains of proteins for gene expression regulation, have certain limitations. For example, cis-riboswitches require upstream sequence modification of target genes. And synthesizing asRNAs requires screening a large number of endogenous regulatory RNAs, which is time-consuming, limiting their application in metabolic engineering and genetic screening. In contrast, the defective nuclease (dCas protein) in the CRISPRa/CRISPRi system, instead of cleaving DNA, fuses with the transcriptional blocking domain, forming an RNA-protein complex. The complex targets the transcription start site (TSS) of a gene to repress its transcription. The complex could also activate transcription of the target gene at the TSS site by binding to scaffold RNA (scRNA), a modified gRNA that encodes both target sequences and RNA hairpins, recruiting transcriptional activators or fusing with transcriptional activation domains.

A comparison of the advantages and disadvantages of the *E. coli* and *B. subtilis* expression systems is shown in Table 4. Compared with the *E. coli* expression system, *B. subtilis* has many advantages. For example, its cell wall contains peptidoglycan and phosphoplasm but does not contain endotoxin, making *B. subtilis* is nonpathogenic. Additionally, *B. subtilis* features an effective signal peptide and molecular chaperone system that prevents inclusion body formation. This characteristic simplifies the purification process of target proteins and eliminates codon bias, coupled with a robust capability for secreting most exogenous proteins [49]. Furthermore, *Bacillus subtilis* boasts a rapid growth cycle, a well-defined genetic background, and modest nutritional requirements.

The applications of *Bacillus subtilis* expression system Teague cloned and expressed the pullulanase gene from *Bacillus naganoensis* (ATCC No. 53909) in the *Bacillus subtilis* expression system and found the collected pullulanase remains excellent properties. Although *B. subtilis*-derived prullanase could be abundantly expressed in *E. coli*, *E. coli* cannot be used in the food processing industry for its cell wall contains lipopolysaccharides which are considered endotoxins. Therefore, studies for the expression of pullulanase in *B. subtilis* is of broader value and significance. Lu applied the shuttle plasmid pWB980 to express pullulanase (PUL) in *B. subtilis* WB600 and obtained a pulullanase with the enzyme activity of 10.94 U mL^{-1}. Subsequently, Song [50] increased the enzyme activity of pullulanase to 24.5 U mL^{-1} by combing different host and promoter during heterologous expression of pullulanase. Deng [51] optimized the expression of pullulanase by

Table 4 Comparison of *E. coli* and *Bacillus subtilis* expression systems

Advantages	Disadvantages
E. coli	
1. Whole-genome sequencing, with a total of 4405 open reading frames 2. Well-established systems for the cloning and expression of genes 3. Rapid reproduction, simple culture, ease of handling, high resistance to contamination and genetic stability 4. Approved by the United States FDA as a safe genetically engineered receptor organism 5. Clearer metabolic pathways and gene expression, with higher levels of expression	1. Lack of complexing function for eukaryotic proteins 2. Lack of modification and processing systems for eukaryotic proteins 3. The periplasm contains a wide variety of endotoxins 4. The tendency to form inclusion bodies, which makes purification difficult 5. Poorly processed posttranslational modifications of the prokaryotic expression system and low biological activity of the expression product
B. subtilis	
1. The composition of the cell wall is simple, and the secreted protein products are not mixed with endotoxins 2. The ability to secrete the target protein directly into the extracellular compartment, with relatively simple recovery and purification 3. Possessing an efficient system of secretory signal peptides and molecular chaperones to accomplish efficient secretory expression of target proteins and avoid the formation of inclusion bodies	1. The intrinsic secretion of proteases that hydrolyze exogenous proteins, which to a large extent affect the accumulation of exogenous proteins 2. Has the instability of the plasmid itself, i.e., the plasmid-attached vector is less stable 3. Strain characteristics that lead to plasmid instability, such as the Bsu M restriction modification system, can affect transformation efficiency and prevent entry of exogenous genes into cells

introducing enhancers into the gene and rationally designing the mRNA 5'-UTR sequence, resulting in enzyme activity of 28.7 U mL^{-1} of pullulanase.

2.1.3 Yeast

At present, commonly used microbial expression systems mainly include *E. coli*, yeast, and filamentous fungi. As one of the model organisms in biological research, *E. coli* has very definite genetic background and gene expression regulation mechanism. Though the *E. coli* expression system includes a variety of expression vectors and hosts, it does not pose the posttranslational modification and processing functions of eukaryotic cells, making it impossible to secrete eukaryotic proteins. The fungal expression system can effectively overcome the shortcomings of *E. coli* for the high growth rate and thriving on simple and cheap media. Besides, it has obvious advantages in terms of the expression and secretion of a wide range of commercially valuable proteins. Taking the typical yeast and *Aspergillus* as examples to introduce fungal chassis cell transformation strategies for high-efficiency expression of enzymes in the following.

In recent years, the yeast has been used as a new choice for expressing recombinant proteins for several advantages such as expressing recombinant proteins at a high level, growing at a high cell density, having a distinct secretion pathway, and could be modified by a variety of molecular biological methods [52]. For example, LipG7, a lipase purified from the Antarctic filamentous fungus *Geomyces* sp. P7, was discovered to be cold adaptable and capable of maintaining activity after heat denaturation. Tomasz Florczak et al. heterologously expressed LipG7 protein in *Saccharomyces cerevisiae* BJ5465 and found that it has the same properties as the natural protein [53].

Metabolic engineering of yeast is a common strategy for increasing the yield of enzymes by heterologous expression, achieving targeted construction of targeted engineering strains and promoting high production of metabolites [54, 55]. The construction of mutant strains is also a strategy to improve the efficient expression of enzymes. The mutation region was determined by comparing the structure differences of enzymes with high and low catalytic efficiency. The amino acid sequence of the mutation site was optimized based on the *Pichia pastoris* codons. The secondary structure stability of the mRNA can be site-directed mutated using overlap or extension PCR [56, 57]. The mutant strains could be screened and heterologously express high-efficient enzymes. Modular pathway engineering has been widely used in metabolic engineering transformation technology in recent years. It usually referred to dividing a microorganism's metabolic pathway into different modules, constructing each module through regulatory elements, assembling it into a metabolic network, and finally adjusting the activity among the modules to obtain the strain with the highest relative activity [58].

However, modular pathway engineering lacks dynamic regulation at different cell growth stages, which would cause the accumulation of intermediate products. The tremendous progress in systems biology and synthetic biology have greatly promoted our understanding of metabolic regulation mechanisms and the creation of artificial gene circuits [59]. In short, metabolic engineering strategies have developed rapidly. Through the introduction of omics analysis and dynamic models, the in-depth study of regulatory mechanisms would provide a deeper understanding of the dynamics of metabolic pathways. We could identify the rate-limiting steps through dynamic metabolic analysis and introduce a reasonably designed synthetic circuit via genetic manipulation to eliminate the rate-limiting steps, increasing the output of exogenous enzymes ultimately.

Increasing the copy number of genes in *P. pastoris* could effectively increase the expression of enzymes. When copies of heterologous genes inserted into the genome of *P. pastoris* was greater, their expression level should be higher. However, if the copies were too large, the protein expression would be restricted in other steps, such as the secretion and transport of the protein. At the same time, excessive protein expression may adversely affect the physiology of the host. In addition, studies have shown that signal peptides could also affect the heterologous expression of enzymes. For example, the alpha maturation factor (α-MF) signal peptide from *S. cerevisiae*, had a strong secretion efficiency, was widely used. The high secretory expression of the enzyme could be achieved by constructing a recombinant expression vector carrying (α-MF) signal peptide.

As a heterologous eukaryotic expression system for efficiency expression of foreign proteins, *P. pastoris* is characterized by containing the AOX1 (alcohol oxidase gene) promoter [60], which can promote expression efficiency of foreign proteins with methanol as the induction. Promoter is a regulatory element that regulates gene transcription, which has a great influence on the level of gene transcription. Promoter transformation methods without changing the starting strength include random mutation (error-prone PCR), truncation mutation (promoter sequence length saturation mutation), hybrid promoter engineering, synthetic promoter engineering, etc. [61]. Random mutation is a powerful technology for improving the function of promoters. For instance, EP-PCR randomly mutates the promoter, easily obtaining a functional mutant promoters library. Hybrid promoter engineering refers to the fusion assembly of enhancer elements and core promoters, which can enhance transcription ability. Hybrid promoter engineering have been applied to *S. cerevisiae* recently to construct a strong promoter library and regulate gene expression (Fig. 2). For example, the endogenous promoters various UAS elements were fused with the core promoter to create a promoter library with a regulatory strength within a dynamic range of 90 times. Additionally, the strong constitutive yeast promoter PGPD was constructed, which increased the level of promoter-regulated mRNA transcription by 2.5 times. A library of lactose-inducible promoters was also created by hybrid promoter engineering. The PGall promoter's strength was increased by 50 times with a dynamic range, and transcription capacity was increased by 15% [62]. These results indicated that endogenous promoters could increase promoter strength by adding upstream activating elements in eukaryotic cells. Synthetic promoter engineering had recently been widely used [63]. For example, the PPFY1 promoter was reconstructed into a synthetic promoter for different protein expression levels in *S. cerevisiae* [64]. In short, a series of methods have been used in promoter modification to realize the efficient regulation and expression of foreign enzymes.

Terminators, like promoters, play an important role in gene expression regulation [65] in *S. cerevisiae*. It is fully proved that choosing different terminators to regulate the expression of foreign proteins would have varying influence on the half-life of mRNA, resulting in an increase or decrease in protein expression. Because the regulation strength of wild-type terminator is often low, artificially synthesized terminators have become hotspots. The synthetic terminator, which has a shorter sequence than the wild-type terminator, could effectively improve the stability of Mrna [61]. Transcription factors are critical regulators of gene expression in the metabolic regulatory network. Many transcription factors, such as the PerR in bacterium and the AFT1 in yeast, had been studied in the regulation of gene expression. By analyzing the putative binding site of AFT1 in *P. pastoris*, it was found that transcription factors were involved in the regulation of many secreted genes. Thus, it was speculated that they played a positive role in enhancing the expression and secretion of enzymes.

Efficient expression of foreign genes requires not only the initiation of transcription by a strong promoter but also the efficient initiation of translation. The intensity

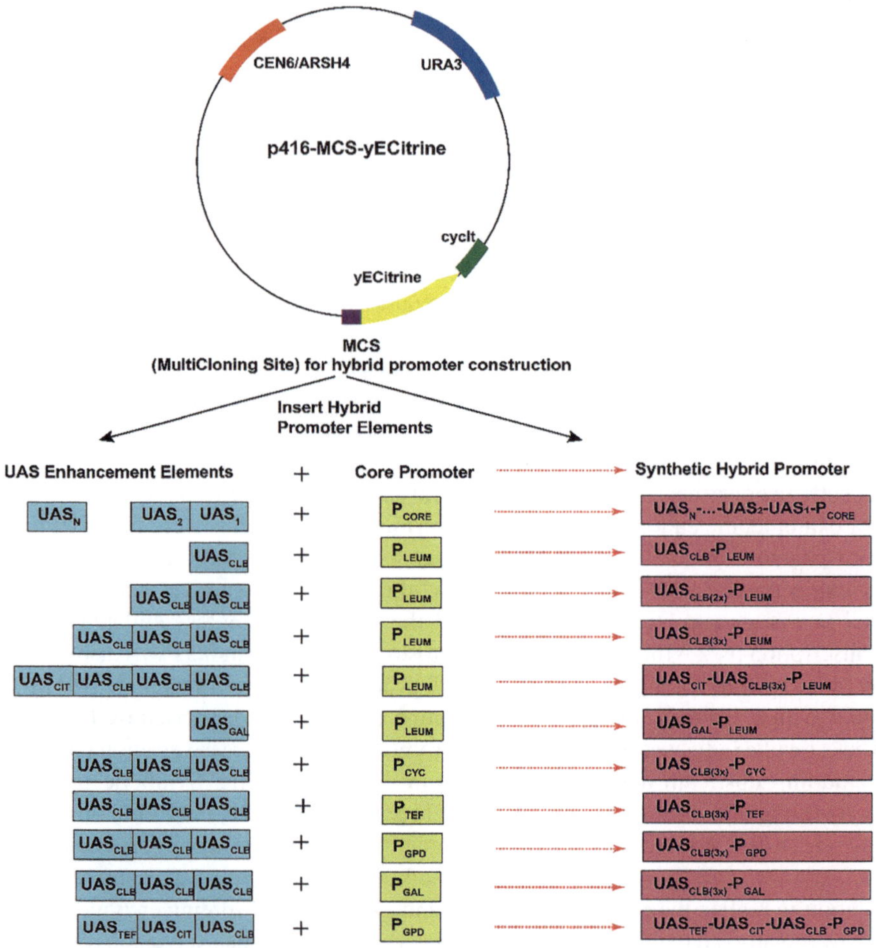

Fig. 2 Schematic diagram of hybrid promoter construction

of gene expression can be controlled in the translation process through ribosome-binding sites. The ratio of the metabolic flux could be optimized by using ribosome-binding sites with varying strengths and different enzymes in the metabolic pathway. The *Pichia pastoris* expression system, a highly efficient eukaryotic mechanism, is extensively utilized for the correct expression and folding of foreign proteins. However, it has been observed that its capability to secrete and express a diverse array of enzymes is not uniformly efficient, representing a significant limitation in the expression of foreign proteins within this system [52]. Enhancements in protein secretion can be achieved by genetically modifying the yeast's secretion pathways [66]. These complex secretory pathways, encompassing protein transport through the endoplasmic reticulum, Golgi apparatus, endosomes, cell membranes, or vacuoles, are regulated by various secretory pathway proteins. Consequently, a comprehensive understanding of these mechanisms is crucial for improving the secretory expression of enzymes.

Currently, numerous high-value proteins have been successfully expressed in *P. pastoris* using molecular biology. However, the *P. pastoris* expression system exhibits several drawbacks, including a complex construction methodology, low conversion rate, and prolonged growth cycles. Additionally, methanol induction presents potential safety risks. The genetic complexity of eukaryotic yeast further complicates matters. Presently, our understanding of the secretion mechanisms of foreign proteins and the functional genes of the genome remains incomplete. Consequently, issues such as inconsistent secretion of products and degradation of expressed products may arise during the expression process [67].

2.1.4 *Aspergillus*

Aspergillus is a significant industrial microorganism and a traditional bacteria used in fermented food production. It is recognized as a food-grade safe strain by both the US Food and Drug Administration and the World Health Organization. Technological advancements have increasingly highlighted the importance of *Aspergillus*, leading to its development into a versatile cell factory capable of multiple expressions. *Aspergillus* surpasses bacteria and yeasts with its robust degrading enzyme system and efficient protein secretion pathways. These features enable rapid growth and reproduction, adaptation to harsh environments, and sustenance in low-value culture mediums and acidic conditions. Furthermore, its highly effective secretion system allows for efficient extracellular secretion of enzymes or other polypeptides, facilitating the separation and purification of recombinant proteins [68]. In recent years, *Aspergillus* has become a primary industrial bacterium, producing a variety of organic acids (such as citric acid, gallic acid, gluconic acid, etc.), industrial enzymes, protein preparations like cellulase, amylase, pectinase, etc. [69].

Enhancing the productivity of recombinants in *Aspergillus* strains necessitates advanced molecular techniques for the efficient transfer and expression of heterologous genes across different hosts. Typically, expression vectors for recombinant proteins are stably integrated into the *Aspergillus* genome, occurring either randomly through nonhomologous/illegal recombination or via homologous recombination (e.g., into mutant marker genes or as tandem repeats of vector copies) [70]. Commonly, multiple insertions ranging from 2 to 10 copies are observed and are often associated with increased protein yields. However, a higher copy number does not invariably lead to enhanced expression in *Aspergillus*. This inconsistency may arise from factors like the titration of endogenous transcription factors or the pleiotropic effect of random integration [65].

Establishing an efficient genetic transformation system is crucial for employing *Aspergillus oryzae* in the expression of exogenous proteins. However, DNA transformation in *Aspergillus* is notably less efficient than in yeast or *E. coli*, largely due to the obstructive fungal cell wall. Techniques such as electro-transformation and knockout processes have demonstrated relatively lower efficiencies [71]. The best results have been achieved by transforming protoplasts, prepared from fungal hyphae or budding conidia, following treatment with cell wall-degrading enzymes [72]. Currently, the transformation of *A. oryzae* predominantly employs the PEG/CaCl2-mediated protoplast transformation method, which is simple and highly efficient. This method is widely used in fungal genetic transformation, achieving success in *S. cerevisiae*, *A. niger*, and *Xylaria*, typically resulting in single-copy integration with a high homologous recombination frequency for efficient DNA transformation [73]. Insights from the sequenced *Aspergillus* genome facilitate the identification of genes that could enhance protein production and secretion. Enhancing protein expression levels is feasible through effective mutagenesis strategies. Selecting suitable marker genes and hosts is a vital aspect of this transformation system. Nutrient-deficient marker genes such as *pyrG*, *argB*, *deA*, *adeB*, *niaD*, *sC*, etc., and genes that induce mutations or establish specific metabolic pathways such as *amdS*, *niaD*, *ptrA*, etc., are commonly used in *Aspergillus* transformations. Employing nutrient-deficient genes as screening markers for targeted *Aspergillus* transformation, with the aim of enhancing enzyme activities, presents a promising approach for industrial applications [74].

Protein expression regulation in *Aspergillus* predominantly occurs at the transcriptional level. The fungal transcriptional control region is crucial in constructing efficient expression vectors. A strong promoter enhances the stability and transcriptional level of mRNA, thereby increasing exogenous protein expression. Prominent examples include the amylase promoters *amyA* and *amyB* from *A. oryzae* and the promoters *glaA* and *glaB* from *A. niger*. Advancements in understanding promoter structure and function have led researchers existing promoters to boost their transcriptional activity. André Fleißner identified several promoters with robust constitutive activity or high inducibility, successfully implementing them in expression systems (Table 5) [75]. Given the complexity of gene expression regulation in *A. oryzae*, a combination of multiple methods is employed to enhance exogenous gene expression levels. Utilizing strong promoters, coupled with the overexpression of multiple target genes in *A. oryzae*, can significantly increase the yield of exogenous proteins.

Table 5 Common promoters and enzyme production systems of *Aspergillus*

Host	Promoter	Enzyme	Reference
A. niger	*alcA/aldA*	Peroxidase	[78]
	amy (taka)	Lipase	[79]
	glaA	Aspartyl protease	[80]
	gpdA	β-glucuronidase	[81]
A. nidulans	*alcA/aldA*	Glycoamylase	[82]
	amyB	Carboxypeptidases OcpC	[83]
A. oryzae	*Tef1*	Polygalacturonase	[84]
	amyA	Endoglucanase	[85]
A. awamori	*exlA*	α-galactosidase	[86]

As a model strain in *Aspergillus*, *A. oryzae* is renowned for its exceptional protein secretion capabilities and safety. It has attracted significant scientific interest due to a plethora of potent promoters and regulatory elements derived from it. Extensive research has led to various strategies to enhance heterologous protein expression in industrial applications of *A. oryzae*, including the use of strong promoters, fusion expression, and knockdown of protease genes [76]. Table 3 illustrates several proteins expressed in *Aspergillus oryzae*. The first heterologous expression of cold-adapted enzymes in filamentous fungi was pioneered by Mao, who developed a novel uracil-deficient *Aspergillus* transformation system. This system, based on the nutrient selection marker for orotate phosphoribosyltransferase (pyrF), facilitated a novel approach in genetic manipulation [77]. A universal, purifiable vector was utilized to insert the gene into the *Aspergillus* host strain. Genomic DNA from the cold-tolerant fungus *Geomyces pannorum* was extracted, containing a novel α-amylase gene. Subsequently, this α-amylase was expressed in the newly devised *A. oryzae* system, achieving an amylase activity of 958 U/ml. This enzyme displayed optimal activity at pH 5 and 40 °C and retained over 20% in the temperature range of 0–20 °C.

The fusion expression strategy is a vital approach to augmenting the expression of exogenous proteins in *A. oryzae*. This method involves attaching the exogenous protein to the 3′ end of *A. oryzae*'s native secretory protein, which serves as a carrier. This attachment ensures the proper positioning of the exogenous protein during secretion and protects it from degradation by endogenous proteases. Furthermore, fusion expression contributes to the stability of mRNA, thereby enhancing the yield of the exogenous protease.

Despite *Aspergillus*' significance as a microbial cell factory in the food and pharmaceutical industries, the range of genetic tools available for the efficient production and secretion of heterologous proteins remains limited. The intricacies of filamentous fungi often obscure the factors influencing the production, stability, and secretion of specific enzymes. Consequently, an in-depth understanding of the physiology and metabolism of these organisms, coupled with the ongoing development of innovative engineering tools, is imperative for devising effective mass production strategies.

2.2 Multi-Enzyme Cascade Catalysis Technologies

Enzymes, characterized by their high efficiency, mild reaction conditions, and notable selectivity and specificity [87], have emerged as a pivotal area in synthetic biology and biocatalysis [88, 89]. They are extensively utilized in the production of chemicals, foods, and pharmaceuticals [90]. They are extensively utilized in the production of chemicals, foods, and pharmaceuticals. They enable reactions unachievable by chemical means, demonstrating higher resource utilization efficiency and reduced environmental impact. Therefore, biological enzyme catalysts are increasingly preferred over traditional chemical catalysts, especially when high stereospecificity, mild reaction conditions, and eco-friendly processes are essential. The development of biocatalyst systems, from design to production, necessitates sourcing from biodiversity and protein engineering. This involves creating robust and efficient multistep enzymatic reaction cascades in a single pot. Enzymes derived from microbes in extreme environments, such as the ocean, provide an expansive protein database for constructing stable and efficient multienzyme cascade systems [91].

2.2.1 Definition of Enzyme Cascade Reaction

The concept of an enzyme cascade refers to the process of generating valuable compounds using enzymes, typically involving multiple steps. Distinct from natural biosynthetic pathways, these cascades have evolved to facilitate multistep synthesis, ranging from substrates to simple metabolites and eventually to complex natural products. Biocatalytic cascade reactions can be categorized into four distinct designs: (i) linear, (ii) orthogonal, (iii) parallel, and (iv) cyclic cascades. In a linear cascade, a single substrate undergoes transformation into a final product via one or more intermediates. This process not only saves time and minimizes waste but also converts unstable or toxic intermediates directly into the final product, thereby enhancing yield and driving the equilibrium toward product formation. An orthogonal cascade involves transforming a substrate into a product with additional steps, such as cofactor regeneration or by-product removal. A parallel cascade simultaneously converts two different substrates into two separate products, each involving distinct enzymatic reactions with cofactors or co-substrates, segregating valuable compounds and discarding by-products. Conversely, a cyclic cascade recycles the material into an intermediate, which is then converted back into the starting material, leading to compound accumulation and providing synergistic benefits by integrating enzymes with chemical processes. The classifications of enzyme cascades are flexible, with new designs often being a hybrid of these various approaches [92].

2.2.2 Strategies for Constructing Multienzyme Complexes

Efficiently directing multistep reactions toward a desired outcome poses a significant challenge. A crucial factor is the controlled transportation of intermediates between reaction stages. Nature addresses this by employing multienzyme complexes to enhance overall catalytic efficiency through three primary methods: (i) direct fusion of genes or peptide conjugation, (ii) compartmentalization techniques, and (iii) immobilization to create enzyme complexes. These approaches facilitate the channeling of intermediates from one stage to another, either through a tunnel or interface, thereby controllably increasing the yield of targeted compounds in a multistep reaction cascade [93].

The integration of genetic genes through direct fusion is a conventional approach to minimize the diffusion distance between enzymes. Typically, peptide linkers are employed to fuse pathway enzymes, creating artificial enzyme complexes. These complexes facilitate continuous reactions that would otherwise require multiple, separate natural enzymes. This fusion process results in a multifunctional chimeric enzyme with several catalytic active centers, significantly reducing the spatial distance between enzymes and synergistically enhancing sequential reactions via substrate channels. Studies have shown that these fusion enzyme complex exhibits kinetic advantages and superior product yield compared to the natural free enzymes [94]. Crucial factors influencing the activity of fusion proteins include the type, length, protease sensitivity, secondary structure, and integration sequence of peptide linkers connecting different proteins. Notably, the length of linker is directly associated with hydrogen production efficiency. Extremely short or long linkers yield lower hydrogen production due to ineffective cis-interaction. Conversely, an optimally sized linker can boost the local concentration of electron transfer proteins, thereby enhancing the electron transfer rate and hydrogen production. Whether the flexible and unstructured N-terminal and C-terminal of protein are fused directly, or the peptide linkers are used as the "bridge" between protein domains, it is necessary to avoid the steric hindrance between proteins. This is essential to ensure adequate space for proper protein folding, preventing negative outcomes like inclusion body aggregation, reduced protein yield, and diminished catalytic activity, etc.

Due to the above limitations of genetic fusion, posttranslational assembly may be more suitable for multicomponent systems. This method decouples protein expression and fusion into two steps, allowing different protein modules to be expressed in optimal hosts and then assembled by chemical/biological pairing, resulting in protein complexes with varying stoichiometries, sizes, geometries, and catalytic efficiency [95]. Key enzymes in a specific metabolic pathway are assembled together to form metabolic pathway nodes, which can increase the yield of specific metabolites. The intermolecular protein conjugation can also be achieved by posttranslational chemical modification of proteins. Bifunctional reagents can be covalently connected to reactive functional groups on proteins (e.g., –SH, $-NH_2$, –COOH) to form intramolecular or intermolecular biological coupling [96].

In the natural metabolic pathway, the regulation of multienzyme reaction by highly complex intracellular microenvironment should not be ignored. Compartmentalization, another method of separating pathways, avoids metabolic crosstalk, and controls flow based on the principle of natural cell compartments. Various multistep cascade reactions could be catalyzed by the precise spatial colocalization of multiple enzymes through encapsulating specific enzymes in a material or protein shell. Various materials such as polymer capsules, phospholipid liposome vesicles, polymersomes, and nano–/micro-sized particles, are employed to form compartment structures, which are extensively utilized in the development of multienzyme compartment systems. Encapsulation of functionally related enzymes in close proximity enables efficient substrate channeling between cascade partners and reduces intermediate loss due to diffusion.

Immobilized enzyme is a kind of technology that binds free enzyme to specific solid supports, enabling their repeated and continuous use while preserving catalytic activity [97]. This technology is emerging as a novel bio-manufacturing platform, offering several advantages for industrial applications. Firstly, it simplifies enzyme separation and enhances recovery and repeatability, thereby reducing costs by allowing enzymes to be used repeatedly. Secondly, immobilized enzymes exhibit greater stability compared to free enzymes. Thirdly, enzyme immobilization technology enhances catalytic efficiency by optimizing the enzyme microenvironment and substrate reaction conditions. Over the years, numerous co-immobilization techniques for enzymes have been developed. These techniques are categorized based on the need for medium materials into carrier immobilization and carrier-free immobilization. And the research on multienzyme immobilization using biological materials, organic materials, and inorganic materials as scaffolds has developed rapidly.

Proteins, as common renewable biomaterials, are extensively utilized in creating protein scaffolds that facilitate the co-localization of multiple enzymes within a single protein complex. The natural interaction domains/ligands, such as cohesin-dockerin domain, the GTPase-binding domain (GBD), the SH3 domain, and the PSD95/DlgA/Zo-1 (PDZ) domain, along with their respective peptide ligands, have been widely used in constructing programmable synthetic scaffolds. According to the precise specificity of Watson-Crick base pairing, a large set of nucleotide sequences can be utilized to design complex and multidimensional DNA nanostructures for directed multienzyme immobilization. The structural programmability and accurate addressability of DNA nanostructures enable the regulation of inter-enzymatic distances with greater precision and flexibility at the nanoscale. Therefore, DNA scaffold-directed assembly of multienzyme reactions is broadly applied to enhance catalytic efficiency, increase reaction selectivity, and investigate the mechanism kinetics of multienzyme reactions. In addition to DNA, RNA has also emerged as a unique, programmable material, capable of being designed and assembled into multidimensional RNA structures [98].

The wide application of biocompatible DNA nanostructures in the construction of multienzyme systems with controllable spatial arrangement and precise positions proves its potential as versatile designer three-dimensional scaffold architectures. However, the complexity and high cost of this enzyme immobilization technology limit its suitability for large-scale industrial applications. Selecting appropriate immobilization carriers remains a primary focus in enzyme immobilization research for effective large-scale catalytic applications. In addition to biological scaffolds, metal organic frameworks, graphene, carbon nanotubes, silica, and other abiotic materials can also be used for enzyme immobilization to improve the stability and reusability of enzymes. Recently, metal and covalent organic frameworks (MOFs and COFs), characterized by their crystalline porous structure, have gained significant attention as solid-phase carriers for enzymes. Their benefits include a large specific surface area, adjustable pore size, exceptional adsorption capacity, marked stability, and a high loading rate. Based on MOF/COF, other materials can be added to improve recovery and avoid leaching or damage of enzymes in the reaction environment. In addition to metal oxides, carbon materials, minerals, and other abiotic materials are also widely used in enzymatic processes. These materials demonstrate enhanced thermal, chemical, and mechanical properties, significantly augmenting the effectiveness and applicability of enzyme catalysts. Abiotic scaffolds typically characterized by high specific surface area, high porosity and adjustable structure, and rich functional groups obtained through surface modification. Different modification methods and the choice of linkers enable the synthesis of a large number of materials with different properties and characteristics, allowing for the design of specific immobilization strategies. These strategies optimize the pore size, shape, volume, and surface area of materials to adapt to different enzyme catalytic processes [99].

Multienzyme co-immobilization is based on the single-enzyme immobilization technology. In order to achieve efficient multienzyme co-immobilization, design of the immobilization carrier or technology of maintaining the enzyme activity and the structure and function of each enzyme is still necessary to carry out. When co-immobilizing multiple enzymes on the same platform, several aspects must be considered. Firstly, the size difference of co-immobilized enzymes is often ignored in the design of combined biocatalysts. Using a porous carrier, the pore size is determined by the larger enzyme, leading to a reduced specific surface area and, consequently, decreased enzyme carrying capacity and mechanical resistance. This reduction may diminish the volumetric activity of the immobilized biocatalyst, potentially resulting in economic drawbacks. Secondly, the rate of enzyme immobilization should be considered. For porous carriers, the immobilization rate and the enzyme's diffusion pathway inside the pores before immobilization influence the immobilization sequence and site. Thirdly, a uniform immobilization strategy for all enzymes in the cascade pathway is essential, requiring all enzyme molecules to be maintained under similar conditions for a certain period if immobilized on the same carrier. Therefore, co-immobilization makes it impossible to improve the properties of all enzymes, unless the optimal immobilization scheme is same for all relevant enzymes by coincidence. At present, no immobilization method or carrier is universally applicable. The appropriate method should be selected according to the advantages and disadvantages of different immobilization methods and enzymes. In addition, a single method is often insufficient for all application needs. Therefore, multiple functional carriers could be used to immobilize enzymes through different mechanisms to obtain immobilized enzymes with better performance, then maximizing the immobilization effect in cascade reactions [100].

2.2.3 Advantages and Challenges of Multienzyme Cascade Systems

In addition to the methods described above, a growing number of emerging construction strategies, such as cell surface displays [101], enzymatic cross-linking [102], etc., are emerging to adapt to a variety of biocatalysis conditions. In multienzyme cascade reactions, enzymes are assembled into macromolecular structures, which modulate reactivity through compartmentalization or spatial organization, facilitating cooperative catalysis among multiple enzymes. These special structures have a major advantage called metabolic channels. This arrangement allows for the direct transfer of products from one enzyme to another as substrates, without requiring equilibration, when the active sites of the enzymes are in close proximity. This phenomenon, known as the substrate channel effect, can effectively increase the overall reaction rate. The influence of this phenomenon significantly enhances the overall reaction rate, by bringing various active sites into closer proximity and assembled into complexes and providing effective control for multifunctional enzymes, multienzyme complexes, or separate enzymes. Additionally, substrate channels protect unstable intermediates and both stable and unstable cofactors, prevent substrate competition between different pathways, inhibit toxic metabolites, and avoid adverse equilibrium. Moreover, some researchers suggest that the enhanced catalytic efficiency in multienzyme cascade reactions is not solely due to the proximity of enzymes. Instead, it is postulated that the immobilized scaffold alters the enzyme properties, creating a conducive microenvironment for enzyme catalysis [103].

Although the multienzyme cascade sequential reactions have significant kinetic advantages in biocatalytic production, the potential challenges in these cascade methods, such as the balance of reaction conditions, the optimization of complex processes, cannot be ignored. The selection of attachment technologies and platforms for enzyme activity maintenance is complicated by the diverse properties of different enzymes. Each enzyme's structure and function must be considered to preserve overall activity, necessitating conditions that ensure high stability and activity for all involved enzymes. In some cases, failure to meet these requirements may completely prevent the possibility of using a one-pot reaction with the required enzymes. Another consideration is that a product of the cascade process such as free radicals, vinyl groups, or hydrogen peroxide may have an enzyme inactivation effect. In a one-pot system, all enzymes are exposed to these harmful compounds, necessitating the selection of enzymes resilient to these harmful reagents, which may limit the range of useful enzymes used in the process. Optimizing a process catalyzed by a single enzyme is intricate, involving not just enzyme reaction conditions, substrates, and intermediates, but often unstable or poorly soluble compounds. Therefore, designing an optimized process for a one-pot tandem reaction will be more complex. From this perspective, the use of different enzymes with various activity/stability under reaction conditions will reduce the range of conditions for process design. Thus, designing an appropriate cascade process for the efficient production of the target product is a critical yet challenging goal in biocatalysts. On one hand, researchers should balance the potential benefits of using several enzymes simultaneously to maximize the kinetic advantages of one-step or one-pot reactions. On the other hand, they need to balance the problems caused by these cascade reactions, such as ensuring enzyme activity and stability and mitigating the adverse effects of harmful compounds [104].

2.2.4 Strategies to Improve the Catalytic Efficiency of Cascade Systems

Beyond the strategies previously discussed, in enzyme cascade catalysis research, the catalytic efficiency can be enhanced through four key approaches including optimizing the stoichiometric ratio, increasing enzyme activity, optimizing temperature conditions, and optimizing pH conditions.

Optimizing the stoichiometric ratio of enzymes is one of the most direct ways to maximize the efficiency of the cascade reaction [105]. Optimization is usually carried out under constraints such as a limited total cost or the number of enzymes in practice. In industrial applications, increasing the use of cheap enzymes to ensure the effectiveness of expensive enzymes is an economically feasible option. The optimization of the enzyme ratio relies on the activity of single enzymes, whether determined manually or through automated means, to ascertain the optimal solution for the enzyme cascade reaction. However, the experimental workload also escalates exponentially with the increasing number of enzymes engaged in the cascade reaction. If the number of enzymes is limited, chemical methods such as immobilization and chemical modification can still be used to improve the catalytic performance of a single enzyme. Studies have shown that some immobilized carriers can aggregate substrate molecules through electrostatic interactions, specific affinity, or other interactions, and the activities of enzymes connected to these carriers increase several times, which is accompanied by a decreased apparent K_m value [106]. Although most enzymes have acceptable activity at mild temperatures (such as physiological temperatures), some of them are psychrophilic or thermophilic, especially extreme enzymes mined from extreme environments. They have evolved various unique mechanisms for maintaining the stability and activity of cell composition to withstand various extreme conditions. Therefore, it is very important to optimize the operating temperature to obtain the maximum catalytic efficiency. Cascade enzymes may operate best under different pH conditions, which may hamper one-pot catalysis. The inherent pH dependence of enzymes is difficult to engineer, because it is mainly controlled by the protonation state of key residues at the active site. However, we can shift the apparent pH-activity profile of the immobilized enzyme to a more acidic or alkaline pH value by attaching the enzyme to charged substrates [107].

The preparation of multienzyme catalysts has attracted more attention of the researchers. And the preparation and connection methods of multienzyme supramolecular complexes are diverse. Realizing complex biocatalytic cascade reactions carried out at industrial scale in a cost-effective manner faces many challenges. The preparation for multienzyme catalysts, the distance and ratio between the multienzymes, etc. also need to be further designed reasonably. Nowadays, there is still a lack of understanding of the molecular transport process and multienzyme coordination mechanism of cascade reactions catalyzed by multienzyme complexes. These challenges and the in-depth exploration of synergy mechanisms can only be solved by increasing interdisciplinary research efforts in bioengineering, biochemistry, chemistry, and process engineering. Advances in synthetic biology, materials science, and protein engineering provide a large number of methods for the design of multienzyme cascade systems, which must be merged to create powerful catalysts and systems to coordinate enzyme cascades and perform synthetic tasks similar to the metabolic networks of living cells.

3 Strategies for Enzyme-Directed Modification

In industrial production, the enzymes are typically required to exhibit high activity, good selectivity, high stability, and strong specificity to natural substrates. However, the characteristics of most natural enzymes usually fall to meet all of these criteria. Directed evolution has become a pivotal approach to address this gap. According to the methods of constructing mutant library, the directional modification strategy of enzyme can be categorized into three strategies including irrational design, semi-rational design, and rational design. By simulating natural evolution, processes involve multiple rounds of mutation; expressing and screening of the target gene are repeated; thereby the evolutions that take thousands of years in nature can be completed in a short time. Finally, proteins with improved performance or new functions are obtained (Fig. 3).

3.1 Irrational Design

Irrational design, also known as directed evolution, was introduced by Frances H. Arnold in 1991 [108]. This method, distinct from approaches reliant on protein molecular structure, is particularly beneficial for the modification of enzyme molecules whose protein structures remain unresolved.

3.1.1 Mutation Strategies In Vivo

Enzyme activity level produced by microbial strains is crucial in the production of enzyme preparations, representing the main constraints for both research and large-scale manufacturing. To overcome this challenge, researchers can employ artificially adopt physical and chemical mutagenesis methods for strain improvement, aiming to achieve high vigor and increased enzyme output. Mutation breeding based on artificially induced gene mutations offers the benefits of rapid, high-efficiency, and simplicity. This approach not only enhances the production capacity of strains but also improves the quality of products and diversifies product varieties. Mutagenesis breeding is categorized into physical mutagenesis and chemical mutagenesis according to the type of mutagens applied.

Ultraviolet mutagenesis and normal pressure and room temperature plasma mutagenesis are commonly used physical mutagenesis methods in basic experiments, which characterized by their simplicity, low equipment costs, and effective

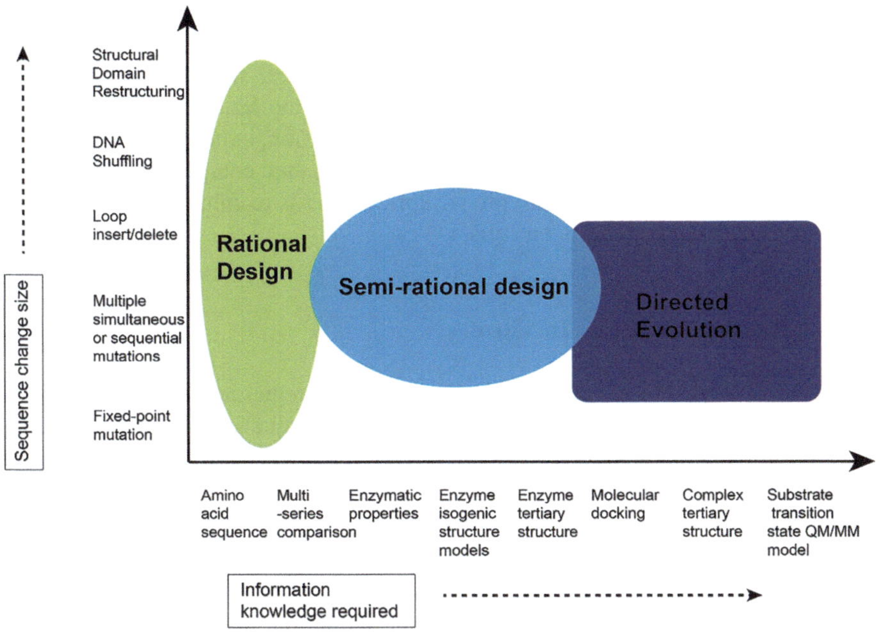

Fig. 3 The relationship between the sequence changes of the three modification strategies and the required knowledge

mutagenic outcomes. This method induces biological effects primarily by causing covalent bonding between two adjacent pyrimidines in the DNA molecule, resulting in pyrimidine dimers. The appearance of the dimer weakens the hydrogen bond between the double bonds, leading to distortions and impediments to proper base pairing, thereby inducing mutations. For instance, *Bacillus atrophicus* FSHM2 was induced to produce an excessive amount of hyperthermic basophilic lipase, tripling enzyme activity compared to the wild type [109]. Atmospheric and room temperature plasma (ARTP) microbial mutagenesis, as a new type of mutagenesis method, mainly uses the high concentration of active particles (excited helium, hydrogen, and oxygen atoms) produced by helium at atmospheric pressure. This interaction can cause DNA single-strand or double-strand breaks, which in turn induces cells to initiate an emergency repair mechanism, prompting microbial mutations. Additionally, these mutations may alter the structure and permeability of microbial cell walls or membranes, offering benefits such as cost-effectiveness, ease of operation, safety, high mutation rates among strains, and environmental sustainability.

Chemical mutagenesis involves treating microorganisms with chemical substances to increase their mutation rate. Chemical inducers currently employed fall into various categories, including alkylating agents, base analogs, frameshift mutagens, inorganic compounds, etc. Due to the diversity of chemical mutagens, each type targets a specific site on the microorganisms' genetic material without affecting other regions. The selection of chemical mutagens varies based on different experimental needs. For instance, alkylating agents like ethyl methylsulfonate (EMS) can alkylate guanine, leading to mutations in base pairing during DNA replication. Base analogs include pyrimidines and purines, such as 5-bromouracil, 2-aminopurine and the like, which inserts into DNA molecules instead of normal bases when cells are undergoing massive DNA replication, inducing structural alterations that result in mutations. Additionally, the inorganic compound lithium chloride (LiCl) is often combined with physical mutagenesis to enhance the destructive effect of physical mutagenesis on cells.

3.1.2 Mutation Strategies In Vitro

As an important part of irrational design, mutation strategies in vitro mainly include error-prone PCR, DNA shuffling, synthetic shuffling, sequence saturation mutation, exon shuffling, recombination extension, etc. It typically involves two stages including constructing a mutation library, followed by screening for the desired mutant enzyme.

Error-prone PCR technology has been widely used in the modification of enzyme molecules. This technology consists of four parts including random mutation of target base, construction of mutation library, mutant screening, and accumulation of beneficial mutations. These mutations are mainly introduced through reasonable methods to introduce base deletions, additions, and substitutions while amplifying the target gene, resulting in random mutations of the target gene to form a mutation library. The main method is to change the standard PCR reaction conditions, such as increasing the concentration of Mg ions, adding Mn ions, changing the relative concentrations of the four dNTPs, and using low-fidelity DNA polymerases. This technology was firstly proposed by Leung et al. in 1989, making it be the earliest method of constructing gene libraries used in directed evolution research. Notably, Cadwell [110] and Kuipers further refined this technique [111]. Mabrouk et al. [112] used error-prone PCR to modify the thermostability of the maltogenic amylase MAUS149 from *Bacillus*, obtaining the mutant enzyme MA-A27, MA-A27, and MA-A27 with four mutation points D46V, P78L, V145A, and K548E. New hydrogen bonds and salt bridges are formed in A27, which improves the thermal stability of the enzyme. However, epPCR's limitations include a target sequence length constraint (500–1000 bases), limited application range, prevalence of neutral mutations, a higher ratio of transition mutations compared to transversions, and codon bias. It has been reported that error-prone PCR usually fails to produce >70% of the beneficial site/amino acid exchange. However, epPCR's limitations include a target sequence length constraint (500–1000 bases), limited application range, prevalence of neutral mutations, a higher ratio of transition mutations compared to transversions, and codon bias [113].

DNA shuffling, a technology surpassing traditional error-prone PCR mutation, not only in accelerating the accumulation of beneficial mutations, but also in integrating two or more optimized enzyme properties (Table 6), has become a pivotal method for directed evolution of enzymes in vitro. Introduced by the American

Table 6 Comparison of various DNA shuffling methods

Mutation technologies	Advantages	Disadvantages
Classic DNA shuffling	Controllable mutation	Parents require high homology
Between nonhomologous sequences DNA shuffling	Does not require parental homology	Need multiple sampling and inactivation
Family shuffling	Can shuffle a series of related genes at the same time, highly beneficial mutations	Parents require high homology
Family shuffling using restriction enzymes	Highly effective reorganization, generating non-repetitive gene fragments	Limited gene diversity
ss-DNA shuffling	Effective reorganization efficiency	Parents require high homology
Stagger extension process(StEP)	Simple operations, does not purify DNA fragments, controllable mutation	Being difficult to grasp the optimal PCR reaction conditions need strictly controlling the PCR extension time and temperature

Stemmer Laboratory in 1994 [114, 115], the DNA shuffling primarily utilizes PCR amplified genes as experimental materials, which are digested with DNaseI and separated into random fragments with a size of 10–50 bp. which are then fragmented using DNaseI into random segments of 10–50 bp, followed by reassembly through primerless and primer PCR to demonstrate DNA recombination in vitro. For example, Zhao [116] et al. applied DNA shuffling to enhance the alkaline tolerance of bacterial CotA laccase, resulting in the 5E29 CotA-laccase mutant with improved functionality under alkaline conditions. Moreover, when applied to a series of related genes, the technique is known as family shuffling, requiring a minimum of 70% homology between genes. To extend the application to sequences with less than 70–80%, Ostermier et al. [117] developed the ITCHY (incremental truncation for the creation of hybrid enzymes) technique, marking a significant advancement. Subsequent innovations of DNA shuffling are included using restriction endonuclease (ER) instead of DNase I to cut double-stranded DNA, and using ss-DNA to replace ds-DNA (double-stranded DNA) as a template occurred. Zhao et al. [118] further refined the DNA shuffling technology by creatively proposing the StEP method, which consolidates annealing and extension phases of PCR, significantly reducing reaction times and bypassing the DNase fragmentation step, thus streamlining the DNA shuffling process. Rabiller et al. [119] combined StEP with random mutagenesis to alter the regioselectivity of the α-galactosidase AgaB from *Bacillus stearothermophilus*, achieving specificity for the 1,3 regions.

Similar to StEP, the Recombination Extension Truncation Technique (RETT) eschews the use of DNase for DNA shuffling, relying instead on the generation of single-stranded DNA (ssDNA) fragments. These fragments are produced either by DNA polymerase activity or through continuous deletion by exonuclease in the presence of random primers. Unidirectionally synthesized ssDNA fragments serve solely as templates in PCR, facilitating the creation of a random recombination gene library from polynucleotides that extend from the primers via template conversion. Yong ChulShin [120] applied RETT to the directed evolution of chitinase variants, achieving a chimeric chitinase with enhanced thermostability compared to the parental enzyme. Exon shuffling, akin to DNA shuffling, involves mutation-containing fragment exchanges but is particularly apt for eukaryotic genomes. Distinct from DNA shuffling, exon shuffling relies on intron homology within the same molecule for recombination, whereas DNA shuffling is unrestricted, applying to entire gene segments. Synthetic shuffling, a further developed protein engineering technique, facilitates recombination among all amino acids from parental sequences, creating a direct path from database sequences to functional libraries through the use of degenerate oligonucleotides.

Recently introduced, sequence saturation mutagenesis is a technique that enables the randomization of a target sequence at each nucleotide position, independent of the inherent mutation bias of DNA polymerase. This method allows for the specific targeting of individual nucleotides or nucleotide types (A, T, G, or C) and grants the ability to tailor the mutation spectrum through the use of universal bases. The unique hybrid base pairing characteristics of universal bases facilitate the generation of random mutations at targeted sites. Research has demonstrated that altering the concentration of sodium chloride and/or sodium hydroxide in the DNA solubilization solution can modulate the mutation frequency [121].

3.2 *Semi-Rational Design*

The essence of the semi-rational design method lies in its ability to identify promising mutation sites for evolutionary modification with limited enzyme information. This approach integrates the unique benefits of both directed evolution and rational design by leveraging bioinformatics tools and site-directed mutagenesis. It employs specialized software to strategically reduce the mutation library size, thereby enhancing mutation and screening efficiencies. This reduction significantly alleviates the task of database construction (Fig. 4).

3.2.1 Semi-Rational Design of Hotspots

Identifying optimal mutation sites is critical in semi-rational design, with significant influence on enzyme performance often attributed to regions known as hotspots. Protein sequences are typically analyzed using bioinformatics tools to gather structural and functional information. The strategy involves employing semi-rational design techniques, including amino acid sequence alignment, protein modeling, molecular dynamics simulations, molecular docking, and calculations of substrate channels, to pinpoint hotspots for saturation mutagenesis and the selection of superior mutants. Subsequent steps may involve conducting combinatory mutations at

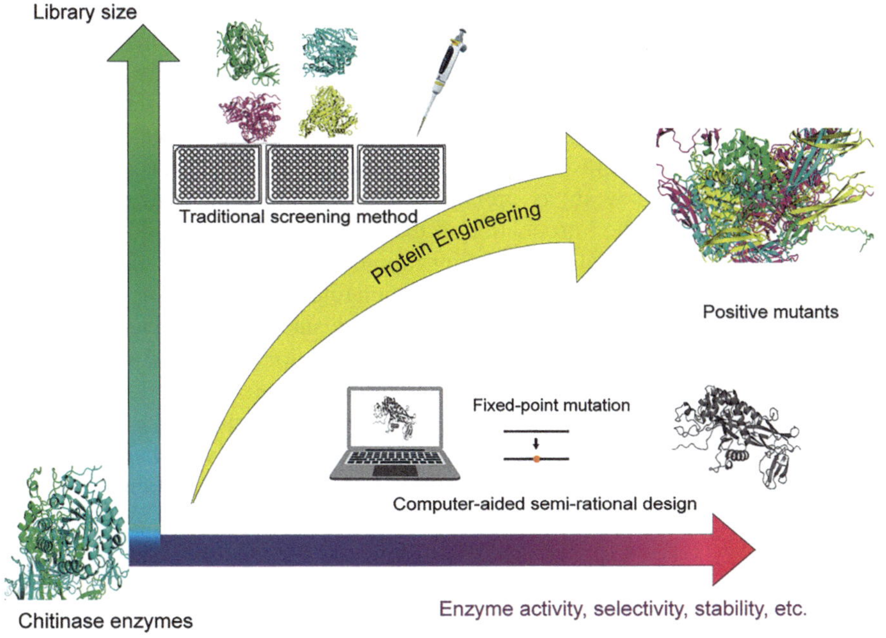

Fig. 4 Overview of semi-rational design

strategic sites such as flexible loops, substrate channels, substrate pockets, or interfaces between subunits to enhance the enzyme's catalytic efficiency. This approach substantially narrows down the mutation library and significantly decreases the experimental workload, reducing the screening process to a manageable number of colonies. *Site-directed mutagenesis in the hotspots of the catalytic active center* Site-directed mutagenesis involves the precise replacement of amino acid residues at specific sites within a protein's primary structure. An advanced form of this technique, site-directed saturation mutagenesis, entails systematically substituting targeted amino acids at certain positions with each of the 19 other natural amino acids, serving as a key semi-rational design strategy. Ozturk et al. enhanced the hydrogen peroxide resistance of α-amylase derived from *Thermotoga maritima* by substituting methionine near the active site with alanine, noting a significant tolerance increase in the M55A mutant. Specifically, after 30 minutes of hydrogen peroxide treatment, M55A maintained 50% of its enzyme activity, in contrast to the wild-type α-amylase, which completely lost its activity under identical conditions [122]. Similarly, Su et al. significantly improved the thermostability of glucuronic acid dehydrogenase from *Agrobacterium tumefaciens* through site-directed mutagenesis within a semi-rational framework, achieving a nearly 400-fold increase in the half-life $t_{1/2}$ (59 °C) at 59 °C with the mutant enzyme (A39P/H99Y/H234K) [123].

Iterative Saturation Mutation in the Hotspots of the Substrate Channel Iterative saturation mutagenesis (ISM) is a potent semi-rational design strategy for enhancing various enzymatic catalytic parameters. When aiming to modify an enzyme's substrate profile, the initial step involves selecting amino acid residues near the binding pocket for saturation mutagenesis to create a mutant library, thereby accumulating beneficial mutations. Pei et al. leveraged this approach to enhance the thermal stability and activity of chitinase PpChi1 from *Paenidenillus paasadenensis* CS0611. The resulting mutant, S244C-I319C/T259P, exhibited increased specific activity at elevated temperatures, with its optimal reaction temperature rising from 45 to 52.5 °C, attributed to the introduction of disulfide bonds and proline substitutions [124]. Similarly, Vojcic et al. applied semi-rational design to the *Subtilisin carlsberg* protease mutant T58A, focusing on the oxidation tolerance critical sites, Trp216 and Met221. Through saturation mutagenesis, four advantageous mutants were identified. Subsequent iterative mutations led to three additional mutants with enhanced oxidation resistance. Moreover, compared to the wild type, mutants T58A/W216M/M221 and T58A/W216L/M221C saw a 1.4-fold increase in their K_{cat} values. Further research has been conducted on applying semi-rational design to trehalose synthase, with recent molecular modification advancements summarized in Table 7.

3.2.2 Codon Saturation Mutation

The semi-rational design approach, employing iterative mutation, sees an exponential increase in screening scale with the addition of pseudo-mutation sites. Codon saturation mutagenesis significantly reduces the screening effort and enhances efficiency by creating a "compact and targeted" mutant library. This method has been effectively used for the directed modification of enzyme parameters, including stereoselectivity, regioselectivity, substrate spectrum expansion, and thermostability.

Single codon saturation mutation A single codon saturation mutation (SCSM) technique employs only one codon to explore the enzyme's catalytic pocket. This approach utilizes the physical and chemical characteristics of the catalytic pocket—like hydrophilicity and hydrophobicity—and existing data to thoughtfully choose a specific amino acid codon as the foundational element for modifying the catalytic pocket. The goal is to enhance or reverse the enzyme's stereoselectivity. For instance, when mutating ten sites simultaneously with 95% coverage, the scale of screening is significantly reduced from approximately 3×10^{15} (for NNK) or 2×10^{11} (for NDT) to roughly 3000. In a model reaction, limonene epoxyhydrolase (LEH) facilitated the production of chiral cyclohexane-1,2-diol from epoxyhexane. Strategic selection of hydrophobic valine (V) and phenylalanine (F) as the building blocks allowed for a combined SCSM scanning and iterative saturation mutations at the chosen ten sites. This process led to screening around 4000 transformants, ultimately yielding mutants that significantly improved stereoselectivity or achieved its reversal. The ee value was raised from 2% in the wild-type (S,S)-configuration to 92%, while the (R,R)- configuration reached 96% [131].

Table 7 Various mutants of TreS by site-directed mutagenesis

Source of microorganisms	Mutant	Enzyme activity	References
	L116M	Maltose activity is 118%, sucrose activity is 78%	
	L116E	Maltose activity is 101%, sucrose activity is 74%	
Thermomonospora curvata	L116G	Maltose activity is 17%, sucrose activity is 144%	[125]
	E330A	Only hydrolase activity, no transglycosidase activity	
	1150F	Transglycosidase activity is 55%, hydrolase activity is 170%	
	G78D/V289L/G322A/ I323L	Transglycosidase activity is 229%, 222%, 195%, 355%	[126]
	N253A	Transglycosidase activity is 11%, hydrolase activity is 180%	
Deinococcus radiodurans	R148A	Transglycosidase activity is 12%, hydrolase activity is 150%	[127]
	Y213A/E320A/E324A	No transglycosidase activity, no hydrolase activity	
	R283G/Y287R/R291G	Transglycosidase activity is 36%	
Thermobaculum terrenum	H534Y	Transglycosidase activity is 50%	[128]
	R283G/Y287R/R291G/ H534Y	Transglycosidase activity is 25%	
	F259A	Transglycosidase activity is 30%	
Pseudomonas putida P06	F223A/R535A	Transglycosidase activity is 100%	[129]
	D294A/E338A/D403A/ D112A/F115A/F255A	No transglycosidase activity, no hydrolase activity	
Pseudomonas putida ATCC47054	V407M/K490L/R680E	Transglycosidase activity is 119%	[130]

Three-codon saturation mutation To significantly reduce the screening workload, Sun et al. employed a strategy that integrated protein sequence analysis (identifying conserved sites through multiple sequence homology comparison) and structural insights (from crystal structures or homology modeling), along with enzyme catalytic characteristics and corroborating experimental data. They strategically chose three amino acid codons as the foundations for saturation mutations and innovatively grouped multiple mutation sites (three to four amino acid residues per group), leading to the development of a novel technique known as Triple Codon Saturation Mutagenesis (TCSM) [132]. Demonstrating this with the mutation of ten sites, they highlighted that the capacity of the mutant library is approximately 3.14×10^6. By organizing these sites into three groups (4 + 3 + 3) and constructing separate mutant libraries for each, the overall library size could be efficiently reduced to 1152 (768 + 192 + 192). Additionally, employing the TCSM method with limonene epoxyhydrolase (LEH) as a case study, they simultaneously improved stereoselectivity, thermal stability, and catalytic activity. Despite a minor reduction in catalytic efficiency, they achieved a notable increase in stereoselectivity to 94% (S,S) and 80% (R,R) and enhanced thermal stability by 5–10 °C, marking the first successful coevolution of these properties [133].

3.3 Rational Design

Compared to irrational evolution and semi-rational design approaches, rational design offers a more expedient and focused methodology that significantly diminishes the screening workload, as well as the associated time and material costs. The precision of computational science coupled with structural enzyme analysis has progressively enabled the rational design and modification of enzymes. Leveraging computer-aided virtual mutations, bioinformatics knowledge facilitates the analysis and prediction of mutations' impacts on enzymes, thus streamlining the selection of promising mutant targets for subsequent tailored modification. Structural calculations guide the identification of potentially beneficial mutations for experimental validation.

3.3.1 Homology Comparison

The homology comparison typically involves a comparative analysis between the enzyme subject to modification and its thermostable homologous counterpart, mostly sourced from thermophilic strains, leading to the foundation of the key amino acids that affects the thermostability and subsequently modifying them. Such an approach is recognized for its precision and efficacy, grounded in evolutionary principles. For conducting homology comparisons, software tools like COBALT and Clustal W are prevalently utilized. Dror et al. [134] successfully increased the thermostability of a lipase from *Geobacillus stearothermophilus* using the sequence consensus method, resulting in a notable enhancement of the Tm value by 4.2 °C.

3.3.2 Surface Charge Optimization

Proteins typically consist of a hydrophobic core and a charged hydrophilic surface, with charged amino acids strategically positioned to facilitate a protective network that mitigates aggregation of partially unfolded proteins, thereby enhancing renaturation efficiency and thermal stability. Optimization of the surface charge distribution—specifically the rearrangement of charged amino acids like glutamic acid, aspartic acid, lysine, arginine, and histidine—aims to improve thermal stability and other protein properties. At present, surface charge optimization has been implemented with the aid of calculation tools such as Modeller, AvNAPSA supercharge, Rosetta supercharge, and so on.

3.3.3 Introduction of Disulfide Bond

The disulfide bond is a stable covalent bond formed between two cysteines in which the hydroxyl group is oxidized to the -S-S- form. The formation of disulfide bonds will quickly close the amino acid residues near cysteine, forming a highly dense area, allowing hydrophobic amino acid residues to gather into a local hydrophobic environment, protecting internal molecular hydrogen bonds, and improving the local stability of protein molecules. In addition, the disulfide bonds can also reduce the self-degradation of proteases and influence protein folding speed and folding pathways. According to the reports, each disulfide bond can increase the enzyme protein's stability by approximately 5 kcal/mol ΔG. At present, calculation tools such as MODIP, BridgeD, Disulfide by Design 2 (DbD2), and SSBOND have been developed to predict the disulfide bonds of proteins. These calculation tools are further utilized alongside thermal stability prediction algorithms such as FoldX and Rosetta, which is a strategic method to identify stable disulfide bond mutants effectively.

3.3.4 Proline Effect

Studies have shown that proline is seldom found in the α-helix structure [135] due to its inability to adopt φ angle necessary for the α-helix formation and the absence of a hydrogen atom on the nitrogen atom, which precludes the formation of hydrogen bonds essential for stabilizing the α-helix. In 1987, Matthews et al. firstly introduced the "proline theory," suggesting that the introduction of proline in non-α-helixes, and specifically the mutation of glycine into proline, can improve protein stability without changing the protein backbone. The introduction of proline is theorized to decrease the entropy associated with protein unfolding and increase stability by augmenting protein rigidity.

3.3.5 Thermal Stability Prediction Algorithm

The Gibbs free energy of protein unfolding (ΔG) is considered to be a general indicator of its thermal stability. Many methods have been used to calculate the Gibbs free energy difference ($\Delta\Delta G$) between the wild type and its mutant to guide the modification of the enzyme. Some of these methods relied on energy functions to calculate ΔG, while others used deep learning methods. The methods of using energy functions are categorized into physical potential approaches (PPA), statistical potential approaches (SPA), and empirical potential approaches (EPA). Notably, FoldX, CUPSAT, and SDM use EPA energy function; dDFIRE, Rosetta ddg_monomer, Eris, etc. employ PPA energy function; and AUTO-MUTE utilize SPA energy function.

3.3.6 SCHEMA Simulation

The SCHEMA simulation method, deriving its principles from the unconventional approach of DNA reorganization, was initially introduced by Voigt et al. in 2002. SCHEMA is a calculation method for identifying sequence blocks. It can analyze the structure of a protein and divide the protein into several independent small fragments based on the structural information of the protein. The algorithm then methodically reassembles these fragments, striving to preserve the integrity of the protein structure. This process enables the incorporation of thermostability-related segments from thermophilic proteins into mesophilic proteins, thus enhancing the structural stability of the latter. Currently, the SCHEMA method has been effectively utilized to enhance the thermostability of cellulase through molecular engineering [136].

3.3.7 Binding of Nonnatural Covalent Bonds

Computer-aided design of nonnatural covalent bond "binding" sites is a recently developed new thermal stability strategy. With advances in genetic code research, it has been realized that nonstandard amino acids (noncanonical amino acids, ncAA) could be directly integrated into proteins through translation, which enhance the thermal stability and chemical resistance of the enzyme, or the endowment of new catalytic functions. Consequently, this area has emerged as a significant focus of contemporary research. For example, Moore et al. [137] directly used this algorithm to design unnatural covalent thioether bonds to stabilize myoglobin. The resulting superior variant, sMb10, exhibited a remarkable increase in its Tm value by 16.8 °C.

3.3.8 Machine Learning Contributed to the Directed Modification of Enzymes

Protein function is fundamentally determined by its three-dimensional (3D) structure, which results from the folding of its amino acid sequence. Protein function is fundamentally determined by its three-dimensional (3D) structure, which results from the folding of its amino acid sequence. AlphaFold2, for instance, has marked a significant breakthrough by accurately predicting the three-dimensional (3D) structure of the protein from their amino acid sequence with atomic precision. This method involves comparing the initial amino acid sequence with homologous sequences to predict the three-dimensional coordinates of all heavy atoms within the protein. Research by P. Bryant and G. Pozzati et al. highlights the application of AlphaFold2 in predicting the structure of heterodimeric protein complexes, achieving an accuracy of 59% [138]. Furthermore, David Baker et al. have developed RoseTTAFold, based on deep learning techniques pioneered during AlphaFold2's development. RoseTTAFold offers accuracy comparable to AlphaFold2 but operates faster and with reduced computational demands [139].

3.4 New Technologies for Continuous Directed Evolution

3.4.1 Continuous Directed Evolution In Vivo

Theoretically, a single living cell can serve as a scalable and experimental system for in vivo continuous evolution. In stressful environments, cells undergo genomic mutations to adapt, with these mutations accumulating as cells proliferate, leading to mutants with enhanced environmental adaptability. However, the presence of numerous essential genes imposes a "mutation rate limit," restricting the speed at which mutations can occur and challenging the feasibility of directed evolution in vivo [140]. In recent years, research on DNA polymerases and mismatch repair systems have facilitated increased genomic or targeted gene mismatch rates, enabling directed evolution within cells. This process, known as in vivo continuous directed evolution, integrates mutation, translation (including evolution non-gene, screening, and replication steps) in a seamless cycle. After completing one evolution cycle, the system automatically progresses to the next, allowing for minimal human intervention and enabling faster, more extensive, and precisely targeted evolutionary trajectories [141]. Various in vivo continuous directed evolution techniques have been explored across different organisms, with rapid advancements leading to widely used methods such as phage-assisted continuous evolution (PACE) and multiplex automated genome engineering (MAGE).

Phages are characterized by their high mutation efficiency and simplicity of manipulation, making them the inaugural models for in vivo continuous evolution [142]. Currently, the PACE invented by Esvelt et al. is now prevalently utilized [143]. This approach integrates the target protein's evolution with the phage's life cycle through automated continuous culture, enabling the perpetual evolution of the target molecule. The life cycle of the phage is short, with progeny phage being generated in just 10 minutes. This system utilizes the proliferation ability of M13 phage and relies on the characteristics of inducible protein III (pIII protein) to construct a selective phage that contains the target gene but is defective in pIII protein. Distinctively, in this system, the target gene is harbored within the M13 phage, whereas both mutagenic and selective factors (mutagenic and helper plasmids, respectively) are hosted in *E. coli*. Upon infection of *E. coli* by the phage, the mutagenic plasmid triggers phage mutation and releases specific mutant phages that induce the production of pIII protein, facilitating the screening process.

In vivo genome modification of bacteria is conducted using multiplex automated genome engineering (MAGE), a technique that enables the continuous evolution of bacteria. MAGE facilitates the alteration of multiple genes or chromosomal sites through various mechanisms, such as insertion, mismatch, or deletion, thereby generating a diverse array of gene mutant strains. This method employs the lambda phage Red recombinase homologous recombination system, introducing synthetic single-stranded DNA fragments containing mutation sites into *E. coli*. The single-stranded DNA fragments can efficiently integrate with the lagging strand under recombinase activity, allowing for the automatic recombination of multiple DNA fragments. The mutant library generated is then subjected to appropriate selection pressure to screen mutants exhibiting superior traits, which can then undergo further rounds of multiple automatic integration and screening. However, compared with traditional directed evolution technology, MAGE achieves a continuous process of translation, screening, and replication, significantly advancing the host cell genome editing and providing a foundation for the advancement of continuous directed evolution technologies.

Based on the in vivo continuous directed evolution method, a highly error-prone orthogonal replication system (OrthoRep) and in vivo continuous evolution technology (ICE) are developed in *S. cerevisiae*. OrthoRep operates independently from the host's genome, exhibiting a mutation rate that is 100,000 times higher than the host's in vivo mutation rate. ICE technology facilitates the efficient acquisition of mutants with enhanced performance across single enzymes, global transcription factors, and multigene metabolic pathways. Despite these advantages, ICE technology's mutations occurring within the genome can also present challenges by potentially impacting cell adaptability and diminishing the effectiveness of the selection process.

In summary, the developments of continuous directed evolution in vivo across phages were showed in Table 8. Each in vivo directed evolution technology presents unique benefits and limitations. The selection of an appropriate evolution technology should be based on the specific evolutionary host, mutation frequency, and the complexity of the screening method.

3.4.2 Application of CRISPR/Cas Technology in Directed Evolution

The base editor uses DNA-modifying enzymes and catalytically defective CRISPR proteins to precisely introduce point mutations. So far, two types of base editors including cytosine base editor (CBE) which could convert C-G to T-A and adenine base editor (ABE) which could convert A-T to G-C have been developed [152]. The cytosine base editing system (CBE) also could realize the conversion from C to T, and the mechanism is that when the protein-RNA complex formed by dCas9 and sgRNA are bind to the target DNA, the double-stranded DNA is opened, and one is replaced with other single-stranded DNA. Then, the target gene PAM sequence NGG is cytosine deaminated at positions -13 to -17 upstream by APOBECI to generate uracil, which will be used to be recognized as T in the subsequent DNA replication process, finally realizing the conversion of target site C to T [153].

In 2016, Nishida et al. [154] fused cytosine deaminase with CRISPR/Cas9 and UGI to form a complex of dCas9-gRNA and PmCDA1, named the Cas9n-AID base

Table 8 The comparison of continuous evolution techniques in vivo

Directed evolution	Technology	Mutation location	Continuous	Mutagenic factor	Mutation rate /bp/ generation	Advantage	Disadvantage	Re
Phage	PACE	Target gene orthogonal	Continuous	M13 polymerase	2.3×10^{-3}	Fast evolution and wide range of applications	Limited to the cytoplasm of *E. coli*	[144]
Prokaryotic cell	MAGE/CREACE	Target gene non-orthogonal	Discontinuous	Chemical synthesis/dnaQ polymerase	1.35×10^{-7} 3×10^{-8}	Automatic integration into the genome	Need to manually design mutation library, discontinuous	[145]
	MMR/ TM-MAGE/ pOPRTMAGE	Genome	Continuous	Knockout methyl repair gene mutS/mutL	1×10^{-8}	Mutation and evolution of the cell genome	Unable to evolve for specific genes	[146]
	Ribosomal switch/ dual function switch	Genome	Continuous	DnaQ polymerase	–	Precise regulation of cell genome	Unable to evolve for specific genes	[147]
	EvolvR	Target genes are not completely orthogonal	Discontinuous	Error-prone polymerase	2.4×10^{-6}	Specific gene evolution	Editable gene length is limited(≤350 bp)	[148]
Eukaryotic cell	CRISPR-X	Target gene non-orthogonal	Discontinuous	Cytidine deaminase	1×10^{-3}	Specific type of base mutation	Can only achieve C to T mutation	[149]
	ICE	Target gene orthogonal	Continuous	Error-prone reverse transcriptase	1.5×10^{-4}	Continuous evolution in vivo	Interfere with cell growth	[150]
	Orthorep	Target gene orthogonal	Continuous	Error-prone DNA polymerase	1×10^{-5}	Continuous evolution in vivo	The system design is complicated and the cell burden is heavy	[151]

editor. In 2016, Ma et al. [155] removed the C-terminus of human cytosine deaminase (hAID) to obtain AIDx and then fused dCas9 with AIDx to develop a targeted AID-mediated mutagenesis (TAM), which could transform C to T, A, and G at bases 5–9 of PAM to generate mutations. Then the human-derived cytosine deaminase mutant was fused to the MS2 protein after removing the nuclear localization signal and recruited into the sgRNA backbone through the hairpin structure of MS2 RNA. After optimization, CRISPR-X was developed through dCas9, which target the highly active cytidine deaminase AID to induce local and diverse point mutations. Active Cas9 introduces mutations mainly through insertions and deletions, or homologous recombination after Cas9 cleavage to generate oligonucleotide mutation libraries. In contrast, CRISPR-X produced local sequence diversification, which simultaneously mutated multiple genomic sites and specifically mutagenized endogenous targets. In 2017, Kim et al. [156] established the YE1-BE3 base editing system through amino acid mutation (W90Y + R126E) in the cytosine deaminase functional domain. The YE1-BE3 base system was edited while maintaining similar editing activity as BE3. The window is more precise. Fusion of Cpf1 and APOBEC1, due to the special structure of Cpf1, could not be designed as a nick targeting chain. And the nontargeting chain is exposed to deaminase. Therefore, similar to TALE-AID, the initial Cpf1-Apobec1 efficiency was not high. In 2018, Li et al. [157] modified the editing editor by introducing the nuclear positioning signal into Cpf1-Apobec1 to form dCpf1-Bes, which successfully increased the editing efficiency of the system to 40%.

In 2019, the research team developed a base editor (phage-assisted continuous evolution of base editors, BE-PACE) that uses phage-assisted continuous evolution to improve its editing efficiency and target sequence compatibility. The efficiency of BE-PACE can carry out dozens of generations of mutations, selection, and copying every day to produce a base editor with improved target sequence compatibility and higher activity. In 2020, Grunewald et al. [158] fused adenosine deaminase (TadA) and cytosine deaminase (PmCDA1) to the N-terminus and C-terminus of Cas9n in order to developed a CRISPR/Cas9-based two-base editor that can simultaneously edit adenine and cytosine. The C-T editing efficiency of the double-base editor is equivalent to that of CBE, and the editing efficiency of A-G is slightly reduced, while the overall base editing efficiency of the double-base editor is higher than that of the CBE and ABE systems alone.

In addition to using the base editor for single-base conversion, the use of the CRISPR system to recruit other mutagenized proteins also can generate local sequence diversification. The system relies on a user-defined guide RNA (gRNA) with a 20 nt spacer sequence to achieve precise positioning. EvolvR involves two important elements, including the variant Cas9 (nickase Cas9 or nCas9), in which the RuvC nuclease domain is mutated and only single-strand breaks occur. There is also an error-prone polymerase (PolI3M) directly fused with nCas9. Firstly, nCas9 locates the target genomic site to create a nick on the single-stranded DNA complementary of the gRNA. After nCas9 is separated, PolI3M bound at the cut position. PolI3M aggregates a new DNA strand, which replaces the old DNA strand and occasionally be introduced wrong bases. Before dissociation of PolI3M, the petaloid endonuclease domain of PolI3M cuts the original replacement strand, leaving a gap for ligation. After the dissociation, the gap repair is completed in the host cell. By introducing semi-random mutations into any small region downstream of the genome or plasmid site that can be targeted by CRISPR/Cas9, EvolvR can relocate and continuously evolve the same site. EvolvR can be used in any bacterial strain but typically only used in *E. coli* in nowadays. Therefore, we need to further optimize its use in other bacteria.

3.4.3 Directed Evolution Based on Ancestor Sequence Reconstruction

The enzymes in nature have evolved over hundreds of millions of years, forming various complex and delicate structures to catalyze all kinds of reactions. Understanding how the current structure and function evolved will help us to explore the early biochemical evolution of the earth, the origin and evolution of species, and the evolution of structure and function. Fortunately, the ancestral sequence reconstruction (ASR) technologies, especially the enzymatic ASR technology, provide new ideas for solving some of these basic problems [159]. The enzymatic ASR technology usually includes six steps, namely, collection of homologous sequence sets, multiple sequence alignment of sequence sets, construction of phylogenetic trees, ancestral sequence analysis based on bioinformation methods, recombinant expression, and characterization of enzymatic properties (Fig. 5) [160]. ASR technology is widely used in enzyme engineering, including rebuilding and resurrecting enzymes in vivo or in vitro to investigate the differences between ancestral and modern enzymes, identify key amino acid that vital to enzyme functions on a time scale, as well as study the relationship among sequence, structure, and functions from a vertical perspective. Meanwhile, the extreme environment of earth during the pre-Cambrian period contributed to the ancestral enzymes a better thermal

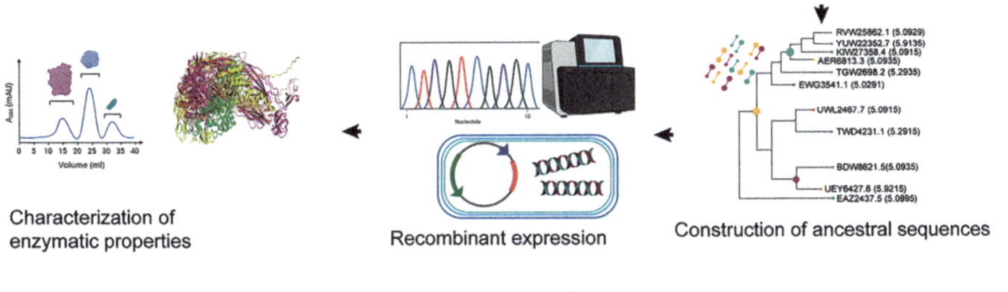

Fig. 5 The processes of ancestor sequence reconstruction

stability, high heterologous expression, low pH activity, catalytic promiscuousness and other characteristics, which are more in line with the demand for industrial enzymes. In addition, the ancestral enzyme also has a better mutation tolerance, which makes it more suitable for a protein scaffold during directed evolution than modern enzymes [161].

In a previous study, a rational design method for thermostable enzymes based on the ASR technology was developed. The thermotolerance of proteins could be improved by adding ancestral amino acid residues to modern protein sequences, which has wide applicability [162]. Based on the probabilistic search of nonconservative functional spaces, new sequences could be generated by the ASR method. Accurate sequence alignment of the input makes each output highly functional. If the input data set is rich and diverse, the obtained ancestral sequence will usually cause a big difference with the existing sequence, even the homology is <30%. This allows researchers to discover beneficial mutations that cannot be obtained by other methods, including coordinated mutation sets [163]. These could change the characteristics determined by the full sequence of the proteins, including stability under heat or other pressures. Therefore, the ancestral enzyme could be employed as a starting point for the evolution of new enzymes (Fig. 6). Previously, the ancestral sequences of several fungal laccases dating back between 500 and 250 million years were reconstructed. The reconstructed laccase was easily secreted by yeast for their

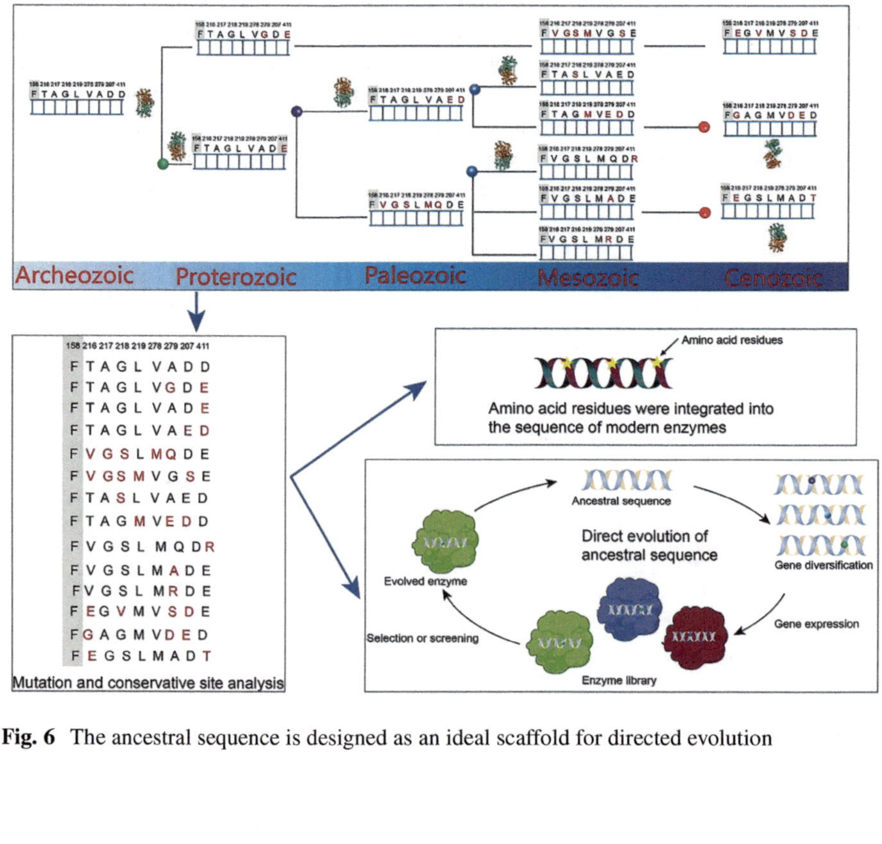

Fig. 6 The ancestral sequence is designed as an ideal scaffold for directed evolution

similar kinetic parameters, broader stability, and pH activity. The ancestor enzyme was further directly evolved to increase its catalytic rate of 1,3-cyclopentadione oxidation [164]. With the development of directed evolution and semi-rational design technologies, the stability of ancestral enzymes, especially mutation stability, has become a unique advantage, which could effectively improve the activity and stability of enzymes.

4 High-Throughput Screening Strategies

4.1 Screening Platforms for Enzyme Engineering with Different Throughputs

Currently, the traditional agar plate screening method and microplate screening method are the two most commonly used approaches. However, both methods suffer from drawbacks such as low-throughput, time-consuming, and challenges in accurate quantification. In order to solve these issues, ultrahigh-throughput screening methods such as fluorescence-activated cell sorting and droplet microfluidic sorting (>10^7) have been developed in recent years for screening of the target mutation from large-capacity mutation libraries during enzyme-directed transformation (Fig. 7) (Table 9) [165–167]. Despite their advantages, most enzyme mutation screenings still rely on the traditional substrate plate or microwell method in nowadays. It is primarily due to the ease of operation and high versatility offered by these two methods. While they have been widely used up to now, their throughput remains limited, typically ranging from 10^4 to 10^7 clones per day.

4.1.1 Fluorescence-Activated Cell Sorting (FACS)

FACS is a fluorescence-activated cell sorting technique that could sort single cells efficiently [168]. FACS can sort cells according to cell size or fluorescence at a rate of up to 10^7 fluxes [169]. Its advantage lies not only in the ability to detect cells possess different fluorescence intensities at an ultrahigh speed of 10^7 per hour but also in the ability to use the function of the flow cytometer to simultaneously sort samples into the flow cytometer. Afterward, the individual cells are wrapped by the sheath fluid stream and separated into individual droplets, arranged in a beam shape and then passed through a laser beam (focused laser beam). At the same time, the detector records and feeds back the information obtained through the scattering (the nature of the cell itself), fluorescence (the enzyme reaction product in the cell), and other signals emitted by excited cells. Subsequently, it manages the sorting of target cells meeting the specified criteria. In one step, the flow cytometer will add different charges to the cells according to the settings. Once the cells enter the high-voltage deflection plate, they will have different paths and finally fall into the corresponding collection tube.

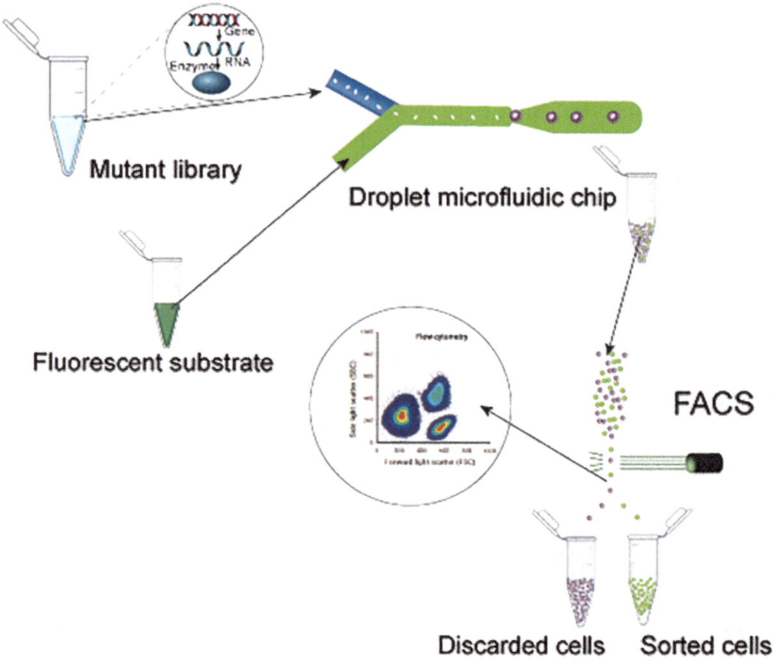

Fig. 7 Ultrahigh-throughput screening systems based on droplet microfluidics

Table 9 Comparison of different screening methods and throughput

Method	Flux (clone/d)
Agar plate screening	1 × 105
Microplate screening	1 × 104
Fluorescence activated cell sorting (FACS)	1 × 107
Droplet Microfluidic Screening (DMFS)	1 × 105

In FACS technology, rational design can be used to make the transporter on cell membrane to recognize the special fluorescent substrate and transport it into the cell for reaction, while the formed fluorescent product cannot be recognized by the specific transmembrane transporter. In the cell, the FACS technology that enriches and induces fluorescence has the characteristics of high sensitivity and high throughput, which enables it to perform high-throughput screening from large-capacity mutation libraries with a very high sorting frequency (10–15 kHz). Thus, the FACS technology has been widely used in discovery of new enzymes in high enzyme activity screening and targeted transformation of cell factories. Since FACS is a screening method at the whole-cell level, strategies such as enriched fluorescent reporter genes that display intracellular fluorescent products on the cell membrane surface must be used to ensure the genotype and expression during the screening process. For example, Tan et al. [170] constructed an ultrahigh-throughput FACS screening method based on the principle of intracellular fluorescent products enrichment and successfully screened the α-1,3-fucosyltransferase mutant with a 14-fold increase in activity. Therefore, FACS screening strategies are special cases of case by case in accumulation with a particularly poor versatility, making it difficult to apply the mutations to other enzymes.

4.1.2 Droplet Microfluidic Screening (DMFS)

Microfluidic technology integrates each step into one or more chips to form a micro-channel, such as sample preparation, sample separation, sample reaction, cell incubation, detection, sorting, etc. This technology is also known as the lab-on-a-chip for the chip can reach the millimeter level, with the microfluidics run through the entire reaction system. On the droplet microfluidic chip, a bacterium or cell is infiltrated into a droplet with a radius of only 2.5–60 μm, and the number of processed cells increased to one million in 24 h, which greatly improves the screening ability. In addition, since the package of the droplets can play the role of concentration, the sensitivity of detection is ensured. The microfluidic technology is currently limited for the screening only relies on fluorescence intensity. However, Gielen et al. [171] proposed an efficient microfluidic absorption-activated droplet sorting (AADS) method in 2016, which could activate droplets by absorbance. The sorting module had a rate of 300 droplets per second. Then the method was successfully used in the improvement of phenylalanine dehydrogenase activity.

The DMFS method realizes the coupling of genotype and phenotype by embedding single cells in droplets with continuous high frequency (>10 kHz) on the chip and performs quantitative analysis and sorting by detecting the substance signals in the droplets [167]. Its screening flux is as high as 10^5, and the water-in-oil droplets provide a nanoliter to skin upgrade reaction compartment, making this screening method not only suitable for screening intracellular enzymes or metabolites but also for screening extracellular secretase or metabolism product. Compared with the conventional microliter to milliliter level reaction system, this method is more than a million times smaller, significantly decreasing the quantity of reagents required. This presents a distinct advantage, particularly when employing costly substrates or reagents. Furthermore, the process permits operations such as the injection, fusion, and splitting of droplets after embedding in a monolayer, enhancing operational flexibility. The water/oil/water double droplets formed after the water-in-oil single droplet is embedded in the water phase could also be screened by FACS, with the throughput notably increased to 10^7. Similarly, the hydrogel droplets produced by the microfluidic chip are amenable to FACS screening. Although the hydrogel droplets offer stability conducive to prolonged cell culture, forming double droplets and hydrogels presents challenges. Hinder the seamless execution of subsequent manipulations, including analytical reagent injections and droplet fusion or splitting, thereby limiting the flexibility of the screening process. DMFS combines sophisticated droplet operation and rapid sorting system and has become a powerful tool for directional modification of extracellular enzyme and metabolite mutation library screening. Since its initial application by Agresti et al. [166] in 2010 for the modification of horseradish peroxidase, DMFS has been extensively applied to the targeted engineering of other enzymes and cell factories. While DMFS offers substantial benefits for the high-throughput screening of extracellular products, the complexity and technical demands of the DMFS process pose challenges for novices and require significant optimization across different targets, necessitating extensive system adjustments.

Fluorescence-activated droplet sorting (FADS) is a method of DMFS on the basis of FACS, through the use of water-oil-water (w/o/w) secondary micro-liquid. This method encapsulates components of the catalytic system—including the enzyme, substrate, and product—within droplets, effectively linking genotype to phenotype without necessitating intricate coupling strategies for diverse enzymes, thereby offering broad applicability. The core concept of FADS involves the integration of microdroplet generation, enzyme protein expression, enzyme reaction, enzyme activity assessment, and microdroplet sorting within a singular microfluidic chip platform. Its principal advantage lies in the generation of uniform droplets via microfluidic chips, allowing for precise quantification of enzyme activity. This significantly enhances screening accuracy, boosts throughput, and lowers costs. Weitz et al. [166] successfully employed FADS for the directed evolution of horseradish peroxidase, enhancing the activity of an already potent commercial enzyme by tenfold through the screening of 10^8 individuals, achieving a catalytic rate that approached the diffusion limit of molecules.

4.2 *Signal Detection for High-Throughput Screening Platforms*

Identifying an effective signal detection strategy is pivotal for developing screening methodologies. The fluorescent signal detection is commonly used in mutation library screening. Although absorbance and Raman spectroscopy have recently been integrated into droplet microfluidic screening systems, these methods require further refinement in terms of sensitivity, operational simplicity, and throughput. Despite these emerging techniques, fluorescence detection remains superior in sensitivity and reliability for quantitative analysis [172]. Its ultrahigh sensitivity, rapid response capability, and the availability of advanced detection technology solidify its status as the most widely utilized detection method in ultrahigh-throughput screening (Fig. 8) [173].

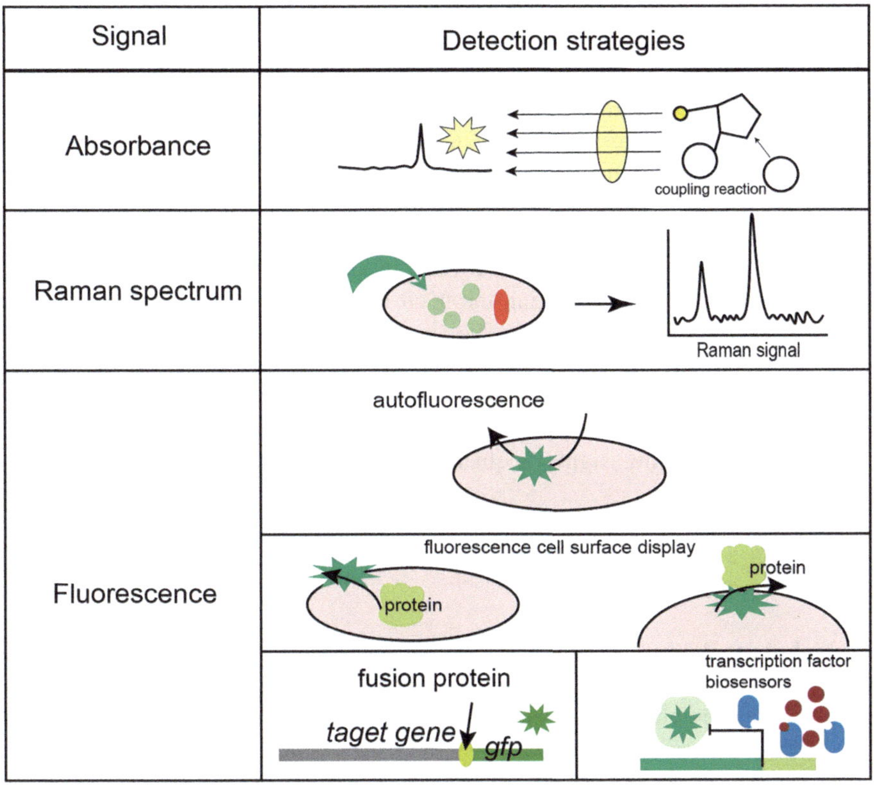

Fig. 8 Signal detection strategies in high-throughput screening process

4.2.1 Spectrophotometric Detection

Spectrophotometry serves as a pivotal technique for both qualitative and quantitative analysis of substances, leveraging the measurement of absorbance at specific wavelengths or within a defined wavelength range. At present, it has been widely used in microplate screening due to its ease of operation. Moreover, ultrahigh-throughput systems utilizing spectrophotometric detection have emerged, notably those employing the enzymatic reaction of phenylalanine dehydrogenase (PheDH) with NAD^+ consumption, coupled with the generation of a blue-violet dye, crystalline formazan, via 1-methoxy-5-methylphenazine methyl sulfate (mPMS) as a redox mediator. This approach significantly enhances the assay's absorbance signal, thereby improving accuracy and sensitivity [171]. However, the sensitivity of absorbance-based detection is considerably lower than that of fluorescence-based methods, by approximately three to four orders of magnitude. A further limitation is the requirement for larger droplets to extend the optical path for adequate signal collection. Despite these challenges, advancements in spectrophotometric detection technology have notably broadened its applications in droplet-based microfluidic fluorescence-activated cell sorting (DMFS), enriching the field [171].

4.2.2 Raman Spectroscopy Detection

Raman spectroscopy, an innovative droplet-based microfluidic technology, is frequently employed for screening when direct labeling of the target product is impractical. The inaugural integrated Raman droplet microfluidic sorting system, designed specifically for astaxanthin screening, was introduced in 2018. This system seamlessly combines Raman signal acquisition and analysis within an automated framework for embedding and sorting in microalgae cells [172]. It operates by acquiring Raman spectral maps from individual cells and comparing these to a database for target cell identification, obviating the need for cellular labeling and thereby preserving cell integrity. This noninvasive approach holds significant potential across various applications. However, the acquisition time is extended due to the weak signal, resulting in significantly lower throughput (260 droplets/min) compared to fluorescence sorting (600 droplets/s). Enhancing the sensitivity of Raman signal detection and reducing signal acquisition time in the future could broaden its application.

4.2.3 Fluorescence Detection

For high-throughput screening using droplet microfluidic technology, the primary challenge is converting the biological activity that needs to be analyzed into a detectable fluorescent signal, which is crucial for the development of droplet microfluidic screening systems. Fluorescence detection, characterized by its high sensitivity, strong responsiveness, and well-established detectors, has emerged as the predominant method in the screening of mutation libraries for enzyme transformation and cell factory applications. Currently, strategies to couple fluorescent signals primarily rely on various fluorescence generation methods. Direct detection is feasible for targets with inherent fluorescence, such as riboflavin, betaine, camptothecin, etc. Unfortunately, most targets, like enzymes, lack autofluorescence and require linkage to fluorescent markers for detection.

For substrates lacking autofluorescence, the fluorescent intracellular enrichment method is employed. This technique correlates enzyme activity with fluorescence intensity by designing a suitable substrate that could permeate the cell membrane, interact with the enzyme inside the cell, and produce a fluorescent product unable to exit the cell. Fluorophore-labeled substrates facilitate fluorescent signal generation through enzyme-catalyzed release of fluorophores, thereby enabling the detection of the target enzyme's activity based on fluorescent signal intensity. While most fluorophore-labeled substrates are applicable to hydrolases and oxidoreductases, only a select few are used for other enzymes, such as ligases, cleavage enzymes, and transferases. Screening aromatic sulfatase mutant libraries with dimethylphosphoryl difluoride-fluorescein as a substrate has resulted in mutants with a sixfold increase in total enzyme activity and expression [174].

The fluorescence cell surface display technique, another widely used method, involves expressing the enzyme on the cell surface, significantly facilitating FACS applications. This approach eliminates the need for substrates to penetrate the cell membrane, as enzymatic reactions occur directly on the cell surface with the enzyme. The most pivotal aspect of the fluorescence cell surface display is linking the fluorescent reaction product to the cell, preventing its diffusion. The genotype-phenotype linkage is achieved by designing substrates that immobilize the corresponding fluorescent product on the cell membrane through specific physicochemical mechanisms. In 2000, the feasibility of immobilizing fluorescent products on cell membranes was demonstrated with OmpT, a serine protease naturally expressed on the surface of *Escherichia coli*, which was firstly evolved using FACS method. A fluorescence resonance energy transfer (FRET) peptide substrate was designed, featuring a fluorescent moiety on the positively charged part (i.e., the part bound to the cell membrane) and a corresponding quenched moiety on the other end. The surface-displayed protease acts on the substrate between these two groups, ensuring that upon cleavage of the quenched group, the fluorescent part remains adsorbed on the cell surface. Using a different design of the amino acid sequence between the "fluorescence-quenching" groups, the hydrolytic activity of OmpT for the unnatural site Arg-Val was successfully increased by 60-fold. With the similar principles, counterselection substrates could be designed to reduce the natural activity of the enzyme and further improve enzyme specificity for nonnatural substrates [175–177].

Beyond the aforementioned methods that use electric charge to immobilize fluorescent products on the cell surfaces [178], a distinct approach employing covalent binding for the adsorption of fluorescent products was introduced in 2007, which successfully implemented directed evolution of horseradish peroxidase (HRP) stereoselectivity. Compared to methods that rely on charge, covalent binding enhances the precision of screenings owing to its stability, which mitigates the cross-interference of signals among cells. This approach has been employed to screen the stereoselective esterase (EstA) derived from *Pseudomonas aeruginosa*. Nonetheless, the technique of fluorescence cell surface display sometimes encounters issues with accuracy, and in some instances, the fluorescence intensity does not accurately reflect enzyme activity.

Fluorescent reporter proteins, genetically encoded by fluorescent chromophores, have been developed extensively since the discovery of the green fluorescent protein (GFP) in the luminescent jellyfish *Aequorea victoria* [179, 180] successively. GFP from different sources, as well as their derivatives, has facilitated the directed evolution of recombinant enzymes by serving as a reporter gene. Cre recombinase, a promising tool for genetic engineering and even synthetic biology, enables sequence recombination at specific LoxP sites, allowing for mutations, gene substitutions, or the creation of genetic switches. Directed evolution that changes the specificity of Cre for DNA sequence recognition would greatly expand its practical applications [181]. A FACS-based approach, which uses GFP as a reporter protein, has also been developed for screening Cre recombinase specificity. In this system, genes for two heterodimers of GFP (GFPUV and EYFP) are positioned between two specific LoxP (LoxP Variants) sites, with GFPUV oriented oppositely to EYFP and T7/Lac promoters. The reversible nature of recombination permits the production of distinct fluorescent signals, contingent on the Cre mutant's ability to recognize altered

LoxP sites, enabling the screening of Cre recombinases for varied specificities through simple modifications of the LoxP sequence. Similarly, this system has been applied to evolve the substrate specificity of amido-tRNA synthetase [182].

When it comes to another methods, the nucleic acid sensors, notably transcription factor biosensors [183] and riboswitch biosensors [184], have been also widely used. Transcription factor biosensors involve the introduction of heterologous transcription factors into a new host, where they attach to the promoter or enhancer regions of a gene. The transcription factor could interact with substrate, product, or metabolite of the target enzyme, thereby modulating the transcriptional process involving RNA polymerase. This modulation allows for the transcription and expression of the fluorescent reporter gene to be correlated with the target enzyme activity or the concentration of metabolites, which thus could be detected. Riboswitch biosensors turn on or off the transcription and expression of the reporter gene through changing the conformation of RNA riboswitch when binding the target product. The yield of the target product can be detected based on the fluorescence intensity, making the fluorescent protein sensor suitable for ultrahigh-throughput screening systems. A strategy which combined with the FACS function of flow cytometry, making strains with strong fluorescent signals and high yields of target products could be sorted out, has played an important role in catalytic element performance enhancement [185, 186] as well as in high-yield strain identification [187, 188]. For example, an arginine-responsive biosensor [185] was used to improve the activity and affinity of arginine deiminase for its substrate, enabling the mutant to catalyze arginine at very low concentrations in physiological conditions, which significantly inhibited the tumor cell growth. Similarly, a SoxR sensor, capable of responding to NADPH, was used to establish a screening system for NADPH-dependent enzymes, markedly improve the catalytic activity of NADPH-dependent ethanol dehydrogenase [186].

In general, fluorescent protein-based detection methods, characterized by easy to operate, have a wide range of application prospects. Nevertheless, several limitations are still need to be broken through. For instance, as a fusion protein, the fluorescent protein may affect the expression of the target protein due to its large molecular weight. In addition, when the fluorescent protein used as a transcription factor or ribosome switch organism, enhancing the biosensor's sensitivity through condition optimization is essential to tailor it for the specific detection range required for product analysis [189].

References

1. Crameri A, Raillard S-A, Bermudez E, Stemmer WPC (1998) DNA shuffling of a family of genes from diverse species accelerates directed evolution. Nature 391(6664):288–291. https://doi.org/10.1038/34663
2. Turner NJ (2003) Directed evolution of enzymes for applied biocatalysis. Trends Biotechnol 21(11):474–478. https://doi.org/10.1016/j.tibtech.2003.09.001

3. Leemhuis H, Kelly RM, Dijkhuizen L (2009) Directed evolution of enzymes: library screening strategies. IUBMB Life 61(3):222–228. https://doi.org/10.1002/iub.165
4. Lewis WH, Tahon G, Geesink P, Sousa DZ, Ettema TJG (2021) Innovations to culturing the uncultured microbial majority. Nat Rev Microbiol 19(4):225–240. https://doi.org/10.1038/s41579-020-00458-8
5. Rappé MS, Connon SA, Vergin KL, Giovannoni SJ (2002) Cultivation of the ubiquitous SAR11 marine bacterioplankton clade. Nature 418(6898):630–633. https://doi.org/10.1038/nature00917
6. Kato C, Bartlett DH (1997) The molecular biology of barophilic bacteria. Extremophiles 1(3):111–116. https://doi.org/10.1007/s007920050023
7. Kaeberlein T, Lewis K, Epstein SS (2002) Isolating "uncultivable" microorganisms in pure culture in a simulated natural environment. Science 296(5570):1127–1129. https://doi.org/10.1126/science.1070633
8. Zengler K, Toledo G, Rappé M, Elkins J, Mathur EJ, Short JM, Keller M (2002) Cultivating the uncultured. Proc Natl Acad Sci 99(24):15681–15686. https://doi.org/10.1073/pnas.252630999
9. Lagier JC, Hugon P, Khelaifia S, Fournier PE, La Scola B, Raoult D (2015) The rebirth of culture in microbiology through the example of culturomics to study human gut microbiota. Clin Microbiol Rev 28(1):237–264. https://doi.org/10.1128/cmr.00014-14
10. Hugon P, Dufour J-C, Colson P, Fournier P-E, Sallah K, Raoult D (2015) A comprehensive repertoire of prokaryotic species identified in human beings. Lancet Infect Dis 15(10):1211–1219. https://doi.org/10.1016/S1473-3099(15)00293-5
11. Lagier J-C, Khelaifia S, Alou MT, Ndongo S, Dione N, Hugon P, Caputo A, Cadoret F, Traore SI, Seck EH, Dubourg G, Durand G, Mourembou G, Guilhot E, Togo A, Bellali S, Bachar D, Cassir N, Bittar F, Delerce J, Mailhe M, Ricaboni D, Bilen M, Dangui Nieko NPM, Dia Badiane NM, Valles C, Mouelhi D, Diop K, Million M, Musso D, Abrahão J, Azhar EI, Bibi F, Yasir M, Diallo A, Sokhna C, Djossou F, Vitton V, Robert C, Rolain JM, La Scola B, Fournier P-E, Levasseur A, Raoult D (2016) Culture of previously uncultured members of the human gut microbiota by culturomics. Nat Microbiol 1(12):16203. https://doi.org/10.1038/nmicrobiol.2016.203
12. Bertrand H, Poly F, Van VT, Lombard N, Nalin R, Vogel TM, Simonet P (2005) High molecular weight DNA recovery from soils prerequisite for biotechnological metagenomic library construction. J Microbiol Methods 62(1):1–11. https://doi.org/10.1016/j.mimet.2005.01.003
13. Schmidt TM, DeLong EF, Pace NR (1991) Analysis of a marine picoplankton community by 16S rRNA gene cloning and sequencing. J Bacteriol 173(14):4371–4378. https://doi.org/10.1128/jb.173.14.4371-4378.1991
14. Béjà O (2004) To BAC or not to BAC: marine ecogenomics. Curr Opin Biotechnol 15(3):187–190. https://doi.org/10.1016/j.copbio.2004.03.005
15. Galperin MY, Koonin EV (2010) From complete genome sequence to 'complete' understanding? Trends Biotechnol 28(8):398–406. https://doi.org/10.1016/j.tibtech.2010.05.006
16. Jiang X, Xu X, Huo Y, Wu Y, Zhu X, Zhang X, Wu M (2012) Identification and characterization of novel esterases from a deep-sea sediment metagenome. Arch Microbiol 194(3):207–214. https://doi.org/10.1007/s00203-011-0745-2
17. Ajito K, Torimitsu K (2002) Laser trapping and Raman spectroscopy of single cellular organelles in the nanometer range. Lab Chip 2(1):11–14. https://doi.org/10.1039/B108744B
18. Hansen C, Quake SR (2003) Microfluidics in structural biology: smaller, faster better. Curr Opin Struct Biol 13(5):538–544. https://doi.org/10.1016/j.sbi.2003.09.010
19. Zaparucha A, de Berardinis V, Vaxelaire-Vergne C (2018) Genome mining for enzyme discovery, in modern biocatalysis: advances towards synthetic biological system. https://doi.org/10.1039/9781788010450-00001
20. Schnoes AM, Brown SD, Dodevski I, Babbitt PC (2009) Annotation error in public databases: misannotation of molecular function in enzyme superfamilies. PLoS Comput Biol 5(12):e1000605. https://doi.org/10.1371/journal.pcbi.1000605

21. Rodríguez Benítez A, Narayan ARH (2019) Frontiers in biocatalysis: profiling function across sequence space. ACS Central Sci 5(11):1747–1749. https://doi.org/10.1021/acscentsci.9b01112
22. Sillitoe I, Lewis TE, Cuff A, Das S, Ashford P, Dawson NL, Furnham N, Laskowski RA, Lee D, Lees JG, Lehtinen S, Studer RA, Thornton J, Orengo CA (2014) CATH: comprehensive structural and functional annotations for genome sequences. Nucleic Acids Res 43(D1):D376–D381. https://doi.org/10.1093/nar/gku947
23. Musil M, Konegger H, Hon J, Bednar D, Damborsky J (2019) Computational design of stable and soluble biocatalysts. ACS Catal 9(2):1033–1054. https://doi.org/10.1021/acscatal.8b03613
24. Jeffryes JG, Colastani RL, Elbadawi-Sidhu M, Kind T, Niehaus TD, Broadbelt LJ, Hanson AD, Fiehn O, Tyo KEJ, Henry CS (2015) MINEs: open access databases of computationally predicted enzyme promiscuity products for untargeted metabolomics. J Cheminform 7(1):44. https://doi.org/10.1186/s13321-015-0087-1
25. Pertusi DA, Moura ME, Jeffryes JG, Prabhu S, Walters Biggs B, Tyo KEJ (2017) Predicting novel substrates for enzymes with minimal experimental effort with active learning. Metab Eng 44:171–181. https://doi.org/10.1016/j.ymben.2017.09.016
26. Ferrario V, Siragusa L, Ebert C, Baroni M, Foscato M, Cruciani G, Gardossi L (2014) BioGPS descriptors for rational engineering of enzyme promiscuity and structure based bioinformatic analysis. PLoS One 9(10):e109354. https://doi.org/10.1371/journal.pone.0109354
27. Kumar A, Wang L, Ng CY, Maranas CD (2018) Pathway design using de novo steps through uncharted biochemical spaces. Nat Commun 9(1):184. https://doi.org/10.1038/s41467-017-02362-x
28. Venter JC, Remington K, Heidelberg JF, Halpern AL, Rusch D, Eisen JA, Wu D, Paulsen I, Nelson KE, Nelson W, Fouts DE, Levy S, Knap AH, Lomas MW, Nealson K, White O, Peterson J, Hoffman J, Parsons R, Baden-Tillson H, Pfannkoch C, Rogers YH, Smith HO (2004) Environmental genome shotgun sequencing of the Sargasso Sea. Science 304(5667):66–74. https://doi.org/10.1126/science.1093857
29. Ausubel JH, Trew Crist D, Waggoner PE (2010) First census of marine life 2010: highlights of a decade of discovery. First census of marine life 2010: highlights of a decade of discovery
30. Meyer F, Paarmann D, D'Souza M, Olson R, Glass EM, Kubal M, Paczian T, Rodriguez A, Stevens R, Wilke A, Wilkening J, Edwards RA (2008) The metagenomics RAST server – a public resource for the automatic phylogenetic and functional analysis of metagenomes. BMC Bioinformatics 9(1):386. https://doi.org/10.1186/1471-2105-9-386
31. Hunter S, Corbett M, Denise H, Fraser M, Gonzalez-Beltran A, Hunter C, Jones P, Leinonen R, McAnulla C, Maguire E, Maslen J, Mitchell A, Nuka G, Oisel A, Pesseat S, Radhakrishnan R, Rocca-Serra P, Scheremetjew M, Sterk P, Vaughan D, Cochrane G, Field D, Sansone S-A (2013) EBI metagenomics—a new resource for the analysis and archiving of metagenomic data. Nucleic Acids Res 42(D1):D600–D606. https://doi.org/10.1093/nar/gkt961
32. Lingner T, Aßhauer KP, Schreiber F, Meinicke P (2011) CoMet—a web server for comparative functional profiling of metagenomes. Nucleic Acids Res 39 (suppl_2):W518-W523:W518. https://doi.org/10.1093/nar/gkr388
33. Adrio JL, Demain AL (2014) Microbial enzymes: tools for biotechnological processes. Biomolecules 4(1):117–139. https://doi.org/10.3390/biom4010117
34. Connaris H, West SM, Hough DW, Danson MJ (1998) Cloning and overexpression in *Escherichia coli* of the gene encoding citrate synthase from the hyperthermophilic Archaeon *Sulfolobus solfataricus*. Extremophiles 2(2):61–66. https://doi.org/10.1007/s007920050043
35. Wang HM, Shih YP, Hu SM, Lo WT, Lin HM, Ding SS, Liao HC, Liang PH (2009) Parallel gene cloning and protein production in multiple expression systems. Biotechnol Prog 25(6):1582–1586. https://doi.org/10.1002/btpr.274
36. Suzuki SM, Stevens RC, Richter RJ, Cole TB, Park S, Otto TC, Cerasoli DM, Lenz DE, Furlong CE (2010) Engineering human PON1 in an *E. coli* expression system. Adv Exp Med Biol 660:37–45. https://doi.org/10.1007/978-1-60761-350-3_5

37. Ou JH, Yamada T, Nagahisa K, Hirasawa T, Furusawa C, Yomo T, Shimizu H (2008) Dynamic change in promoter activation during lysine biosynthesis in *Escherichia coli* cells. Mol BioSyst 4(2):128–134. https://doi.org/10.1039/b711035a
38. Sharrocks AD (1994) A T7 expression vector for producing N- and C-terminal fusion proteins with glutathione S-transferase. Gene 138(1–2):105–108. https://doi.org/10.1016/0378-1119(94)90789-7
39. Stanton BC, Nielsen AAK, Tamsir A, Clancy K, Peterson T, Voigt CA (2014) Genomic mining of prokaryotic repressors for orthogonal logic gates. Nat Chem Biol 10(2):99–105. https://doi.org/10.1038/nchembio.1411
40. Cho S, Choe D, Lee E, Kim SC, Palsson B, Cho BK (2018) High-level dCas9 expression induces abnormal cell morphology in *Escherichia coli*. Acs Synth Biol 7(4):1085–+. https://doi.org/10.1021/acssynbio.7b00462
41. Hess JM, Tchernajenko V, Vieille C, Zeikus JG, Kelly RM (1998) *Thermotoga neapolitana* homotetrameric xylose isomerase is expressed as a catalytically active and thermostable dimer in *Escherichia coli*. Appl Environ Microb 64(7):2357–2360. https://doi.org/10.1016/j.physe.2010.01.004
42. Buchanan CL, Connaris H, Danson MJ, Reeve CD, Hough DW (1999) An extremely thermostable aldolase from *Sulfolobus solfataricus* with specificity for non-phosphorylated substrates. Biochem J 343:563–570. https://doi.org/10.1042/0264-6021:3430563
43. Hough DW, Danson MJ (1999) Extremozymes. Curr Opin Chem Biol 3(1):39–46. https://doi.org/10.1016/S1367-5931(99)80008-8
44. Cendrin F, Chroboczek J, Zaccai G, Eisenberg H, Mevarech M (1993) Cloning, sequencing, and expression in *Escherichia coli* of the gene coding for malate-dehydrogenase of the extremely halophilic Archaebacterium *Haloarcula Marismortui*. Biochemistry-Us 32(16):4308–4313. https://doi.org/10.1021/bi00067a020
45. Kulakova L, Galkin A, Kurihara T, Yoshimura T, Esaki N (1999) Cold-active serine alkaline protease from the psychrotrophic bacterium *Shewanella* strain Ac10: gene cloning and enzyme purification and characterization. Appl Environ Microb 65(2):611–617. https://doi.org/10.1128/AEM.65.2.611-617.1999
46. Ara K, Ozaki K, Nakamura K, Yamane K, Sekiguchi J, Ogasawara N (2007) *Bacillus* minimum genome factory: effective utilization of microbial genome information. Biotechnol Appl Bioc 46:169–178. https://doi.org/10.1042/Ba20060111
47. Westers H, Dorenbos R, van Dijl JM, Kabel J, Flanagan T, Devine KM, Jude F, Seror SJ, Beekman AC, Darmon E, Eschevins C, de Jong A, Bron S, Kuipers OP, Albertini AM, Antelmann H, Hecker M, Zamboni N, Sauer U, Bruand C, Ehrlich DS, Alonso JC, Salas M, Quax WJ (2003) Genome engineering reveals large dispensable regions in *Bacillus subtilis*. Mol Biol Evol 20(12):2076–2090. https://doi.org/10.1093/molbev/msg219
48. Jakociunas T, Jensen MK, Keasling JD (2016) CRISPR/Cas9 advances engineering of microbial cell factories. Metab Eng 34:44–59. https://doi.org/10.1016/j.ymben.2015.12.003
49. Westers L, Westers H, Quax WJ (2004) *Bacillus subtilis* as cell factory for pharmaceutical proteins: a biotechnological approach to optimize the host organism. BBA-Mol Cell Res 1694(1–3):299–310. https://doi.org/10.1016/j.bbamcr.2004.02.011
50. Song W, Nie Y, Mu XQ, Xu Y (2016) Enhancement of extracellular expression of *Bacillus naganoensis* pullulanase from recombinant *Bacillus subtilis*: effects of promoter and host. Protein Expres Purif 124:23–31. https://doi.org/10.1016/j.pep.2016.04.008
51. Deng Y, Nie Y, Zhang Y, Wan Y, Xu Y (2018) Improved inducible expression of *Bacillus naganoensis* pullulanase from recombinant *Bacillus subtilis* by enhancer regulation. Protein Expres Purif 148:9–15. https://doi.org/10.1016/j.pep.2018.03.012
52. Mattanovich D, Branduardi P, Dato L, Gasser B, Sauer M, Porro D (2012) Recombinant protein production in yeasts. Methods Mol Biol 824:329–358. https://doi.org/10.1007/978-1-61779-433-9_17
53. Florczak T, Daroch M, Wilkinson MC, Bialkowska A, Bates AD, Turkiewicz M, Iwanejko LA (2013) Purification, characterisation and expression in Saccharomyces cerevisiae of LipG7

an enantioselective, cold-adapted lipase from the Antarctic filamentous fungus Geomyces sp. P7 with unusual thermostability characteristics. Enzym Microb Technol 53 (1):18-24:18. https://doi.org/10.1016/j.enzmictec.2013.03.021
54. Chen Y, Nielsen J (2013) Advances in metabolic pathway and strain engineering paving the way for sustainable production of chemical building blocks. Curr Opin Biotechnol 24(6):965–972. https://doi.org/10.1016/j.copbio.2013.03.008
55. Paddon CJ, Keasling JD (2014) Semi-synthetic artemisinin: a model for the use of synthetic biology in pharmaceutical development. Nat Rev Microbiol 12(5):355–367. https://doi.org/10.1038/nrmicro3240
56. Carrigan PE, Ballar P, Tuzmen S (2011) Site-directed mutagenesis. In: Disease gene identification. Springer, pp 107–124
57. Ho SN, Hunt HD, Horton RM, Pullen JK, Pease LRJG (1989) Site-directed mutagenesis by overlap extension using the polymerase chain reaction. 77(1):51–59
58. Na D, Yoo SM, Chung H, Park H, Park JH, Lee SY (2013) Metabolic engineering of Escherichia coli using synthetic small regulatory RNAs. Nat Biotechnol 31(2):170–174. https://doi.org/10.1038/nbt.2461
59. Lee JW, Na D, Park JM, Lee J, Choi S, Lee SY (2012) Systems metabolic engineering of microorganisms for natural and non-natural chemicals. Nat Chem Biol 8(6):536–546. https://doi.org/10.1038/nchembio.970
60. Cereghino GPL, Cereghino JL, Sunga AJ, Johnson MA, Lim M, Gleeson MAG, Cregg JM (2001) New selectable marker/auxotrophic host strain combinations for molecular genetic manipulation of Pichia pastoris. Gene 263(1–2):159–169. https://doi.org/10.1016/S0378-1119(00)00576-X
61. Deaner M, Alper HSJSBME (2016) Promoter and terminator discovery and engineering. pp 21–44
62. Blazeck J, Garg R, Reed B, Alper HS (2012) Controlling promoter strength and regulation in *Saccharomyces cerevisiae* using synthetic hybrid promoters. Biotechnol Bioeng 109(11):2884–2895. https://doi.org/10.1002/bit.24552
63. Garg A, Lohmueller JJ, Silver PA, Armel TZ (2012) Engineering synthetic TAL effectors with orthogonal target sites. Nucleic Acids Res 40(15):7584–7595. https://doi.org/10.1093/nar/gks404
64. Blount BA, Weenink T, Vasylechko S, Ellis T (2012) Rational diversification of a promoter providing fine-tuned expression and orthogonal regulation for synthetic biology. PLoS One 7(3):e33279. https://doi.org/10.1371/journal.pone.0033279
65. Du L, Villarreal S, Forster AC (2012) Multigene expression in vivo: supremacy of large versus small terminators for T7 RNA polymerase. Biotechnol Bioeng 109(4):1043–1050. https://doi.org/10.1002/bit.24379
66. Lambertz C, Garvey M, Klinger J, Heesel D, Klose H, Fischer R, Commandeur U (2014) Challenges and advances in the heterologous expression of cellulolytic enzymes: a review. Biotechnol Biofuels 7(1):135. https://doi.org/10.1186/s13068-014-0135-5
67. Macauley-Patrick S, Fazenda ML, McNeil B, Harvey LM (2005) Heterologous protein production using the *Pichia pastoris* expression system. Yeast 22(4):249–270. https://doi.org/10.1002/yea.1208
68. Ward OP, Qin WM, Dhanjoon J, Ye J, Singh A (2005) Physiology and biotechnology of Aspergillus. Adv Appl Microbiol 58C:1–75. https://doi.org/10.1016/S0065-2164(05)58001-8
69. Lubertozzi D, Keasling JD (2009) Developing *Aspergillus* as a host for heterologous expression. Biotechnol Adv 27(1):53–75. https://doi.org/10.1016/j.biotechadv.2008.09.001
70. Weld RJ, Plummer KM, Carpenter MA, Ridgway HJ (2006) Approaches to functional genomics in filamentous fungi. Cell Res 16(1):31–44. https://doi.org/10.1038/sj.cr.7310006
71. Meyer V, Mueller D, Strowig T, Stahl U (2003) Comparison of different transformation methods for *Aspergillus giganteus*. Curr Genet 43(5):371–377. https://doi.org/10.1007/s00294-003-0406-3

72. de Bekker C, Wiebenga A, Aguilar G, Wosten HA (2009) An enzyme cocktail for efficient protoplast formation in *Aspergillus niger*. J Microbiol Methods 76(3):305–306. https://doi.org/10.1016/j.mimet.2008.11.001
73. Michielse CB, Hooykaas PJ, van den Hondel CA, Ram AF (2008) Agrobacterium-mediated transformation of the filamentous fungus *Aspergillus awamori*. Nat Protoc 3(10):1671–1678. https://doi.org/10.1038/nprot.2008.154
74. Skory CD, Ibrahim AS (2007) Native and modified lactate dehydrogenase expression in a fumaric acid producing isolate Rhizopus oryzae 99-880. Curr Genet 52(1):23–33. https://doi.org/10.1007/s00294-007-0135-0
75. Fleissner A, Dersch P (2010) Expression and export: recombinant protein production systems for Aspergillus. Appl Microbiol Biotechnol 87(4):1255–1270. https://doi.org/10.1007/s00253-010-2672-6
76. Minetoki T, Kumagai C, Gomi K, Kitamoto K, Takahashi K (1998) Improvement of promoter activity by the introduction of multiple copies of the conserved region III sequence, involved in the efficient expression of *Aspergillus oryzae* amylase-encoding genes. Appl Microbiol Biotechnol 50(4):459–467. https://doi.org/10.1007/s002530051321
77. Mao Y, Yin Y, Zhang L, Alias SA, Gao B, Wei D (2015) Development of a novel *Aspergillus* uracil deficient expression system and its application in expressing a cold-adapted α-amylase gene from Antarctic fungi Geomyces pannorum. Process Biochem 50(10):1581–1590. https://doi.org/10.1016/j.procbio.2015.06.016
78. Bleichrodt R, Vinck A, Krijgsheld P, van Leeuwen MR, Dijksterhuis J, Wosten HA (2013) Cytosolic streaming in vegetative mycelium and aerial structures of *Aspergillus niger*. Stud Mycol 74(1):31–46. https://doi.org/10.3114/sim0007
79. Singh B, Satyanarayana T (2006) Phytase production by thermophilic mold Sporotrichum thermophile in solid-state fermentation and its application in dephytinization of sesame oil cake. Appl Biochem Biotechnol 133(3):239–250. https://doi.org/10.1385/abab:133:3:239
80. MacKenzie DA, Jeenes DJ, Gou X, Archer DB (2000) Molecular basis of glucoamylase overproduction by a mutagenised industrial strain of *Aspergillus niger*. Enzym Microb Technol 26(2–4):193–200. https://doi.org/10.1016/s0141-0229(99)00145-3
81. Xiao X, Ouyang L, Qi J, Chu J (2021) Exploration and characterization of hypoxia-inducible endogenous promoters in *Aspergillus niger*. Appl Microbiol Biotechnol 105(13):5529–5539. https://doi.org/10.1007/s00253-021-11417-5
82. Devchand M, Gwynne DI (1991) Expression of heterologous proteins in *Aspergillus*. J Biotechnol 17(1):3–9. https://doi.org/10.1016/0168-1656(91)90022-n
83. Morita H, Abo H, Okamoto A, Maeda H, Yamagata Y, Kusumoto KI, Amano H, Ishida H, Takeuchi M (2011) Enzymatic properties of the recombinant serine-type carboxypeptidase OcpC, which is unique to *Aspergillus oryzae*. Biosci Biotech Bioch 75(4):662–668. https://doi.org/10.1271/bbb.100749
84. Yoshino-Yasuda S, Kato M, Kitamoto N (2011) Sequence analysis and heterologous expression of Polygalacturonase gene (AspecA) from a Shoyu Koji Mold, Aspergillus sojae KBN1340. Food Sci Technol Res 17(6):579–584. https://doi.org/10.3136/fstr.17.579
85. Lin H, Wang Q, Shen Q, Ma JW, Fu JR, Zhao YH (2014) Engineering *Aspergillus oryzae* A-4 through the chromosomal insertion of foreign cellulase expression cassette to improve conversion of cellulosic biomass into lipids. PLoS One 9(9):e108442. https://doi.org/10.1371/journal.pone.0108442
86. Koushki MM, Rouhani H, Farsi M (2011) Genetic manipulation of fungal strains for the improvement of heterologous genes expression (a mini-review). Afr J Biotechnol 10(41):7939–7948. https://doi.org/10.5897/AJB11.041
87. France SP, Hepworth LJ, Turner NJ, Flitsch SL (2017) Constructing biocatalytic cascades: *in vitro* and *in vivo* approaches to de novo multi-enzyme pathways. ACS Catal 7(1):710–724. https://doi.org/10.1021/acscatal.6b02979
88. Sheldon RA, Brady D (2018) The limits to biocatalysis: pushing the envelope. Chem Commun 54(48):6088–6104. https://doi.org/10.1039/C8CC02463D

89. Bornscheuer UT (2018) The fourth wave of biocatalysis is approaching. Philos Trans R Soc A Math Phys Eng Sci 376(2110):20170063. https://doi.org/10.1098/rsta.2017.0063
90. Reetz MT (2013) Biocatalysis in organic chemistry and biotechnology: past, present, and future. J Am Chem Soc 135(34):12480–12496. https://doi.org/10.1021/ja405051f
91. Bornscheuer UT (2016) Biocatalysis: successfully crossing boundaries. Angew Chem Int Ed 55(14):4372–4373. https://doi.org/10.1002/anie.201510042
92. Hwang ET, Lee S (2019) Multienzymatic cascade reactions via enzyme complex by immobilization. ACS Catal 9(5):4402–4425. https://doi.org/10.1021/acscatal.8b04921
93. Wheeldon I, Minteer SD, Banta S, Barton SC, Atanassov P, Sigman M (2016) Substrate channelling as an approach to cascade reactions. Nat Chem 8(4):299–309. https://doi.org/10.1038/nchem.2459
94. Mullaney JA, Rehm BHA (2010) Design of a single-chain multi-enzyme fusion protein establishing the polyhydroxybutyrate biosynthesis pathway. J Biotechnol 147(1):31–36. https://doi.org/10.1016/j.jbiotec.2010.02.021
95. Kang W, Ma T, Liu M, Qu J, Liu Z, Zhang H, Shi B, Fu S, Ma J, Lai LTF, He S, Qu J, Wing-Ngor AS, Ho Kang B, Yu Lau WC, Deng Z, Xia J, Liu T (2019) Modular enzyme assembly for enhanced cascade biocatalysis and metabolic flux. Nat Commun 10(1):4248. https://doi.org/10.1038/s41467-019-12247-w
96. Baslé E, Joubert N, Pucheault M (2010) Protein chemical modification on endogenous amino acids. Chem Biol 17(3):213–227. https://doi.org/10.1016/j.chembiol.2010.02.008
97. Jia F, Narasimhan B, Mallapragada S (2014) Materials-based strategies for multi-enzyme immobilization and co-localization: a review. Biotechnol Bioeng 111(2):209–222. https://doi.org/10.1002/bit.25136
98. Fu J, Wang Z, Liang XH, Oh SW, St Iago-McRae E, Zhang T (2020) DNA-scaffolded proximity assembly and confinement of multienzyme reactions. Top Curr Chem (Cham) 378(3):38–38. https://doi.org/10.1007/s41061-020-0299-3
99. Xu K, Chen X, Zheng R, Zheng Y (2020) Immobilization of multi-enzymes on support materials for efficient biocatalysis. Front Bioeng Biotechnol 8(660). https://doi.org/10.3389/fbioe.2020.00660
100. Arana-Peña S, Carballares D, Morellon-Sterlling R, Berenguer-Murcia Á, Alcántara AR, Rodrigues RC, Fernandez-Lafuente R (2021) Enzyme co-immobilization: always the biocatalyst designers' choic or not? Biotechnol Adv 51:107584. https://doi.org/10.1016/j.biotechadv.2020.107584
101. Jiang L, Song X, Li Y, Xu Q, Pu J, Huang H, Zhong C (2018) Programming integrative extracellular and intracellular biocatalysis for rapid, robust, and recyclable synthesis of trehalose. ACS Catal 8(3):1837–1842. https://doi.org/10.1021/acscatal.7b03445
102. Katyal P, Mahmoudinobar F, Montclare JK (2020) Recent trends in peptide and protein-based hydrogels. Curr Opin Struct Biol 63:97–105. https://doi.org/10.1016/j.sbi.2020.04.007
103. Ellis GA, Klein WP, Lasarte-Aragonés G, Thakur M, Walper SA, Medintz IL (2019) Artificial multienzyme scaffolds: pursuing *in vitro* substrate channeling with an overview of current progress. ACS Catal 9(12):10812–10869. https://doi.org/10.1021/acscatal.9b02413
104. Zhang Y, Hess H (2017) Toward rational design of high-efficiency enzyme cascades. ACS Catal 7(9):6018–6027. https://doi.org/10.1021/acscatal.7b01766
105. Bao J, Liu N, Zhu L, Xu Q, Huang H, Jiang L (2018) Programming a biofilm-mediated multienzyme-assembly-cascade system for the biocatalytic production of glucosamine from chitin. J Agric Food Chem 66(30):8061–8068. https://doi.org/10.1021/acs.jafc.8b02142
106. Lin J-L, Wheeldon I (2013) Kinetic enhancements in DNA–enzyme nanostructures mimic the sabatier principle. ACS Catal 3(4):560–564. https://doi.org/10.1021/cs300766d
107. Zhang Y, Wang Q, Hess H (2017) Increasing enzyme cascade throughput by pH-engineering the microenvironment of individual enzymes. ACS Catal 7(3):2047–2051. https://doi.org/10.1021/acscatal.6b03431

108. Chen KQ, Arnold FH (1991) Enzyme engineering for nonaqueous solvents-random mutagenesis to enhance activity of subtilisin. Bio/Technology 9(11):1073–1077. https://doi.org/10.1038/nbt1191-1073
109. Ameri A, Shakibaie M, Soleimani-Kermani M, Faramarzi MA, Forootanfar H (2019) Overproduction of thermoalkalophilic lipase secreted by Bacillus atrophaeus FSHM2 using UV-induced mutagenesis and statistical optimization of medium components. Prep Biochem Biotechnol 49(2):184–191. https://doi.org/10.1080/10826068.2019.1566148
110. Cadwell RC, Joyce GF (1992) Randomization of genes by PCR mutagenesis. PCR Methods Appl 2(1):28–33. https://doi.org/10.1101/gr.2.1.28
111. Kuipers OPJMMB (1996) Random mutagenesis by using mixtures of dNTP and dITP in PCR. 57:351
112. Ben Mabrouk S, Ayadi DZ, Ben Hlima H, Bejar S (2013) Thermostability improvement of maltogenic amylase MAUS149 by error prone PCR. J Biotechnol 168(4):601–606. https://doi.org/10.1016/j.jbiotec.2013.08.026
113. Yang J, Ruff AJ, Arlt M, Schwaneberg U (2017) Casting epPCR (cepPCR): a simple random mutagenesis method to generate high quality mutant libraries. Biotechnol Bioeng 114(9):1921–1927. https://doi.org/10.1002/bit.26327
114. Stemmer WPC (1994) Rapid evolution of a protein in-vitro by DNA shuffling. Nature 370(6488):389–391. https://doi.org/10.1038/370389a0
115. Stemmer WPC (1994) DNA shuffling by random fragmentation and reassembly- in vitro recombination for molecular evolution. Proc Natl Acad Sci USA 91(22):10747–10751. https://doi.org/10.1073/pnas.91.22.10747
116. Ouyang F, Zhao M (2019) Enhanced catalytic efficiency of CotA-laccase by DNA shuffling. Bioengineered 10(1):182–189. https://doi.org/10.1080/21655979.2019.1621134
117. Ostermeier M, Shim JH, Benkovic SJ (1999) A combinatorial approach to hybrid enzymes independent of DNA homology. Nat Biotechnol 17(12):1205–1209. https://doi.org/10.1038/70754
118. Zhao HM, Giver L, Shao ZX, Affholter JA, Arnold FH (1998) Molecular evolution by staggered extension process (StEP) in vitro recombination. Nat Biotechnol 16(3):258–261. https://doi.org/10.1038/nbt0398-258
119. Dion M, Nisole A, Spangenberg P, Andre C, Glottin-Fleury A, Mattes R, Tellier C, Rabiller C (2001) Modulation of the regioselectivity of a Bacillus alpha-galactosidase by directed evolution. Glycoconj J 18(3):215–223. https://doi.org/10.1023/a:1012448522187
120. Lee SH, Ryu EJ, Kang MJ, Wang ES, Piao Z, Choi YJ, Jung KH, Jeon JYJ, Shin YC (2003) A new approach to directed gene evolution by recombined extension on truncated templates (RETT). J Mol Catal B Enzym 26(3–6):119–129. https://doi.org/10.1016/j.molcatb.2003.05.001
121. Wong TS, Tee KL, Hauer B, Schwaneberg U (2005) Sequence saturation mutagenesis with tunable mutation frequencies. Anal Biochem 341(1):187–189. https://doi.org/10.1016/j.ab.2005.03.023
122. Ozturk H, Ece S, Gundeger E, Evran S (2013) Site-directed mutagenesis of methionine residues for improving the oxidative stability of alpha-amylase from *Thermotoga maritima*. J Biosci Bioeng 116(4):449–451. https://doi.org/10.1016/j.jbiosc.2013.04.018
123. Su H-H, Peng F, Xu P, Wu X-L, Zong M-H, Yang J-G, Lou W-Y (2019) Enhancing the thermostability and activity of uronate dehydrogenase from *Agrobacterium tumefaciens* LBA4404 by semi-rational engineering. Bioresour Bioprocess 6(1). https://doi.org/10.1186/s40643-019-0267-3
124. Xu P, Ni Z-F, Zong M-H, Ou X-Y, Yang J-G, Lou W-Y (2020) Improving the thermostability and activity of *Paenibacillus pasadenensis* chitinase through semi-rational design. Int J Biol Macromol 150:9–15. https://doi.org/10.1016/j.ijbiomac.2020.02.033
125. Liang J, Huang R, Huang Y, Wang X, Du L, Wei Y (2013) Cloning, expression, properties, and functional amino acid residues of new trehalose synthase from *Thermomonospora curvata* DSM 43183. J Mol Catal B Enzym 90:26–32. https://doi.org/10.1016/j.molcatb.2013.01.014

126. Chen Y, Zhao Y, Zhou X, Liu N, Ming D, Zhu L, Jiang L (2021) Improving the thermostability of trehalose synthase from Thermomonospora curvata by covalent cyclization using peptide tags and investigation of the underlying molecular mechanism. Int J Biol Macromol 168:13–21. https://doi.org/10.1016/j.ijbiomac.2020.11.195
127. Wang Y-L, Chow S-Y, Lin Y-T, Hsieh Y-C, Lee G-C, Liaw S-H (2014) Structures of trehalose synthase from *Deinococcus radiodurans* reveal that a closed conformation is involved in catalysis of the intramolecular isomerization. Acta Crystallogr D Biol Crystallogr 70:3144–3154. https://doi.org/10.1107/s1399004714022500
128. Wang J, Ren X, Wang R, Su J, Wang F (2017) Structural characteristics and function of a new kind of thermostable trehalose synthase from *Thermobaculum terrenum*. J Agric Food Chem 65(35):7726–7735. https://doi.org/10.1021/acs.jafc.7b02732
129. Su J, Wang T, Ma C, Li Z, Li Z, Wang R (2014) Homology modeling and function of trehalose synthase from *Pseudomonas putida* P06. Biotechnol Lett 36(5):1009–1013. https://doi.org/10.1007/s10529-013-1450-9
130. Liu H, Wang X, Yang S, Wang R, Wang T (2019) Saturation mutagenesis and self-inducible expression of trehalose synthase in *Bacillus subtilis*. Biotechnol Prog 35(4):e2826. https://doi.org/10.1002/btpr.2826
131. Sun Z, Lonsdale R, Kong X-D, Xu J-H, Zhou J, Reetz MT (2015) Reshaping an enzyme binding pocket for enhanced and inverted stereoselectivity: use of smallest amino acid alphabets in directed evolution. Angew Chem Int Ed Engl 54(42):12410–12415. https://doi.org/10.1002/anie.201501809
132. Sun Z, Lonsdale R, Ilie A, Li G, Zhou J, Reetz MT (2016) Catalytic asymmetric reduction of difficult-to-reduce ketones: triple-code saturation mutagenesis of an alcohol dehydrogenase. ACS Catal 6(3):1598–1605. https://doi.org/10.1021/acscatal.5b02752
133. Li G, Zhang H, Sun Z, Liu X, Reetz MT (2016) Multiparameter optimization in directed evolution: engineering thermostability, enantioselectivity, and activity of an epoxide hydrolase. ACS Catal 6(6):3679–3687. https://doi.org/10.1021/acscatal.6b01113
134. Dror A, Shemesh E, Dayan N, Fishman A (2014) Protein engineering by random mutagenesis and structure-guided consensus of *Geobacillus stearothermophilus* lipase T6 for enhanced stability in methanol. Appl Environ Microbiol 80(4):1515–1527. https://doi.org/10.1128/aem.03371-13
135. Clark SE, Muslin EH, Henson CA (2004) Effect of adding and removing N-glycosylation recognition sites on the thermostability of barley alpha-glucosidase. Protein Eng Des Sel 17(3):245–249. https://doi.org/10.1093/protein/gzh028
136. Heinzelman P, Komor R, Kanaan A, Romero P, Yu X, Mohler S, Snow C, Arnold F (2010) Efficient screening of fungal cellobiohydrolase class I enzymes for thermostabilizing sequence blocks by SCHEMA structure-guided recombination. Protein Eng Des Sel 23(11):871–880. https://doi.org/10.1093/protein/gzq063
137. Moore E, Fasan R (2018) Enzyme stabilization via computationally guided protein stapling. Abstr Pap Am Chem Soc 255
138. Bryant P, Pozzati G, Elofsson A (2021) Improved prediction of protein-protein interactions using AlphaFold2. bioRxiv – Bioinformatics. https://doi.org/10.1101/2021.09.15.460468
139. Baek M, DiMaio F, Anishchenko I, Dauparas J, Ovchinnikov S, Lee GR, Wang J, Cong Q, Kinch LN, Schaeffer RD, Millan C, Park H, Adams C, Glassman CR, DeGiovanni A, Pereira JH, Rodrigues AV, van Dijk AA, Ebrecht AC, Opperman DJ, Sagmeister T, Buhlheller C, Pavkov-Keller T, Rathinaswamy MK, Dalwadi U, Yip CK, Burke JE, Garcia KC, Grishin NV, Adams PD, Read RJ, Baker D (2021) Accurate prediction of protein structures and interactions using a three-track neural network. Science 373 (6557):871-+:871. https://doi.org/10.1126/science.abj8754
140. Lang GI, Murray AW (2008) Estimating the per-base-pair mutation rate in the yeast *Saccharomyces cerevisiae*. Genetics 178(1):67–82. https://doi.org/10.1534/genetics.107.071506

141. Camps M, Naukkarinen J, Johnson BP, Loeb LA (2003) Targeted gene evolution in *Escherichia coli* using a highly error-prone DNA polymerase I. Proc Natl Acad Sci USA 100(17):9727–9732. https://doi.org/10.1073/pnas.1333928100
142. Husimi (1989) Selection and evolution of bacteriophages in cellstat. Adv Biophys 25:1–43
143. Esvelt KM, Carlson JC, Liu DR (2011) A system for the continuous directed evolution of biomolecules. Nature 472(7344):499–503. https://doi.org/10.1038/nature09929
144. Wang T, Badran AH, Huang TP, Liu DR (2018) Continuous directed evolution of proteins with improved soluble expression. Nat Chem Biol 14(10):972–980. https://doi.org/10.1038/s41589-018-0121-5
145. Wang HH, Isaacs FJ, Carr PA, Sun ZZ, Xu G, Forest CR, Church GM (2009) Programming cells by multiplex genome engineering and accelerated evolution. Nature 460(7257):894–898. https://doi.org/10.1038/nature08187
146. Luan GD, Cai Z, Gong FY, Dong HJ, Lin Z, Zhang YP, Li Y (2013) Developing controllable hypermutable Clostridium cells through manipulating its methyl-directed mismatch repair system. Protein Cell 4(11):854–862. https://doi.org/10.1007/s13238-013-3079-9
147. Pham HL, Wong A, Chua N, Teo WS, Yew WS, Chang MW (2017) Engineering a riboswitch-based genetic platform for the self-directed evolution of acid-tolerant phenotypes. Nat Commun 8(1):411. https://doi.org/10.1038/s41467-017-00511-w
148. Halperin SO, Tou CJ, Wong EB, Modavi C, Schaffer DV, Dueber JE (2018) CRISPR-guided DNA polymerases enable diversification of all nucleotides in a tunable window. Nature 560(7717):248-+. https://doi.org/10.1038/s41586-018-0384-8
149. Hess GT, Fresard L, Han K, Lee CH, Li A, Cimprich KA, Montgomery SB, Bassik MC (2016) Directed evolution using dCas9-targeted somatic hypermutation in mammalian cells. Nat Methods 13(12):1036−+. https://doi.org/10.1038/nmeth.4038
150. Crook N, Abatemarco J, Sun J, Wagner JM, Schmitz A, Alper HS (2016) In vivo continuous evolution of genes and pathways in yeast. Nat Commun 7:13051. https://doi.org/10.1038/ncomms13051
151. Ravikumar A, Arzumanyan GA, Obadi MKA, Javanpour AA, Liu CC (2018) Scalable, continuous evolution of genes at mutation rates above genomic error thresholds. Cell 175(7):1946−+. https://doi.org/10.1016/j.cell.2018.10.021
152. Rees HA, Liu DR (2018) Base editing: precision chemistry on the genome and transcriptome of living cells (vol 19, pg 770, 2018). Nat Rev Genet 19(12):801. https://doi.org/10.1038/s41576-018-0068-0
153. Jansen R, van Embden JDA, Gaastra W, Schouls LM (2002) Identification of genes that are associated with DNA repeats in prokaryotes. Mol Microbiol 43(6):1565–1575. https://doi.org/10.1046/j.1365-2958.2002.02839.x
154. Nishida K, Arazoe T, Yachie N, Banno S, Kakimoto M, Tabata M, Mochizuki M, Miyabe A, Araki M, Hara KY, Shimatani Z, Kondo A (2016) Targeted nucleotide editing using hybrid prokaryotic and vertebrate adaptive immune systems. Science 353(6305):aaf8729. https://doi.org/10.1126/science.aaf8729
155. Ma Y, Zhang J, Yin W, Zhang Z, Song Y, Chang X (2016) Targeted AID-mediated mutagenesis (TAM) enables efficient genomic diversification in mammalian cells. Nat Methods 13(12):1029–1035. https://doi.org/10.1038/nmeth.4027
156. Kim YB, Komor AC, Levy JM, Packer MS, Zhao KT, Liu DR (2017) Increasing the genome-targeting scope and precision of base editing with engineered Cas9-cytidine deaminase fusions. Nat Biotechnol 35(4):371–376. https://doi.org/10.1038/nbt.3803
157. Li X, Wang Y, Liu Y, Yang B, Wang X, Wei J, Lu Z, Zhang Y, Wu J, Huang X, Yang L, Chen J (2018) Base editing with a Cpf1-cytidine deaminase fusion. Nat Biotechnol 36(4):324–327. https://doi.org/10.1038/nbt.4102
158. Grunewald J, Zhou R, Lareau CA, Garcia SP, Iyer S, Miller BR, Langner LM, Hsu JY, Aryee MJ, Joung JK (2020) A dual-deaminase CRISPR base editor enables concurrent adenine and cytosine editing. Nat Biotechnol 38(7):861–864. https://doi.org/10.1038/s41587-020-0535-y

159. Schriever K, Saenz-Mendez P, Rudraraju RS, Hendrikse NM, Hudson EP, Biundo A, Schnell R, Syrén P-O (2021) Engineering of ancestors as a tool to elucidate structure, mechanism, and specificity of extant terpene cyclase. J Am Chem Soc 143:3794. https://doi.org/10.1021/jacs.0c10214
160. Merkl R, Sterner R (2016) Ancestral protein reconstruction: techniques and applications. Biol Chem 397(1):1–21. https://doi.org/10.1515/hsz-2015-0158
161. Alcalde M (2017) When directed evolution met ancestral enzyme resurrection. Microb Biotechnol 10(1):22–24. https://doi.org/10.1111/1751-7915.12452
162. Miyazaki J, Nakaya S, Suzuki T, Tamakoshi M, Oshima T, Yamagishi A (2001) Ancestral residues stabilizing 3-Isopropylmalate dehydrogenase of an extreme thermophile: experimental evidence supporting the thermophilic common ancestor hypothesis. J Biochem 129(5):777–782. https://doi.org/10.1093/oxfordjournals.jbchem.a002919
163. Gumulya Y, Baek J-M, Wun S-J, Thomson RES, Harris KL, Hunter DJB, Behrendorff JBYH, Kulig J, Zheng S, Wu X, Wu B, Stok JE, De Voss JJ, Schenk G, Jurva U, Andersson S, Isin EM, Bodén M, Guddat L, Gillam EMJ (2018) Engineering highly functional thermostable proteins using ancestral sequence reconstruction. Nat Catal 1(11):878–888. https://doi.org/10.1038/s41929-018-0159-5
164. Gomez-Fernandez BJ, Risso VA, Rueda A, Sanchez-Ruiz JM, Alcalde M, Master ER (2020) Ancestral resurrection and directed evolution of fungal mesozoic laccases. Appl Environ Microbiol 86(14):e00778–e00720. https://doi.org/10.1128/AEM.00778-20
165. Ruff AJ, Dennig A, Wirtz G, Blanusa M, Schwaneberg U (2012) Flow cytometer-based high-throughput screening system for accelerated directed evolution of P450 monooxygenases. ACS Catal 2(12):2724–2728. https://doi.org/10.1021/cs300115d
166. Agresti JJ, Antipov E, Abate AR, Ahn K, Rowat AC, Baret JC, Marquez M, Klibanov AM, Griffiths AD, Weitz DA (2010) Ultrahigh-throughput screening in drop-based microfluidics for directed evolution. Proc Natl Acad Sci USA 107(9):4004–4009. https://doi.org/10.1073/pnas.0910781107
167. Kaminski TS, Scheler O, Garstecki P (2016) Droplet microfluidics for microbiology: techniques, applications and challenges. Lab Chip 16(12):2168–2187. https://doi.org/10.1039/c6lc00367b
168. Yang GY, Withers SG (2009) Ultrahigh-throughput FACS-based screening for directed enzyme evolution. Chembiochem 10(17):2704–2715. https://doi.org/10.1002/cbic.200900384
169. Ma FQ, Feng Y, Yang GY (2012) Ultrahigh-throughput enzymatic screening method based on fluorescence-activated cell sorting and its applications. Prog Biochem Biophys 39(4):299–306. https://doi.org/10.3724/sp.J.1206.2011.00281
170. Tan YM, Zhang Y, Han YB, Liu H, Chen HF, Ma FQ, Withers SG, Feng Y, Yang GY (2019) Directed evolution of an alpha 1,3-fucosyltransferase using a single-cell ultrahigh-throughput screening method. Sci Adv 5(10):11. https://doi.org/10.1126/sciadv.aaw8451
171. Gielen F, Hours R, Emond S, Fischlechner M, Schell U, Hollfelder F (2016) Ultrahigh-throughput-directed enzyme evolution by absorbance-activated droplet sorting (AADS). Proc Natl Acad Sci USA 113(47):E7383–E7389. https://doi.org/10.1073/pnas.1606927113
172. Wang XX, Ren LH, Su YT, Ji YT, Liu YP, Li CY, Li XR, Zhang Y, Wang W, Hu Q, Han DX, Xu J, Ma B (2017) Raman-activated droplet sorting (RADS) for label-free high-throughput screening of microalgal single-cells. Anal Chem 89(22):12569–12577. https://doi.org/10.1021/acs.analchem.7b03884
173. Baret JC, Miller OJ, Taly V, Ryckelynck M, El-Harrak A, Frenz L, Rick C, Samuels ML, Hutchison JB, Agresti JJ, Link DR, Weitz DA, Griffiths AD (2009) Fluorescence-activated droplet sorting (FADS): efficient microfluidic cell sorting based on enzymatic activity. Lab Chip 9(13):1850–1858. https://doi.org/10.1039/b902504a
174. Kintses B, Hein C, Mohamed MF, Fischlechner M, Courtois F, Leine C, Hollfelder F (2012) Picoliter cell lysate assays in microfluidic droplet compartments for directed enzyme evolution. Chem Biol 19(8):1001–1009. https://doi.org/10.1016/j.chembiol.2012.06.009

175. Varadarajan N, Gam J, Olsen MJ, Georgiou G, Iverson BL (2005) Engineering of protease variants exhibiting high catalytic activity and exquisite substrate selectivity. Proc Natl Acad Sci USA 102(19):6855–6860. https://doi.org/10.1073/pnas.0500063102
176. Varadarajan N, Rodriguez S, Hwang BY, Georgiou G, Iverson BL (2008) Highly active and selective endopeptidases with programmed substrate specificities. Nat Chem Biol 4(5):290–294. https://doi.org/10.1038/nchembio.80
177. Varadarajan N, Georgiou G, Iverson BL (2008) An engineered protease that cleaves specifically after sulfated tyrosine. Angew Chem Int Ed Engl 47(41):7861–7863. https://doi.org/10.1002/anie.200800736
178. Lipovsek D, Antipov E, Armstrong KA, Olsen MJ, Klibanov AM, Tidor B, Wittrup KD (2007) Selection of horseradish peroxidase variants with enhanced enantioselectivity by yeast surface display. Chem Biol 14(10):1176–1185. https://doi.org/10.1016/j.chembiol.2007.09.008
179. Rodriguez EA, Campbell RE, Lin JY, Lin MZ, Miyawaki A, Palmer AE, Shu XK, Zhang J, Tsien RY (2017) The growing and glowing toolbox of fluorescent and photoactive proteins. Trends Biochem Sci 42(2):111–129. https://doi.org/10.1016/j.tibs.2016.09.010
180. Kremers GJ, Gilbert SG, Cranfill PJ, Davidson MW, Piston DW (2011) Fluorescent proteins at a glance (vol 124, pg 157, 2011). J Cell Sci 124(15):2676–2676. https://doi.org/10.1242/jcs.095059
181. Santoro SW, Schultz PG (2002) Directed evolution of the site specificity of Cre recombinase. Proc Natl Acad Sci USA 99(7):4185–4190. https://doi.org/10.1073/pnas.022039799
182. Santoro SW, Wang L, Herberich B, King DS, Schultz PG (2002) An efficient system for the evolution of aminoacyl-tRNA synthetase specificity. Nat Biotechnol 20(10):1044–1048. https://doi.org/10.1038/nbt742
183. Aharoni A, Amitai G, Bernath K, Magdassi S, Tawfik DS (2005) High-throughput screening of enzyme libraries: thiolactonases evolved by fluorescence-activated sorting of single cells in emulsion compartments. Chem Biol 12(12):1281–1289. https://doi.org/10.1016/j.chembiol.2005.09.012
184. Gupta RD, Goldsmith M, Ashani Y, Simo Y, Mullokandov G, Bar H, Ben-David M, Leader H, Margalit R, Silman I, Sussman JL, Tawfik DS (2011) Directed evolution of hydrolases for prevention of G-type nerve agent intoxication. Nat Chem Biol 7(2):120–125. https://doi.org/10.1038/nchembio.510
185. Cheng F, Kardashliev T, Pitzler C, Shehzad A, Lue HQ, Bernhagen J, Zhu LL, Schwaneberg U (2015) A competitive flow cytometry screening system for directed evolution of therapeutic enzyme. ACS Synth Biol 4(7):768–775. https://doi.org/10.1021/sb500343g
186. Siedler S, Schendzielorz G, Binder S, Eggeling L, Bringer S, Bott M (2014) SoxR as a single-cell biosensor for NADPH-consuming enzymes in Escherichia coli. ACS Synth Biol 3(1):41–47. https://doi.org/10.1021/sb400110j
187. Schallmey M, Frunzke J, Eggeling L, Marienhagen J (2014) Looking for the pick of the bunch: high-throughput screening of producing microorganisms with biosensors. Curr Opin Biotechnol 26:148–154. https://doi.org/10.1016/j.copbio.2014.01.005
188. Binder S, Schendzielorz G, Stabler N, Krumbach K, Hoffmann K, Bott M, Eggeling L (2012) A high-throughput approach to identify genomic variants of bacterial metabolite producers at the single-cell level. Genome Biol 13(5):R40. https://doi.org/10.1186/gb-2012-13-5-r40
189. Woronoff G, Ryckelynck M, Wessel J, Schicke O, Griffiths AD, Soumillion P (2015) Activity-fed translation (AFT) assay: a new high-throughput screening strategy for enzymes in droplets. Chembiochem 16(9):1343–1349. https://doi.org/10.1002/cbic.201500087

Chitinase

Shaoqing Yang

Abbreviations

CBM	Carbohydrate-binding module
DHB	2, 5-dihydroxybenzoic acid
DP	Degree of polymerization
GC-MS	Gas chromatography-mass spectrometer
GH	Glycoside hydrolases
GlcNAc	*N*-acetyl-D-glucosamine
HPAEC	High-performance anion exchange chromatography
HPLC	High-performance liquid chromatography
LC-MS	Liquid chromatograph-mass spectrometer
MALDI-TOF	Matrix-assisted laser desorption-ionization
N-acetyl COSs	Chitin oligosaccharides
PAD	Pulsed amperometric detection
RTIL	Room temperature ionic liquid
SEC	Size-exclusion chromatography
TLC	Thin-layer chromatography

S. Yang (✉)
College of Food Science and Nutritional Engineering, China Agricultural University, Beijing, China
e-mail: ysq@cau.edu.cn

X. Mao (ed.), *Degrading Enzymes for Marine Polysaccharides and Their Applications*, https://doi.org/10.1007/978-981-96-9517-1_4

1 Introduction of Chitin

1.1 The Structural Characteristics of Chitin

Chitin, a structural biopolymer, functions in a manner comparable to that of cellulose in terrestrial plants and collagen in higher animals. Plants produce cellulose for their cell walls, whereas insects and crustaceans produce chitin for their exoskeletons. As a result, the essential, fundamentally interlaced polysaccharides cellulose and chitin emerge, giving plants and animals their primary respectability and defense mechanisms. Chitin structures make translucent microfibrils, which are necessary for the exoskeletons of arthropods and the cell walls of organisms and yeast, in their usual environments. In shellfish, chitin appears as strong material weaved inside a six-deserted protein helix. Chitin is like changed cellulose, because it has an acetamido bunch in place of the hydroxyl bunch at position C-2. Polymerized monosaccharide structures incorporate chitin, which is comprised of units of - (1-4)- 2-acetamido-2-deoxy-D-glucose, and cellulose, which is comprised of units of - (1-4)- 2-deoxy-D-glucopyranose (Fig. 1). Consequently, chitin is described by poly(-(1-4)-N-acetyl-D-glucosamine) (Fig. 2). Additionally, chitin can be found in three distinct polymorphic structures (α, β, and γ), with ongoing research demonstrating that the γ-structure is a variant of the α-structure. These polymorphic assortments in chitin result from contrasts in the blueprint and limit of neighboring chains in moderate sheets. In α-chitin, the chains are changed nonequivalent, while in β-chitin, all chains are equivalent. Chitin's subatomic strategy is determined by its physiological capabilities and tissue properties. Unadulterated chitin, for example, is found in the getting a handle on spines of sagitta, which give the unbending nature expected to prey catch, and in the driven diatom *Thalassiosira*. It has been seen that treating α-chitin with 20% NaOH followed by water washing switches it over totally to β-chitin, showing the novel association between chitin's polymorphic designs and external effects.

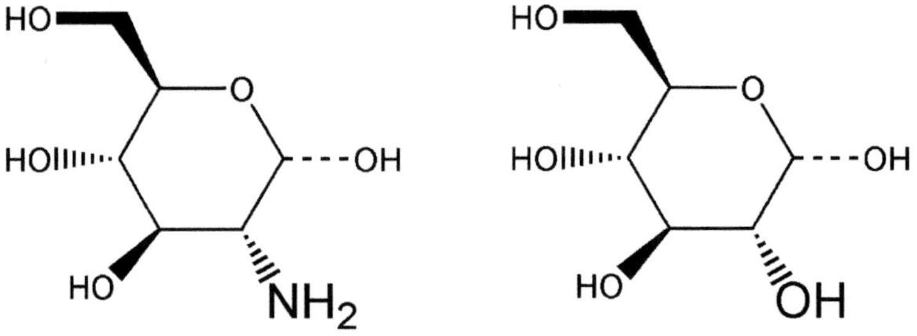

Fig. 1 Structure of glucosamine (monomer of chitosan) and glucose (monomer of cellulose)

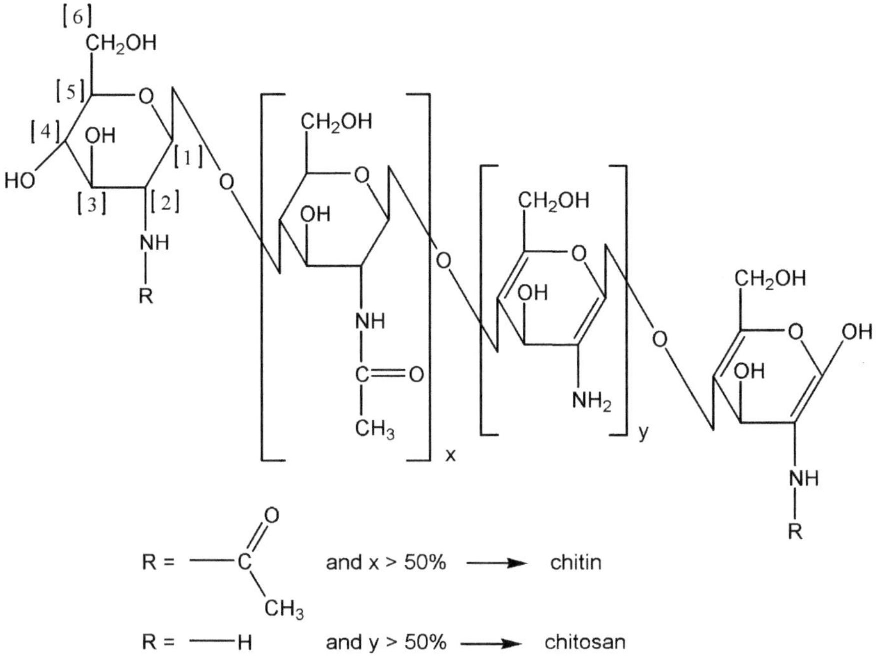

Fig. 2 Structure of chitin and chitosan

1.2 Functions of Chitin Oligosaccharides

Chitin is an *N*-acetyl-D-glucosamine (GlcNAc) unit-linked linear insoluble polymer, making it the second most common polysaccharide after cellulose [31]. Over 100 billion tons of chitin are produced annually worldwide, with much of it going unutilized. The disposal of this excess chitin in natural habitats not only wastes valuable biomass resources but also raises concerns about the environment [57, 60]. Chitin oligosaccharides (*N*-acetyl COSs) and GlcNAc are results of the incomplete or complete debasement of chitin, accomplished through synthetic, physical, or enzymatic techniques. As of late, *N*-acetyl chitooligosaccharides (*N*-acetyl COSs), particularly those with a level of polymerization (DP) going from two to six, have drawn in critical interest because of their remarkable properties, for example, water solvency, steadiness across a scope of temperatures and pH levels, and showed organic exercises, including antimicrobial, hostile to growth, and hypoglycemic impacts [4, 62]. In addition, *N*-acetyl COSs promote bifidus proliferation by acting as plant elicitors and potent angiogenesis inhibitors [58]. *N*-acetyl COSs are extremely adaptable and useful in the pharmaceutical, food, and agricultural sectors because of these properties [4, 44, 61]. As a result, the production of *N*-acetyl chitooligosaccharides (*N*-acetyl COSs) is a good way to deal with and reuse a lot of chitin waste.

1.3 Chemical Methods for Preparation of Chitin Oligosaccharides

By and large, the business creation of *N*-acetyl chitooligosaccharides (COSs) has relied upon an interaction in view of concentrated corrosive hydrolysis of unrefined chitin materials [18]. Normally, chitin debasement by means of substance strategies includes the use areas of strength for of hydrolysis, especially utilizing hydrochloric corrosive. The ideal operating temperature and acid concentration are crucial factors that must be carefully selected and controlled. These boundaries should be sufficiently high to work with chitin debasement without exorbitantly corrupting the subsequent *N*-acetyl COSs items. Hydrochloric acid concentrations between 15% and 36% (w/v) and temperatures between 40 and 80 °C [7] are favorable reaction conditions. The cycle includes starting crushing and sieving of rough chitin, trailed by brooding in 12 M concentrated hydrochloric corrosive. Using thin-layer chromatography (TLC), the resulting solution is filtered, desalted, and qualitatively analyzed. The solution is then neutralized with sodium hydroxide. The hydrolysis items are hence freeze-dried to yield an end result wealthy in *N*-acetyl COSs. Then again, phosphoric corrosive or hydrofluoric corrosive can be utilized instead of hydrochloric corrosive. One detailed strategy includes suspending chitin in 37% (w/v) hydrochloric corrosive, blending for 2 h at 4 °C, cooling to 0 °C, and changing the pH to 7 utilizing half sodium hydroxide. Centrifugation of the suspension combination isolates the supernatant, which is then sifted utilizing a glass channel and thought with a revolving evaporator. The example goes through additional fractionation on a Bio-Gel P2 acrylamide segment and quantitative examination by means of elite execution fluid chromatography (HPLC). A number of *N*-acetyl COSs derivatives, including GlcNAc, $(GlcNAc)_2$, $(GlcNAc)_3$, $(GlcNAc)_4$, and $(GlcNAc)_6$, are produced by the fractions.

An alternative method for synthesizing GlcNAc involves using GlcN as the substrate. Initially, chitin is dissolved in concentrated hydrochloric acid and then subjected to 3 h of heating in boiling water. These stringent conditions ensure complete chitin depolymerization and facilitate the removal of the acetyl group from GlcNAc. Consequently, an additional *N*-acetylation reaction is required to produce GlcNAc. Specifically, GlcN is dissolved in a 10% (w/v) methanol solution and undergoes *N*-acetylation in the presence of acetic anhydride and an ion exchange resin (Dowex1 in carbonate form). Subsequent filtration and crystallization yield an overall GlcNAc production efficiency of 43%. An enhanced *N*-acetylation procedure described by Zhan et al. involves GlcN undergoing *N*-acetylation using pyridine as a solvent, with acetic anhydride and tributylamine. This optimized method results in a maximum yield of 70% GlcNAc and achieves purity levels up to 99% following a series of purification steps.

The products that result from the chemical synthesis of *N*-acetyl COSs and GlcNAc are regarded as chemically modified rather than natural materials, despite the fact that these synthesis methods are both simple and economically viable [37]. Furthermore, beside *N*-acetylated items, follow measures of O-acetylated and di-acetylated items have been recognized utilizing fluid chromatography-mass spectrometry (LC-MS). Also, the items were found to contain hints of both dissolvable and tributylamine, as uncovered by gas chromatography-mass spectrometry (GC-MS) investigation [59]. Chemically synthesized *N*-acetyl COSs are limited in their use across industries, particularly the food industry, because these residual substances contribute to their bitter taste [10, 63].

1.4 Physical Methods for Preparation of Chitin Oligosaccharides

Ball-milling, ultrasound, and microwaves are just a few of the alternative methods that have shown promise in recent years for creating high-energy microenvironments that are favorable to efficient biomass conversion [45]. The limited warm and mechanical circumstances delivered during mechanochemical processes offer a proficient means to start reactivity in natural materials, including chitin and cellulose. This approach kills the requirement for solvents and really upsets the glass-like construction of polysaccharides by debilitating the intermolecular hydrogen bonds. Particularly important for the conversion of chitin is the reduction of crystallinity.

Within frameworks for biomass valorization, mechanochemical processes have demonstrated significant potential for environmentally friendly and sustainable chitin processing. Chitin's crystallinity can be reduced by 50% using a controlled 2-h ball-milling process, according to X-ray diffraction analysis (Fig. 3). Additionally,

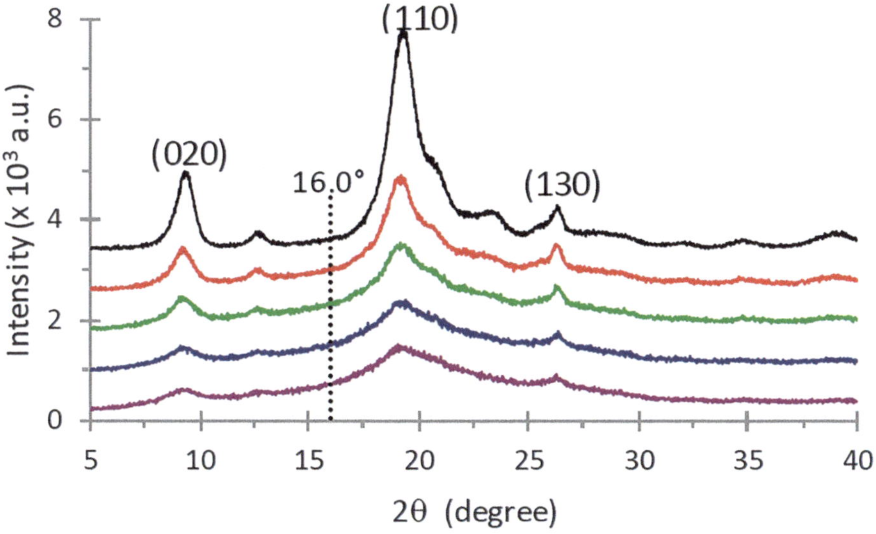

Fig. 3 XRD analysis of untreated (black signal) and milled α-chitin by 2 × 0.5″ balls for 30 (red), 60 (green), 90 (blue), and 120 (pink) min, respectively [29]

N-acetyl groups were retained despite a decrease in glycosidic linkage content as determined by spectral analysis. The solubility of the milled materials significantly increases when kaolinite or natural clay is included in the ball-milling process. More specifically, over 75% of the products are water soluble with the addition of kaolinite, whereas only 35% are water soluble without it. In this method, the items are described as *N*-acetyl chitooligosaccharides (COSs) with levels of polymerization going from 1 to 5 utilizing a clever quantitative framework helped laser desorption-ionization (MALDI-TOF) mass spectrometry strategy [29]. A colorimetric evaluation of reducing termini and size-exclusion chromatography (SEC) are in line with these outcomes. GlcNAc and $(GlcNAc)_2$ yields of 5.1% and 3.9%, respectively, were achieved within 6 h, matching the yields of glucose and cellobiose obtained from cellulose degradation using the same method.

1.5 Enzymatic Methods for Preparation of Chitin Oligosaccharides

The production of *N*-acetyl COSs can be accomplished through chemical and physical methods. However, the utilization of concentrated acids in chemical procedures poses various environmental challenges [60]. Moreover, the yields obtained through physical methods remain insufficient for scalability, notwithstanding the significant energy consumption involved. Consequently, the enzymatic approach emerges as a preferred method for *N*-acetyl COSs production, characterized by its consistency in product output, environmental friendliness, and alignment with principles of sustainable development [33]. The selection of appropriate hydrolytic enzymes is pivotal for the enzymatic conversion of chitin. To date, a variety of enzymes, including chitinase enzymes and lipases, have been utilized for the hydrolysis of chitin materials to produce *N*-acetyl chitooligosaccharides (COSs). Chitinolytic enzymes are predominantly used for the production of *N*-acetyl chitooligosaccharides (COSs) from chitin materials. These enzymes are classified into two main categories based on their mechanisms of action: (1) chitinases (EC 3.2.1.14), which catalyze the random cleavage of glycosidic bonds within internal sites of chitin, and (2) β-*N*-acetylhexosaminidases (EC 3.2.1.52), which sequentially remove GlcNAc residues from the nonreducing end of the chitin chain backbone [2]. Extensive research has focused on identifying, characterizing, and applying various chitinolytic enzymes to enhance the hydrolysis efficiency of chitin for the production of *N*-acetyl COSs and GlcNAc. Additionally, numerous commercially available non-chitinase hydrolytic enzymes, such as lipase, lysozyme, pectinase, papain, glucanase, and tannase, have been documented as capable of chitin hydrolysis for the production of *N*-acetyl chitooligosaccharides (COSs) and GlcNAc [18].

1.6 The Pretreatments of Chitin

To facilitate efficient bioconversion, pretreatments play a crucial role as initial steps aimed at altering the supramolecular structure of biopolymers. Analogous to pretreatment strategies employed in lignocellulosic biomass processing, these procedures typically involve an incubation phase. Although pretreatments are extensively documented for cellulosic biomass, their application to chitin is crucial due to its robust crystallinity and insolubility in aqueous environments. In order to make it easier for chemicals to break down chitin, a tough material found in nature, scientists often use different methods to weaken its structure. This can include treating it with substances like phosphoric acid (to make it bigger) or hydrochloric acid (to make it smaller), causing its crystal structure to be disrupted [18]. There are many different ways that chitin can be treated before being broken down by enzymes to make *N*-acetyl COSs and GlcNAc. These treatments can involve grinding it up, exposing it to strong acids, or even using extreme temperatures and pressures to break it apart [32]. Recently, researchers have found more environmentally friendly ways to turn chitin into useful substances like *N*-acetyl chitooligosaccharides (COSs) and GlcNAc. These methods can involve using high pressure [60], special liquids [16], or even bacteria to help break down the chitin [68]. High-pressure homogenization has been particularly effective in preparing chitin for conversion into these valuable substances. Wei et al. demonstrated that high-pressure homogenization can induce a transformation in the crayfish shell, resulting in a porous netted structure. A GlcNAc yield of 3.9 g/L (from 25 g/L of crayfish shell) was achieved in a batch enzymatic reaction employing chitinase hydrolysis following pretreatment with five cycles of high-pressure homogenization at 400 bar (Fig. 4). The findings indicate that high-pressure homogenization could potentially serve as an effective approach for the enzymatic synthesis of GlcNAc directly from crayfish shell.

Husson et al. demonstrated that pretreating chitin with a room temperature ionic liquid (RTIL), specifically 1-ethyl-3-methyl imidazolium acetate ([C2mim][OAc]), significantly enhances enzyme accessibility to the substrate. This pretreatment facilitated efficient production of GlcNAc at a rate of 185 mg/g chitin and $(GlcNAc)_2$ at 668 mg/g chitin during enzymatic hydrolysis using chitinases from *Trichoderma viride* and *Streptomyces griseus*, respectively. Additionally, a simultaneous enzymatic hydrolysis approach was developed as a promising alternative. This method involves conducting one-pot enzymatic hydrolysis of chitin in a [C2mim][OAc]-aqueous medium, allowing selective production of GlcNAc or $(GlcNAc)_2$. This approach reduces the amount of [C2mim][OAc] required and simplifies the process by minimizing the number of steps involved. Ultimately, it was demonstrated that the synergistic action of chitinases from *T. viride* and *S. griseus* significantly enhances GlcNAc production, achieving 760 mg/g chitin. This underscores the

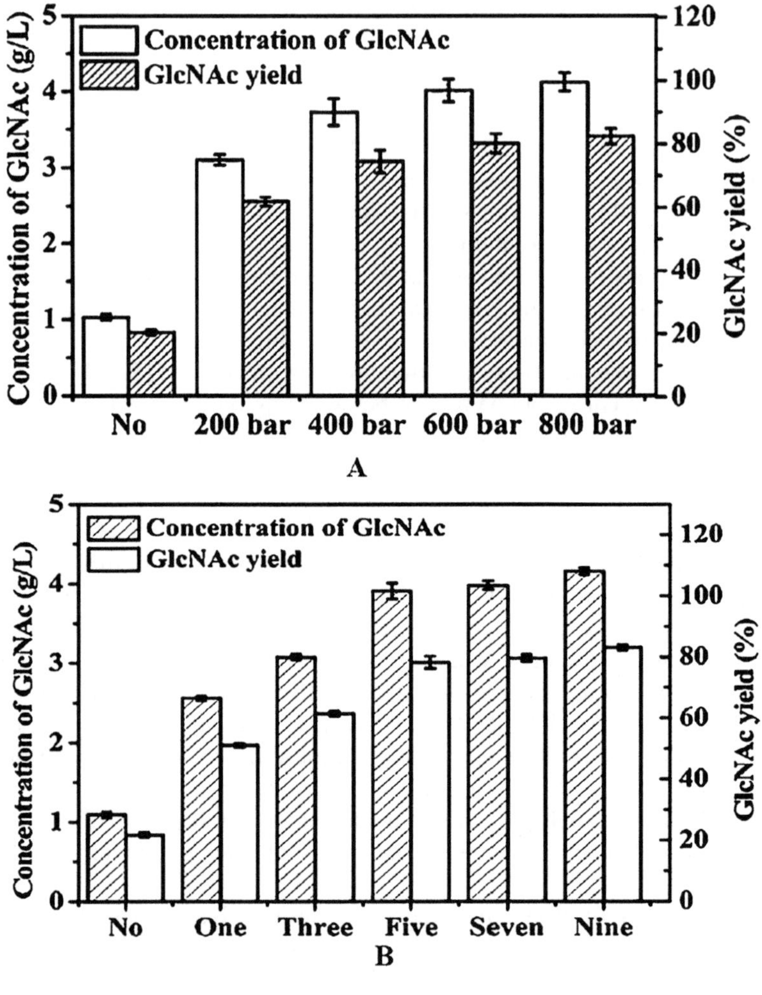

Fig. 4 Effects of high-pressure treatment on the GlcNAc conversion efficiency from crayfish shells. (**a**) Effect of different pressures applied in five cycles; (**b**) effect of cycle numbers treated at 400 bar [60]

importance of employing multiple chitinases to achieve greater enzymatic conversion efficiency (Fig. 5).

Furthermore, bacterial fermentation represents another avenue for enhancing the production of *N*-acetyl COSs and GlcNAc. Zhang et al. documented that pretreatment of chitin powder with bacterial fermentation, specifically employing *Chitinolyticbacter meiyuanensis* SYBC-H1, induced the formation of a visibly

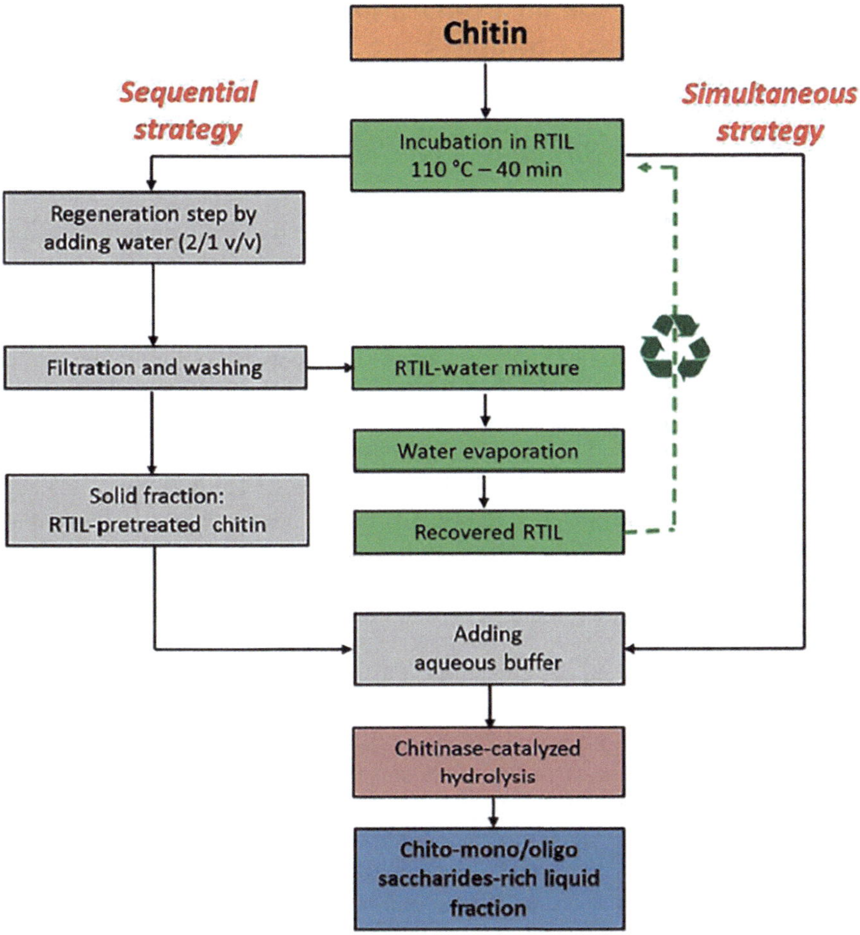

Fig. 5 Skeleton of sequential and simultaneous strategies for the biocatalytic conversion of chitin [16]

fleecy structure. Scanning electron microscopy (SEM) analysis revealed a fiber-like surface structure on the chitin particles, with diameters ranging from 10 to 200 nm. By employing this pretreatment method in conjunction with a subsequent enzymatic reaction lasting 6 h, Zhang et al. achieved the highest GlcNAc yield of 96%, equivalent to 19.2 g/L, from an initial concentration of 20 g/L chitin. The novel method offers an effective, gentle, and eco-friendly approach to decrease the crystalline structure of chitin. It can augment the GlcNAc conversion efficiency devoid of employing acids, oxidizing agents, or alternative chemical reagents.

2 The Enzymes Used for Preparation of Chitin Oligosaccharides: Chitinases and β-*N*-acetylglucosaminidases

2.1 Classification, Structure, and Catalytic Mechanism

Glycoside hydrolases (GH), as extensively documented in the literature, have been systematically compiled and categorized into distinct GH families based on sequence homology. The CAZy database (http://www.cazy.org) offers an updated compilation of these GH families. Chitinases are predominantly grouped into GH families 18, 19, 23, and 48 [11, 50], whereas β-*N*-acetylhexosaminidases are primarily distributed across GH families 3, 20, 84, and 116 [9, 35, 65]. The catalytic domains of enzymes from various GH families generally exhibit a common three-dimensional architecture. For example, members of GH families 20 and 84 possess similar $(\beta/\alpha)_8$ barrel domains [43], while enzymes in GH families 19 and 23 adopt an α+β structural motif [49]. In contrast, GH family 3 enzymes typically display a bipartite structure with a $(\beta/\alpha_{)8}$ barrel followed by an $(\alpha/\beta)_6$ sandwich, although some may lack the $(\alpha/\beta)_6$ sandwich motif. GH family 48 enzymes are characterized by an $(\alpha/\alpha)_6$ barrel structure, featuring six central α-helices surrounded by six external α-helices [2] (Fig. 6).

Chitinases are many times characterized as particular proteins, which normally incorporate helper spaces, for example, the sugar restricting module (CBM) notwithstanding the synergist area [13]. The CBM assumes a fundamental part in enlarging protein movement on insoluble substrates by elevating chemical adherence to the substrate and disturbing the substrate's translucent design to produce free chain closes. Like reactant modules, CBMs are gathered into different homologous protein families, with far-reaching data available in the CAZy data set (http://www.cazy.org). The presence of a deep and narrow substrate-binding cleft lined with aromatic residues is one significant feature that aids chitinases in overcoming

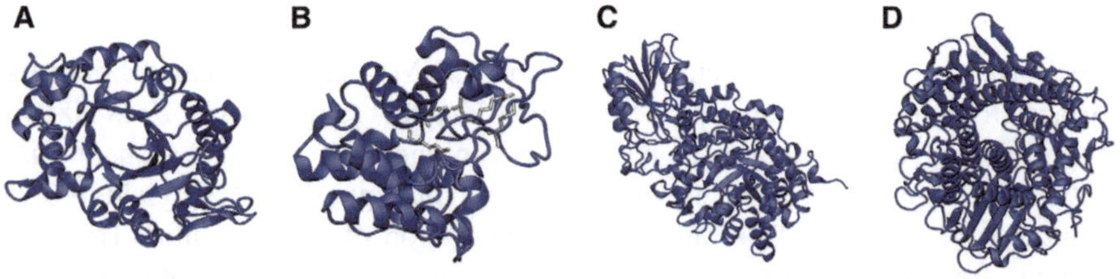

Fig. 6 Three-dimensional structure of several representative chitinases from different GH families. (**a**) The GH family 18 chitinase from **Aspergillus fumigatus**, characterized by a (β/α)8 barrel structure. (PDB ID: 2XVP.) (**b**) The GH family 19 chitinase from **Carica papaya**, which features an α + β structure. (PDB ID: 3CQL) (**c**) The GH family 3 glucohydrolase from **Hordeum vulgare**, comprising a bipartite domain with a (β/α)8 barrel in the lower right and an (α/β)6 sandwich in the upper left. (PDB ID: 1IEQ) (**d**) The GH family 48 cellobiohydrolase from **Clostridium thermocellum**, exhibiting an (α/α)6 barrel structure (PDB ID: 1L1Y) [2]

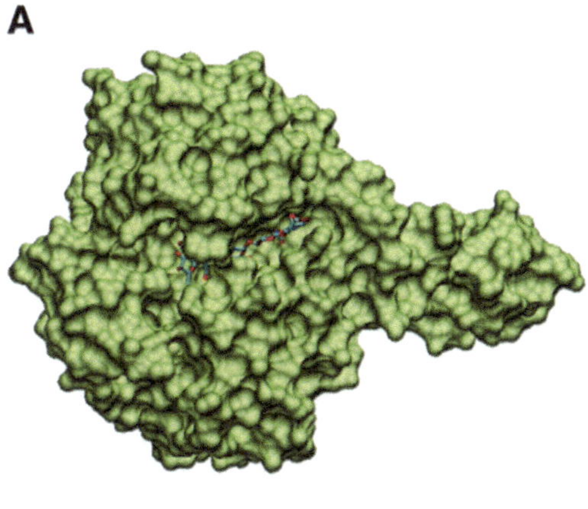

Fig. 7 Illustrated are the three-dimensional structures of the substrate-binding clefts of both processive and non-processive chitinases. (**a**) The chitinase from *Serratia marcescens*, which exhibits processive behavior (PDB ID: 1E6N). (**b**) The chitinase from *Hevea brasiliensis*, characterized by non-processive activity (PDB ID: 1KQY). Substrates in both structures are depicted using stick representation [2]

their limited access to insoluble substrates (Fig. 7) [67]. These chitinases capability in a processive mode, restricting to a substrate chain and guiding it through their reactant split, severing it at numerous hydrolytic locales without delivering the substrate after every cleavage. In order to create a favorable environment for the flexible binding and movement of the substrate within the active site, aromatic residues are essential. While processive catalysis is a successful methodology for hydrolyzing insoluble substrates, it every now and again brings about diminished chemical action on dissolvable or more open polymeric substrates.

Chitinolytic compounds commonly work with substrate depolymerization by means of two unmistakable components: single uprooting and twofold relocation. Two distinct catalytic groups are required for both routes. The first group is typically a conserved glutamate residue in the enzyme's active site and serves as a proton donor. This glutamate can occasionally be replaced by an aspartate residue, as is the case with some GH family 84 chitinases. In either the single-displacement or double-displacement mechanisms, the second catalytic group acts as a base or nucleophile. This gathering can be a carboxyl moiety from a preserved aspartate or glutamate buildup, or the *N*-acetyl gathering of the sugar situated at the - 1 subsite of the enzyme [1]. From −n to +n, the substrate's subsites are designated, with negative numbers denoting the chain's nonreducing end and positive numbers denoting the chain's reducing end. Cleavage happens between the −1 and +1 subsites. The hydrolyzed GlcNAc residue's anomeric configuration is inverted by the single displacement, or inverting, mechanism, while the double displacement, or retaining, mechanism keeps the anomeric configuration.

2.2 *Exploration and Properties of Chitinases and β-N-acetylglucosaminidases*

In recent years, an increasing interest has emerged in the discovery and exploration of novel chitinolytic enzymes. These enzymes, found across a wide range of organisms such as bacteria, fungi, plants, insects, and animals, fulfill various roles including morphogenesis, nutrition, and defense against chitin-containing pathogens. Many of these organisms possess multiple genes encoding chitinolytic enzymes. For instance, filamentous fungi typically harbor between 10 and 20 distinct genes encoding chitinolytic enzymes, while mycoparasitic species may contain as many as 30 or more such genes. These chitinolytic enzymes often work synergistically or successively to degrade chitin. Certain higher organisms, such as *Arabidopsis*, also host a significant number of genes encoding chitinolytic enzymes. Nonetheless, certain genes do not yield active enzymes. For instance, various organisms such as plants, invertebrates, and animals express genes that encode chitinase-like lectins which lack chitinolytic activity, potentially because of alterations in key catalytic residues. Despite retaining the ability to bind chitin, these proteins lack enzymatic function.

Chitinases have a wide range of molecular masses, ranging from 20 to 120 kDa. Similar to plant chitinases (25–40 kDa) and smaller than insect chitinases (40–85 kDa), the molecular masses of microbial chitinases typically range from 20 to 110 kDa. Chitinases that have been isolated from a variety of microorganisms exhibit activity over a wide temperature and pH range [64]. At 28 °C and 80 °C, respectively, the activity of the endochitinases produced by *Streptomyces violaceusniger* and *Streptomyces thermoviolaceus* peaks. In addition, the chitinase chemical from *S. thermoviolaceus* is best inside a pH scope of 8.0–10.0, while the chitinase got from *Stenotrophomonas maltophilia* C3 accomplishes ideal action at a pH scope of 4.5–5. Moreover, chitinases beginning from different sources show a wide cluster of isoelectric focuses (*p*I), traversing from 3.0 to 10.0 [68].

Likewise, microbial β-*N*-acetylglucosaminidases regularly have subatomic masses going from 40 to 90 kDa. Certainly! Here is a cleaned rendition of the sentence: β-*N*-acetylglucosaminidases from different microorganisms, like chitinases, display broad scopes of ideal temperatures (30–70 °C) and pH levels (3.5–10.0). For example, the β-*N*-acetylglucosaminidases from *Streptomyces* sp. Peak activity is seen in NK5 and *Trichoderma reesei* at pH values of 10.0 and 4.0, respectively, with optimal temperatures of 60 °C and 70 °C.

2.3 *Future Prospective of Chitinases and β-*N*-acetylglucosaminidases*

A multitude of efforts have been devoted to enhancing the catalytic efficiencies of chitinases and β-*N*-acetylglucosaminidases through techniques such as directed evolution, site-directed mutagenesis, and domain fusion. These studies have yielded crucial insights into the roles of specific domains, loops, and residues within these enzymes, thus presenting opportunities for optimizing their use in diverse applications. Further investigation into the functional roles of domains promises to enable the customized design of chitinases and β-*N*-acetylglucosaminidases with superior properties. Additionally, investigations into the interactions between chitinases and diverse forms of chitin can shed light on the diverse antifungal properties exhibited by these enzymes.

Furthermore, the predetermination and comparative analysis of the behaviors of diverse chitinases through molecular dynamics simulations have the potential to unveil novel interaction mechanisms. Subsequently, this knowledge can inform the development of chitinases with improved antifungal, hydrolytic, or transglycosylating capabilities through refined design and mutation strategies. Lastly, leveraging the rapidly advancing field of bioinformatics, along with the proliferation of computational tools and databases facilitating the description and prediction of protein structure and function, offers a rational approach to engineering chitinases for enhanced activity and stability tailored to diverse biotechnological applications.

3 Specific Enzymatic Methods

3.1 Preparation of GlcNAc

The complete hydrolysis of chitin is only possible through the cooperation of chitinases and *N*-acetyl-hexosaminidase in the bioproduction of GlcNAc. Various chitinolytic catalysts have been distinguished that can work with chitin hydrolysis to deliver GlcNAc. For instance, *Burkholderia cepacia* TU09's chitinase has demonstrated its ability to hydrolyze both β-chitin and α-chitin, achieving GlcNAc yields of 85% and 80% after 1 day and 7 days, respectively. In a similar vein, the chitinase from *Bacillus licheniformis* SK-1 was able to produce GlcNAc and $(GlcNAc)_2$ from -chitin with a yield of 75% and 20%, respectively, in just 6 days. This chitinase produced a 41% GlcNAc conversion yield when applied to β-chitin [34]. Notably, for the production of GlcNAc, α-chitin from squid pen is generally preferred to -chitin from crab or shrimp shells. Additionally, the enzymatic hydrolysis of colloidal -chitin with a crude enzyme from *Paenibacillus illinoisensis* has been reported. After 24 h of incubation, the highest production yield of 62.2% was achieved, with continuous increases in GlcNAc production rates over time. Simultaneously, the yields of $(GlcNAc)_2$, $(GlcNAc)_3$, $(GlcNAc)_7$, and $(GlcNAc)_8$ were 4.9%, 1.2%, 4.1%, and 9.6%, individually [19].

Additionally, enzymes other than chitinase have been employed for the hydrolysis of chitin to enhance the production of GlcNAc. For example, crude cellulase cultures derived from *Trichoderma viride* and *Acremonium cellulolyticus*, along with the crude enzyme culture from *Trichoderma reesei*, which includes activities of both cellulose and β-glucanase, have been utilized for this purpose. Importantly, a combination of cellulase and lipase enzymes from *Aspergillus niger* has also been used in the hydrolysis of β-chitin [37]. Moreover, flake-type chitin has been identified as an effective substrate for the production of *N*-acetyl COSs, with studies reporting the successful hydrolysis of crab shell α-chitin flakes by crude enzyme extracts from *Aeromonas hydrophila*, achieving GlcNAc yields between 66% and 77% after a 10-day incubation period [38].

Furthermore, *N*-acetyl COSs can serve as substrates for the production of (GlcNAc) through exo-type chitinases. Notably, various enzymes, including exochitinases from *Bacillus thuringiensis* [48] and *Aeromonas hydrophila* [21], as well as *N*-acetyl-β-hexosaminidases from *Stenotrophomonas maltophilia*, have been implicated in this process. While oligomers pose challenges in commercial production due to complexity and cost, they serve as valuable tools for elucidating enzyme properties [25].

Moreover, endeavors have been coordinated toward the particular creation of GlcNAc and $(GlcNAc)_2$ [23]. A crude *Aeromonas* sp. enzyme culture, for instance, GJ-18, which contained both *N*-acetyl-hexosaminidase and *N*-acetyl chitobiase, demonstrated that depending on the temperature of the incubation, selective hydrolysis of -chitin produced either GlcNAc or $(GlcNAc)_2$. After 5 days of incubation, the best yields of GlcNAc (74% and 35%, respectively) were obtained at 45 and 55 °C [24].

3.2 Preparation of Chitin Oligosaccharides

The selective hydrolysis of chitin materials into $(GlcNAc)_2$ was made easier by the regulation of the balance between *N*-acetyl-hexosaminidase and *N*-acetyl chitobiase activities in the culture medium of *Aeromonas* sp. Utilizing crude enzymes that had been pretreated at 50 °C to deactivate *N*-acetyl-hexosaminidase, substantial yields of $(GlcNAc)_2$ were obtained after a cultivation period of 7 days, with 78.9% coming from swollen chitin and 56.6% coming from powdered chitin [22]. Colloidal chitin was distinguished as the most positive substrate for $(GlcNAc)_2$ creation because of its solvency in water and simplicity of hydrolysis. Different endeavors have been made to enzymatically produce $(GlcNAc)_2$ from colloidal chitin. For instance, the marine bacterium *Vibrio anguillarum*'s chitinase hydrolyzed colloidal chitin with efficiency, producing a maximum $(GlcNAc)_2$ yield of 40.3% [46]. Also, business cow-like pepsin was utilized to debase shrimp α-chitin, bringing about yields of 19% for GlcNAc, 75% for $(GlcNAc)_2$, and 9.5% for $(GlcNAc)_3$ [17]. Additionally, the hyperthermophilic archaeon *Thermococcus kodakaraensis* KOD1's chitinase hydrolyzed colloidal chitin effectively, primarily to produce $(GlcNAc)_2$ [47]. Mass delivered unrefined compounds have been utilized to some degree hydrolyze marine biopolymer chitin, bringing about $(GlcNAc)_2$ creation. For instance, after 24 h of incubation, a pepsin with high chitinolytic activity hydrolyzed pepsin-treated chitin at 44 °C and pH 5.4 to yield 71.5% $(GlcNAc)_2$. Moreover, a clever recombinant chitinase (PbChi70) from the marine bacterium *Paenicibacillus barengoltzii* was communicated and sanitized to homogeneity, exhibiting the capacity to hydrolyze

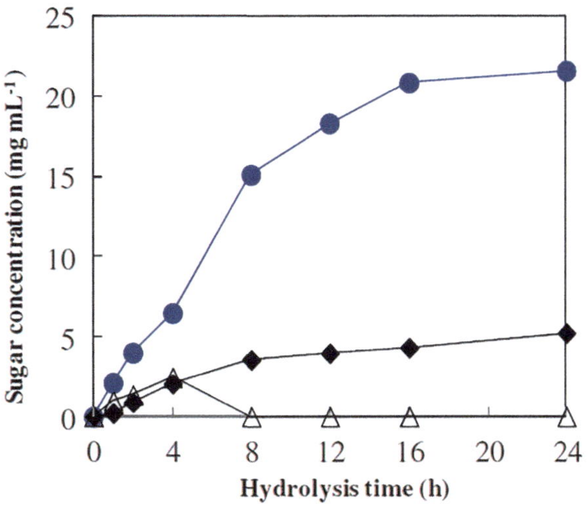

Fig. 8 Effect of incubation time on the production of $(GlcNAc)_2$ from colloidal chitin by PbChi70 from *P. barengoltzii* [64]. Symbols: (◆) GlcNAc; (●) $(GlcNAc)_2$; (△) $(GlcNAc)_3$

colloidal chitin and produce $(GlcNAc)_2$ with a best return of 89.5% [64]. Moreover, $(GlcNAc)_2$ can be blended through a consecutive microbial maturation process, beginning with *Lactobacillus paracasei* aging for demineralization, trailed by *Serratia marcescens* maturation for deproteinization, lastly *Aeromonas* sp. aging for the hydrolysis of the lingering chitin (Fig. 8).

Most chitinolytic catalysts principally catalyze the hydrolysis of chitin, yielding GlcNAc and $(GlcNAc)_2$ as the prevalent items because of their synergist systems. In addition to their hydrolytic function, some chitinolytic enzymes also have transglycosylation activity, allowing them to transfer the freed oligosaccharide moieties to the appropriate acceptors for the formation of new glycosidic bonds. Chitinolytic enzymes possess a significant potential for the synthesis of size- and stereo-specific N-acetyl COSs and their derivatives, thanks to their ability to transglycosylate. Besides, the transglycosylation action displayed by glycolytic chemicals, including chitosanase, chitinase, and different glycosidases, gives a technique to create N-acetyl chitooligosaccharides (COSs) with high levels of polymerization (DP). For example, *Trichoderma reesei* chitinase works with the transglycosylation of $(GlcNAc)_4$, bringing about the creation of $(GlcNAc)_2$ (55.7%) and $(GlcNAc)_6$ (39.6%). Akiyama et al. exhibited the blend of *N*-acetyl COSs with DP going from 4 to 12 by means of a transglycosylation response catalyzed by lysozyme, utilizing N,N′,N″-triacetylchitotriose and N,N′,N″-tri(monochloro)acetylchitotriose as substrates, trailed by the evacuation of N-monochloroacetyl bunches through a base-catalyzed response. In a similar fashion, Hattori et al. used lysozyme-catalyzed transglycosylation of $(GlcNAc)_3$ as a starting material in an aqueous solution containing 30% $(NH_4)_2SO_4$ [18] to produce *N*-acetyl COSs with high DPs (6-15). Table 1 provides comprehensive information on the partial production of GlcNAc and *N*-acetyl COSs by microbial enzymes.

The predominant catalytic mechanisms of most chitinolytic enzymes result in the production of chitin oligosaccharides primarily comprising $(GlcNAc)_2$ and

GlcNAc, facilitating their ease of purification. However, the limited diversity of chitin oligosaccharides, primarily consisting of $(GlcNAc)_2$ and GlcNAc, hampers their commercial viability. Although certain chitinases with transglycosylation activity can produce chitin oligosaccharides with higher degrees of polymerization (DP), their yields remain modest. Consequently, addressing these limitations by identifying chitinases with heightened transglycosylation activity or exploring alternative synthetic strategies for oligosaccharide production from chitin holds considerable promise for enhancing the commercial utility of these compounds.

4 Purification and Characterization of Chitin Oligosaccharides

A variety of oligomers are formed when chitin is hydrolyzed using either chemicals or enzymes. *N*-acetyl chitooligosaccharides (*N*-acetyl COSs) can be separated and purified using a variety of methods that are widely used. Gel filtration [41], ultrafiltration [26], ion-exchange chromatography [14], thin-layer chromatography (TLC), and high-performance liquid chromatography (HPLC) are some of these methods. Commonly, size avoidance chromatography (SEC) is utilized for the preparative partition of *N*-acetyl COSs in view of their subatomic sizes. A SEC framework outfitted with SuperdexTM30 (GE Medical services) segments in series has demonstrated to be proficient in isolating *N*-acetyl COSs with comparative levels of polymerization (DP) values going from 2 to 20.

N-acetyl chitooligosaccharides (COSs) can be separated using a variety of alternative methods. For example, a functioning charcoal segment can be utilized to segregate $(GlcNAc)_2$. At first, the hydrolysate is brought into the section, and the sugars bound to it are eluted utilizing an ethanol arrangement with a slope going from 0% to 40% (v/v) at a stream pace of 1 mL min^{-1}. The purity of the collected fractions is then assessed using TLC and HPLC [64]. A straightforward and speedy method for separating and purifying high-molecular-weight *N*-acetyl COSs has been proposed by Katano et al. More specifically, 100 L of 1 M hydrochloric acid is used to treat 10 mg of a commercially available *N*-acetyl COSs sample containing $(GlcNAc)_{n\ (n\,=\,1–7)}$. Following this, all $(GlcNAc)_n$ compounds are broken up in the fluid arrangement and joined with 900 μL of ethyl liquor. Centrifugation is used to collect the precipitate, which is then analyzed qualitatively. The hasten prevalently contains $(GlcNAc)_n$ ($n \geq 5$), showing remarkable solvency contrasts between short-chain $(GlcNAc)_n$ compounds (with $n \leq 3$) and longer ones. Long-chain $(GlcNAc)_n$ samples of high purity can be obtained by iteratively repeating this procedure (n≥5). Given their acknowledged effectiveness as elicitors, these long-chain $(GlcNAc)_n$ compounds hold promise for applications in plant pathology and chitin chemistry [20].

Techniques for separating and analyzing native *N*-acetyl COSs at low concentrations are still limited, despite significant advancements in the separation of *N*-acetyl chitooligosaccharides (*N*-acetyl COSs) over the past decade. A considerable lot of

Table 1 Partial enzymatic production of GlcNAc and *N*-acetyl COSs

Chitin source	Enzyme	Enzyme source	Mol. Wt.	Condition	Product and yield	Analysis method
Swollen chitin	Crude enzyme	*Aeromonas* sp. GJ-18	–	40 °C, 9 days	GlcNAc 94.9%	HPLC (NH2P-50 4E)
Swollen chitin	Crude enzyme	*Aeromonas* sp. GJ-18	–	45 °C, 5 days	GlcNAc 74%, $(GlcNAc)_2$ 4.8%	HPLC (NH2P-50 4E)
				55 °C, 5 days	GlcNAc 3.9%, $(GlcNAc)_2$ 34.7%	
α-chitin, β-chitin	Crude enzyme	*A. hydrophila* H-2330	–	17 °C, 10 days	GlcNAc 64 ~ 77%	HPLC (NH2P-50)
Chitin	Chitinase	*A. hydrophila* H-2330	–	37 °C,12 h	GlcNAc, $(GlcNAc)_2$ ~ $(GlcNAc)_5$	TLC
Colloidal chitin	Exochitinase	*Bacillus thuringiensis* subsp. *pakistani*	66/60/47/32 kDa	37 °C, 24 h	GlcNAc	TLC
Swollen chitin	Chitinase	*Paenibacillus illinoisensis* KJA-424	38/54/63 kDa	37 °C, 24 h	GlcNAc 62.2%	HPLC
α-Chitin	Cellulase	*Trichoderma viride* *Acremonium cellulolyticus*	–	37 °C, 3 days	GlcNAc 16% GlcNAc 22%	HPLC/NMR
Chitin	Chitinase	*Thermococcus kodakaraensis* KOD1	90 kDa	70 °C, 3 h	$(GlcNAc)_2$	TLC
Chitin	Exochitinase	*Vibrio anguillarum* E-383a	–	–	$(GlcNAc)_2$ 40.3%	HPLC

the ongoing techniques are semiquantitative, require derivatization, or have generally high location limits, which request specific hardware. Underivatized carbohydrates can be sensitively detected, quantified, and separated using high-performance anion exchange chromatography (HPAEC) and pulsed amperometric detection (PAD). This approach takes into consideration the evaluation and partition of a different scope of non-named sugars. Munster et al. fostered a quick and proficient technique for the partition and delicate recognition of non-marked *N*-acetyl COSs with levels of polymerization (DP) from 1 to 6. This strategy used a portable stage made out of 25 mM sodium hydroxide with 22.2 mM sodium acetic acid derivation, accomplishing the compelling division of $(GlcNAc)_{1\text{-}6}$ in 6 min or less. In particular, the CarboPac Dad 1 segment was kept up with at 25 °C, utilizing arrangements A, B, and C, which involved 0.1 M sodium hydroxide, 0.1 M sodium hydroxide with 0.6 M sodium acetic acid derivation, and ultrapure water, separately [56].

The column was subjected to the following washing and equilibration steps prior to sample injection: (1) rinsing with solution A at a flow rate of 1 ml min^{-1} for 10 min; (2) flushing with 100% solution B for 10 min; (3) washing with a mixture of 3.7% B and 96.3% A for 10 min; (4) flushing with a solution composed of 3.7% B, 21.3% A, and 75% C for 20 min; and (5) equilibration. To summarize, an autosampler equipped with a sample loop was used to introduce a volume of 25 L, either of $(GlcNAc)_{1\text{-}6}$ standards or samples containing the internal standard (IS) (10 M L-fucose). After that, the column was washed with solution C to separate the samples. In this way, 225 mM sodium hydroxide was controlled at a stream pace of 2 ml min^{-1} and joined in the blending loop for a concise period. Location of sugars happened 16 min post-infusion, using a standard fourfold likely waveform. By comparing the ratios of the peak areas of standards made with predetermined amounts of *N*-acetyl COSs, the concentrations of *N*-acetyl COS in the samples were determined.

Nuclear magnetic resonance (NMR) spectroscopy stands out as a traditional and dependable approach for the subsequent structural elucidation of *N*-acetyl chitooligosaccharides (*N*-acetyl COSs) that have been separated. Simultaneously, the recently referenced methods, like slender layer chromatography (attention) (see Fig. 9), superior execution fluid chromatography (HPLC) (allude to Fig. 10a), and superior execution anion trade chromatography combined with beat amperometric location (HPAEC-Cushion) (delineated in Fig. 11), give clear and proficient strategies to the underlying portrayal of *N*-acetyl COSs. Matrix-assisted laser desorption/ionization time-of-flight mass spectrometry (MALDI-TOF MS) is still the most common method for determining the molecular weight of *N*-acetyl COSs (Fig. 10c). In a nutshell, the *N*-acetyl COSs sample is mixed with an equal volume of 2,5-dihydroxybenzoic acid (DHB) solution after being dissolved in ultrapure water. After that, 0.75 ml of the mixture are put on and left to dry at room temperature. A spectrometer with a 337 nm pulsed nitrogen laser is used to collect all mass spectrometric data [69].

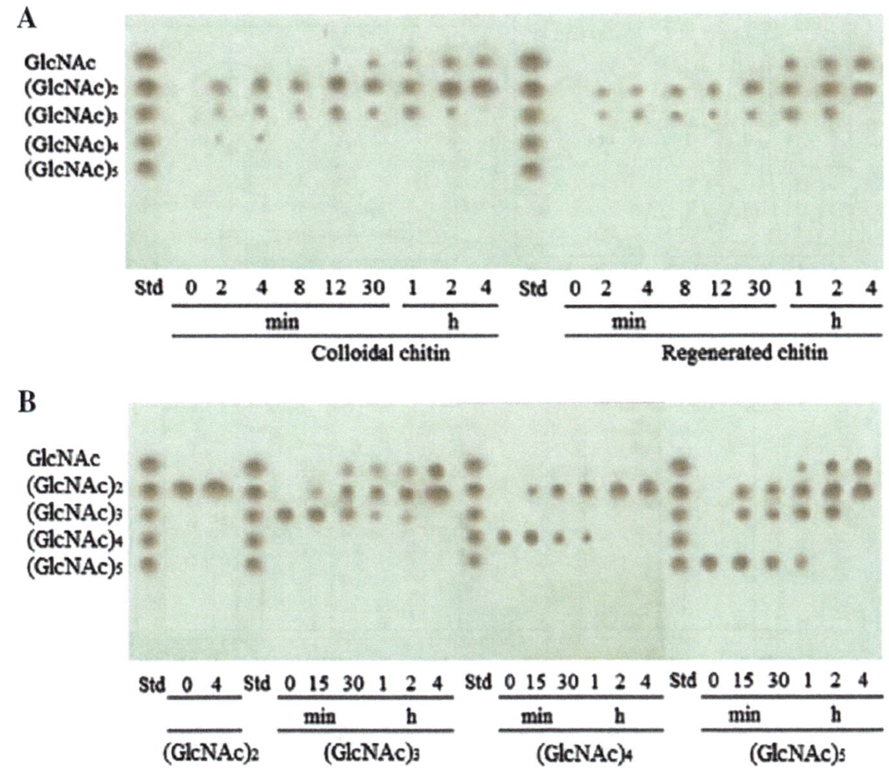

Fig. 9 TLC analysis [64] of hydrolysis products of colloidal chitin (**a**) and *N*-acetyl COSs (DP 2–5) (**b**) by the chitinase from *P. barengoltzii*

5 Enzymatic Synthesis of Chitin Oligosaccharide

5.1 *Preparation of Chitin Oligosaccharide by Reverse Hydrolysis and Transglycosylation Reaction*

The chemical synthesis of chitin oligosaccharides requires multiple protection and deprotection steps, which make it a less desirable method for producing bioactive materials. This process is characterized by a low yield rate and the potential introduction of toxic by-products. In contrast, the biosynthesis of chitin oligosaccharides has attracted considerable interest due to its more efficient and potentially safer production pathway. In this process, two types of enzymes, namely, glycosyltransferases and glycoside hydrolases, play pivotal roles. While glycosyltransferases are utilized in this biosynthesis, certain glycoside hydrolases offer distinct advantages, including broad substrate specificity, accessibility, and the utilization of more cost-effective substrates. In regular frameworks, glycoside hydrolases for the most part

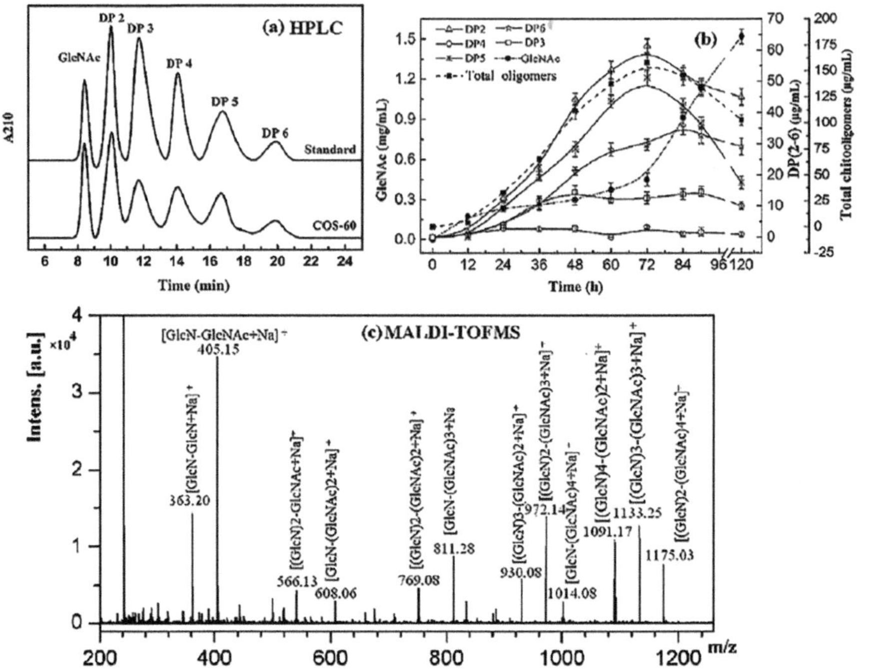

Fig. 10 The detection and identification of hydrolyzed chitin products catalyzed by chitinolytic enzymes can be illustrated as follows: (**a**) The HPLC chromatogram depicts the *N*-acetyl chitooligosaccharides (COSs) obtained from the hydrolysis of HLP by chitinolytic enzymes. (**b**) The impact of varying hydrolysis durations on the concentration of *N*-acetyl COSs derived from HLP through the action of chitinolytic enzymes is analyzed. (**c**) The MALDI-TOF MS spectrum shows the *N*-acetyl COSs resulting from the hydrolysis of HLP by chitinolytic enzymes [69]

catalyze the hydrolysis of glycosidic bonds, with the oligosaccharide working as the leaving gathering and water going about as the nucleophile (glycosyl acceptor). In any case, monosaccharides can likewise supplant water as nucleophiles in these responses. This capacity acquaints the potential with shift glycosidase-catalyzed responses from hydrolysis to union. Reverse hydrolysis and transglycosylation are the two primary mechanisms by which glycoside hydrolases synthesize oligosaccharides.

5.1.1 Reverse Hydrolysis Reaction

Chitin oligosaccharides are made by using *N*-acetylglucosamine and -N-acetylglucosaminidase in reverse hydrolysis. Because they make it easier for the leaving group to leave, modified substrates are frequently used in reverse hydrolysis reactions. Notably, pNP-GlcNAc is frequently used as the donor in

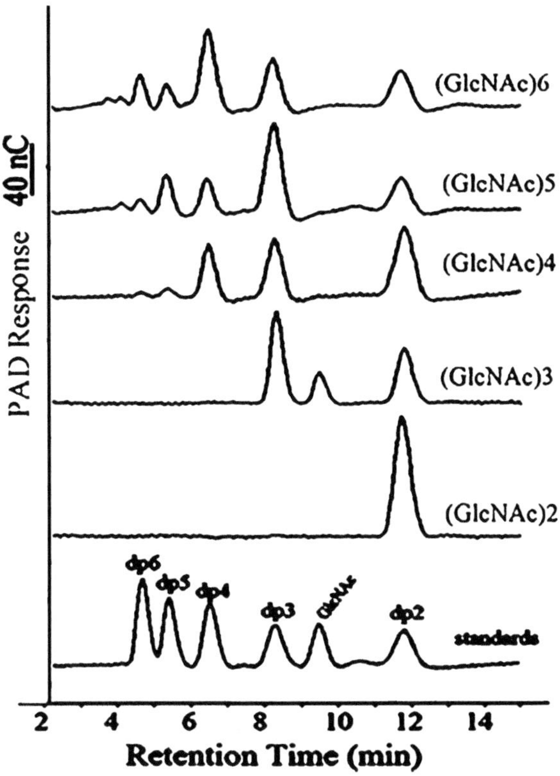

Fig. 11 HPAEC-PAD analysis of the hydrolysis products from *N*-acetyl COSs by a chitinase ChiEn1 [56]

the -N-acetylhexosaminidase-mediated reverse hydrolysis reaction due to its straightforward release as a leaving group. Conversely, the immediate utilization of N-acetylglucosamine as a substrate is less habitually revealed in academic distributions. In 2004, Rauvolfova and partners blended GlcNAc (β-1,3) GlcNAc, GlcNAc (β-1,4) GlcNAc, and GlcNAc (β-1,6) GlcNAc utilizing β-N-acetylhexosaminidase from *Penicillium funiculosum* as the substrate, accomplishing yields of 3.8%, 1.7%, and 10.0%, separately [36]. Thermodynamics is primarily in charge of the reverse hydrolysis reaction, with equilibrium limiting the final yield. (I) Substrate concentration: Higher concentrations of *N*-acetylglucosamine can facilitate the reaction toward synthesis and influence isomer formation. Reverse hydrolysis is influenced by several factors. For instance, the yields of GlcNAc (-1,3) and GlcNAc (-1,6) tripled when the concentration of GlcNAc was increased from 0.5 to 1.0 M. However, enzyme activity may be impeded by excessive substrate concentrations. (II) Concentration of the enzyme: Within a certain range, incremental adjustments to the quantity of -N-acetylglucosaminidase can improve conversion efficiency. In any case, unnecessary chemical levels might animate hydrolysis as opposed to amalgamation. (III) Water action: Each glycosidase shows a basic water action level. Substandard water movement levels can diminish protein action however favor more significant returns. Conversely, high levels of water activity may result in decreased yields while also accelerating reaction kinetics. It has been demonstrated that the addition of salts like $(NH_4)_2SO_4$ or $MgSO_4$ boosts the regioselectivity and yield of -*N*-acetylhexosaminidase. (IV) Product concentration: The equilibrium-restricted nature of the reaction allows for selective adsorption of products to boost yield and promote synthesis. (V) Enzyme source: Although the synthesis of amino oligosaccharides by -N-acetylglucosaminidases from fungal sources and glycoside hydrolase family 20 has been demonstrated, the synthesis of amino oligosaccharides by bacterial sources has not been investigated in this context [40].

5.1.2 Transglycosylation Reaction

New glycosidic links between donor and acceptor sugar molecules are made in the transglycosylation (TG) reaction. A glycosyl-enzyme intermediate is produced during the reaction that is sparked by glucoside hydrolases that have been retained. This middle can go through two pathways: either assault by water prompts hydrolysis or assault by another sugar atom brings about transglycosylation. The transglycosylation response empowers the blend of longer chitin oligosaccharides utilizing chitinases and chitin oligosaccharides as substrates. Most of transglycosylation responses reported to date have utilized chitin oligosaccharides with a level of polymerization (DP) somewhere in the range of 3 and 6. For instance, chitinase-D from *Serratia marcescens* showed transglycosylation movement on $(GlcNAc)_{3\text{-}6}$, bringing about the arrangement of longer-chain chitin oligosaccharides [3]. Also, the endochitinase from *Flavobacterium johnsoniae* showed transglycosylation action utilizing $(GlcNAc)_5$ and $(GlcNAc)_6$ as substrates [54]. Interestingly, only Asaki Uehara et al. have reported that *Stenotrophomonas maltophilia* chitinase was able to synthesize 40% trisaccharide in 3 h using 10% $(GlcNAc)_2$ as a substrate [51]. For the most part, noticed patterns incorporate higher item yields and more limited response times in transglycosylation responses contrasted with turnaround hydrolysis responses, which regularly require over 4 days. Also, chitinases display changing substrate particularity in transglycosylation responses.

The kinetics of transglycosylation reactions dictate that the donor concentration typically exceeds that of the acceptor by two to four times. The transglycosylation yield hinges on the equilibrium between donor synthesis and hydrolysis rates, regulated primarily by product hydrolysis [39]. As reactants become depleted and product concentrations peak during transglycosylation, chitinases, predominantly sourced from glycoside hydrolase families 18 and 19, emerge as pivotal factors influencing the reaction. Chitinases play a crucial role in transglycosylation reactions, with a wide array of sources exhibiting transglycosylation activity, including organisms such as *Serratia marcescens*, *Vibrio carchariae*, *Stenotrophomonas maltophilia*, *Cycas revoluta*, and *Enterobacter cloacae* [5, 28, 42, 53, 55].

Upgrading transglycosylation movement through site-explicit amino corrosive transformations in chitinases addresses a promising technique. It is believed that the transglycosylation activity of GH18 chitinases is heavily reliant on aromatic residues located at glycoside binding sites [30]. In chitinase mutants, it has been demonstrated that increasing transglycosylation activity can be accomplished by altering tryptophan (Trp) side chains, either by removing them or adding them. Madhuprakash et al. performed molecular dynamics simulations on mutants like G119W and discovered that the indoles of adjacent tryptophan residues prevent DP4 products from moving toward the catalytic site [27]. Nathan et al. examined moderated amino corrosive changes inside the limiting depressions of chitinase obtained from *Serratia marcescens* [3]. Their examination uncovered that the freak W167A showed an especially expanded transglycosylation movement, creating 45% $(GlcNAc)_3$ as a substrate, rather than the wild kind, which displayed just 8% action. These systems are pointed toward dragging out substrate home time inside the chemical's dynamic site. In addition, the role of chitinase nucleophiles in transglycosylation has been highlighted in research. The glycosides near the glycosidic cleavage site was mutated into glutamic. For example, the transformation D246E in AfChiB prompted a decline in chitinolytic movement however an expansion in transglycosylation action. In contrast, the mutation W246E maintained hydrolytic activity but completely eliminated transglycosylation activity [3]. These discoveries propose that upsetting nucleophile Asp buildups might improve transglycosylation movement.

5.2 Preparation of Chitin Oligosaccharide by Glycosidic Synthase and Glycosyltransferases

5.2.1 Glycosidic Synthase

Glycosidases possess dual catalytic capabilities, allowing them to both cleave glycosidic bonds and form new glycosidic linkages. The process of transglycosylation, a prevalent reaction mechanism, is predominantly governed by kinetic factors and is supported by a regenerative exoglycosidase utilizing an activated glycoside. However, the overall yield is often compromised due to substrate breakdown and the formation of transglycosylation products, typically ranging from 20 to 40%. In recent years, conventional enzymatic glycoside synthesis has evolved toward specialized applications, necessitating the exploration of innovative techniques such as substrate and enzyme modification [15].

A glycosidic synthase is generated via the method of site-directed mutagenesis applied to a nucleophilic residue within a glycoside hydrolase. This specific alteration converts the nucleophilic attacking component of the glycoside hydrolase into an amino acid devoid of nucleophilic properties, thereby disrupting the initial hydrolysis step and eliminating natural hydrolytic activity. The simultaneous use of -D-fluorinated sugar as a glycosyl donor, which is distinguished by its capacity for self-capping, facilitates the transglycosylation reaction at the enzyme's active site. This change changes over the glycoside hydrolase into a glycoside synthase, subsequently capably catalyzing the blend of glycosidic bonds. The existing glycoside synthases can be categorized according to the various glycosidic synthesis mechanisms [12]. Utilizing the catalytic mechanism of retained glycoside hydrolysis, retained glycoside synthases are engineered variants of glycoside hydrolases. Although modification of inverted glycoside hydrolases also represents a significant source of glycoside synthases, the majority of glycoside synthase enzymes are currently derived from mutated retained glycoside hydrolases [8].

Endo-M, or endo-*N*-acetylglucosaminidase from *Mucor hiemalis*, is a member of the 85-member glycoside hydrolase (GH) family [52]. Umekawa found that this enzyme not only makes hydrolysis easier, but it also initiates transglycosylation reactions, which involve transferring sugar chains released during hydrolysis to form new glycosidic bonds with receptors other than water. The hydrolytic activity is eliminated by mutating Asn 175 to behave like a glycoside synthase, and the mutant enzyme can transglycosylate with highly reactive glucoazolamide as a donor substrate. This addresses the main detailed example of a sugar synthase starting from endo-β-N-acetylglucosaminidase and working by means of a substrate-helped system. Martinez demonstrated that chitinases from two GH families could function as sugar synthases by binding oxazoline-activated donors to fermentation-derived chitin oligosaccharide receptors due to mutations in crucial catalytic amino acids [30]. Individual mutations on three conserved carboxylic acids were carried out on chitinase A1 from *Bacillus circulans* (BcChiA1) and chitinase 42 from *Trichoderma harzianum* (ThChit42). According to the findings, catalysis requires all amino acid residues, including the general acid/base glutamate and the two aspartates that were identified as the proposed stabilizer and its assistant. Due to their capacity to catalyze natural chitin oligosaccharide transglycosylation reactions despite lacking hydrolytic activity, these mutants can be categorized as chitin transglycosylases. In her exploration, Slámová effectively changed a GH family 20 β-*N*-acetylhexosaminidase from a parasitic source, delivering three freaks that displayed transglycosidase action through single amino corrosive modifications at the chemical's dynamic site [40]. Specific substitutions included asparagine, phenylalanine, and histidine in place of tyrosine 470, which interacts with the heterocyclic reaction intermediate. With minimal hydrolysis of the products, these mutants were successfully used to synthesize two chitooligosaccharides with distinct aglycone moieties. The synthesis method proved to be simple and scalable easily. While different common β-*N*-acetyl-aminohexosaminease substrates were tried as glycosyl contributors, just *p*NP-GlcNAc demonstrated reasonable for creating higher oligomeric oligomers [6]. Remarkably, chitotriose, a characteristic hexosaminidase substrate, yielded higher oligomers, denoting the underlying report of an exoglycosidase progressing into a transglycosidase through a substrate-helped synergist system. In addition, in contrast to previous research that made use of mutant chitinase and endo-N-acetylglucosaminidase, the products of the transglycosylation reaction were purified, characterized, and identified using NMR.

Despite these obstacles, scientists have increasingly relied on glycosidic synthases to modify target molecules via glycosylation, which enhances water solubility and promotes structural and biological diversity. Currently, research on glycosidic synthases is in the phase of principle verification and faces challenges such as poor stability of sugar donors, low enzyme vitality, and limited suitability for industrial applications. The best glycosidic synthases are characterized by their significant enzyme stability, superior catalytic efficiency, extensive substrate specificity, and availability of stable sugar donors for use in industrial applications. Nonetheless, the advancement direction of each modern compound requires consistent refinement and streamlining. As a result, the development of ideal glycosidic synthases depends on developments in relevant science and technology and the persistent efforts of scientists.

5.2.2 Chemical Enzymatic Synthesis (Based on Glycosyltransferases)

The availability of glycosyltransferase is often constrained by its source, posing challenges in its isolation, purification, and heterogeneous expression. The enzymatic synthesis catalyzed by this enzyme exhibits remarkable specificity and selectivity toward both donor and acceptor substrates. Typically, nucleotide sugars such as UDP-glucoside serve as donors for glycosylation reactions; however, their utilization is hindered by their high cost, difficulty in preparation, and limited availability in large quantities. In response to these limitations, a novel strategy termed chemical enzymatic synthesis, based on glycosyltransferases, has garnered attention in recent years. This approach represents a highly efficient synthesis technology that integrates enzyme-catalyzed synthesis with chemical methods. Notably, enzymatic synthesis offers advantages including high selectivity, specificity, and synthesis yield, obviating the need for protective group manipulations and thereby reducing time and cost investments. Moreover, the wet reaction conditions associated with this method mitigate safety risks and environmental pollution. By amalgamating the strengths of chemical and enzymatic methodologies, chemical enzymatic synthesis emerges as a promising new approach for oligosaccharide synthesis, anticipated to address current challenges and offer novel solutions in the field.

Successful research in the field includes the chemical enzymatic synthesis of heparin oligosaccharides. In this endeavor, researchers employed a multistep approach wherein they simulated the in vivo synthesis pathway of heparin, identified pivotal enzymes involved in heparin synthesis, and engineered bacteria capable of producing these enzymes. Following the identification of enzymes and the engineering of bacteria, heparin oligosaccharides were efficiently synthesized under meticulously controlled in vitro conditions. The synthetic procedure began with the creation of both a glycosyl donor and a sulfate-based donor. Subsequently, a di-hexasaccharide skeleton was assembled using a commercially available *p*-nitrophenyl-β-D-glucuronide substrate. This di-hexasaccharide scaffold underwent chemical modifications to produce various structural alterations. Given the structural resemblances between the backbone compositions of chitin and heparin oligosaccharides, it is postulated that similar methodologies could be applied to synthesize chitin oligosaccharides.

During the glycosyl donor synthesis process, a variety of starting materials, including GlcNAc (*N*-acetylglucosamine) and GlcNTFA (N-trifluoroacetylglucosamine), along with ATP (adenosine triphosphate) and UTP (uridine triphosphate), were employed. This synthesis was facilitated by the action of three recombinant enzymes: NahK (N-acetyl hexosamine 1-position kinase), GlmU (1-phospho-N-acetylglucosamine uridine transaminase), and PPA (pyrophosphorylase). Concurrently, commercially available nitrobenzene-β-D-glucuronide (GlcA-pNP) served as the starting material for the synthesis of sugar skeletons ranging from two to six units, with glycosyltransferases KfiA and PmHS2 catalyzing the reactions.

The chemical-enzymatic approach showcases high efficiency, specificity, and simplicity, characterized by straightforward reaction steps. Following an uncomplicated purification process, gram-scale glycosyl donors can be readily prepared, achieving product purity levels of up to 98%. Moreover, the method ensures a high yield of approximately 99% for the oligosaccharide skeleton. Such attributes position this methodology as a promising avenue for advancing research in chitin oligosaccharide synthesis.

6 Perspectives

The environmental benefits and ease of controlling product composition of the bio-production pathway make it more promising, even though the efficiency and economic feasibility of producing *N*-acetyl COS through microbial fermentation or enzymatic bio-production are significantly lower than those of traditional chemical methods. For the manufacture of oligosaccharides, it is favored from a commercial and industrial standpoint to employ crude enzyme combinations, such as those including protease, pepsin, lysozyme, chitosanase, chitinase, and cellulase. Furthermore, synthesis by chitinolytic enzyme transglycosylation activity is showing promise as a means of generating *N*-acetyl COSs and their derivatives, especially those with targeted or desired functions.

A significant obstacle lies in the enzymatic production of *N*-acetyl chitooligosaccharides (COSs) directly from various chitin sources, such as the shells of crabs and shrimp and microbial cell walls. Be that as it may, it is feasible to produce oligosaccharides from chitin and crab shells through a cycle including mechano-chemical crushing pretreatment followed by enzymatic hydrolysis. The glass-like construction of chitin and crab shells extraordinarily restricts the capacity of chitinolytic catalysts to get to the substrates, bringing about low effectiveness in enzymatic change. Consequently, improving chitin materials' enzymatic degradation efficiency necessitates disrupting this crystalline structure. Solubilization is the favored strategy for upsetting chitin crystallinity. However, because the reactions must take place in organic or heterogeneous solvent systems, this method requires specific enzymes to hydrolyze the substrate. Therefore, identifying new enzymes that demonstrate high stability in extreme conditions holds significant promise for the enzymatic synthesis of *N*-acetyl COSs. For the enzymatic synthesis of *N*-acetyl COSs, therefore, the discovery of new enzymes with high stability in extreme conditions holds significant promise.

In spite of the regular utilization of enzymatic techniques for blending chito/chitin oligosaccharides, challenges emerge in getting to chitin polysaccharide sources because of natural contamination during assembling and restricted accessibility of marine chitin assets. Thusly, ongoing years have seen the investigation of novel biosynthetic pathways. The enzymatic synthesis of monosaccharide chito/chitin oligosaccharides into chito/chitin oligosaccharides holds a lot of promise for future developments. Glycosyl transferases, transglycosidases/transglycosylases, and glycosynthases are important enzymes in this enzymatic synthesis.

References

1. Aam BB, Heggset EB, Norberg AL et al (2010) Production of chitooligosaccharides and their potential applications in medicine. Mar Drugs 8:1482–1517
2. Adrangi S, Faramarzi MA (2013) From bacteria to human: a journey into the world of chitinases. Biotechnol Adv 31:1786–1795
3. Aronson JN, Halloran BA, Alexeyev MF et al (2006) Mutation of a conserved tryptophan in the chitin-binding cleft of *Serratia marcescens* chitinase A enhances transglycosylation. Biosci Biotechnol Biochem 70:243–251
4. Bhattacharya D, Nagpure A, Gupta RK (2007) Bacterial chitinases: properties and potential. Crit Rev Biotechnol 27:21–28
5. Bhuvanachandra B, Madhuprakash J, Podile AR (2018) Active-site mutations improved the transglycosylation activity of, *Stenotrophomonas maltophilia*, chitinase A. Biochim Biophys Acta Proteins Proteom 1866:407–414
6. Bojarová P, Slámová K, Křenek K et al (2011) Charged hexosaminides as new substrates for β-*N*-acetylhexosaminidase-catalyzed synthesis of immunomodulatory disaccharides. Adv Synth Catal 353:2409–2420
7. Chen JK, Shen CR, Liu CL (2010) *N*-acetylglucosamine: production and applications. Mar Drugs 8:2493–2516
8. Cobucci-Ponzano B, Perugino G, Strazzulli A et al (2012) Chapter fifteen-thermophilic glycosynthases for oligosaccharides synthesis. Methods Enzymol 510:273–300

9. Ferrara MC, Cobucci-Ponzano B, Carpentieri A et al (2014) The identification and molecular characterization of the first archaeal bifunctional exo-β-glucosidase/*N*-acetyl-β-glucosaminidase demonstrate that family GH116 is made of three functionally distinct subfamilies. Biochim Biophys Acta Mol basis Dis 1840:367–377
10. Fu X, Yan QJ, Wang J et al (2016) Purification and biochemical characterization of novel acidic chitinase from *Paenicibacillus barengoltzii*. Int J Biol Macromol 91:973–979
11. Fujita K, Shimomura K, Yamamoto K et al (2006) A chitinase structurally related to the glycoside hydrolase family 48 is indispensable for the hormonally induced diapause termination in a beetle. Biochem Biophys Res Commun 345:502–507
12. Gloster TM, Turkenburg JP, Potts JR et al (2008) Divergence of catalytic mechanism within a glycosidase family provides insight into evolution of carbohydrate metabolism by human gut flora. Chem Biol 15:1058–1067
13. Guillen D, Sanchez S, Rodriguez-Sanoja R (2010) Carbohydrate-binding domains: multiplicity of biological roles. Appl Microbiol Biotechnol 85:1241–1249
14. Haebel S, Bahrke S, Peter MG (2007) Quantitative sequencing of complex mixtures of heterochitooligosaccharides by MALDI-linear ion trap mass spectrometry. Anal Chem 79:5557–5566
15. Honda Y, Kitaoka M (2006) The first glycosynthase derived from an inverting glycoside hydrolase. J Biol Chem 281:1426–1431
16. Husson E, Hadad C, Huet G et al (2017) Effect of room temperature ionic liquids on selective biocatalytic hydrolysis of chitin via sequential or simultaneous strategies. Green Chem 19:4122–4131
17. Ilankovan P, Hein S, Ng CH et al (2006) Production of *N*-acetyl chitobiose from various chitin substrates using commercial enzymes. Carbohydr Polym 63:245–250
18. Jung WJ, Park RD (2014) Bioproduction of chitooligosaccharides: present and perspectives. Mar Drugs 12:5328–5356
19. Jung WJ, Jo GY, Kuk JH et al (2007) Production of chitin from red crab shell waste by successive fermentation with *Lactobacillus paracasei* KCTC-3074 and *Serratia marcescens* FS-3. Carbohydr Polym 68:746–750
20. Katano H, Noba S, Sato K et al (2017) Solubility-based separation and purification of long-chain chitin oligosaccharides with an organic-water mixed solvent. Anal Sci 33:639–642
21. Katta S, Ankati S, Podile AR (2013) Chitooligosaccharides are converted to *N*-acetylglucosamine by *N*-acetyl-β-hexosaminidase from *Stenotrophomonas maltophilia*. FEMS Microbiol Lett 348:19–25
22. Kuk JH, Jung WJ, Jo GH et al (2006) Production of *N,N′*-diacetylchitobiose from chitin using temperature-sensitive chitinolytic enzyme preparations of *Aeromonas* sp. GJ-18. World J Microbiol Biotechnol 22:135–139
23. Kuk JH, Jung WJ, Jo GH et al (2005a) Production of *N*-acetyl-β-D-glucosamine from chitin by *Aeromonas* sp. GJ-18 crude enzyme. Appl Microbiol Biotechnol 68:384–389
24. Kuk JH, Jung WJ, Jo GH et al (2005b) Selective preparation of *N*-acetyl-D-glucosamine and *N,N′*-diacetylchitobiose from chitin using a crude enzyme preparation from *Aeromonas* sp. Biotechnol Lett 27:7–11
25. Lan X, Ozawa N, Nishiwaki N et al (2004) Purification, cloning, and sequence analysis of β-*N*-acetylglucosaminidase from the chitinolytic bacterium *Aeromonas hydrophila* strain SUWA-9. Biosci Biotechnol Biochem 68:1082–1090
26. Lopatin SA, Derbeneva MS, Kulikov SN et al (2009) Fractionation of chitosan by ultrafiltration. J Anal Chem 64:648–651
27. Madhuprakash J, Tanneeru K, Karlapudi B et al (2014) Mutagenesis and molecular dynamics simulations revealed the chitooligosaccharide entry and exit points for chitinase D from *Serratia proteamaculans*. BBA-Gen Subjects 1840:2685–2694
28. Mallakuntla MK, Vaikuntapu PR, Bhuvanachandra B et al (2017) Transglycosylation by a chitinase from *Enterobacter cloacae* subsp. cloacae generates longer chitin oligosaccharides. Sci Rep 7(1):5113

29. Margoutidis G, Parsons V, Bottaro CS et al (2018) Mechanochemical amorphization of α-chitin and conversion into oligomers of *N*-acetyl-D-glucosamine. ACS Sustain Chem Eng 6:1662–1669
30. Martinez EA, Boer H, Koivula A et al (2012) Engineering chitinases for the synthesis of chitin oligosaccharides: catalytic amino acid mutations convert the GH-18 family glycoside hydrolases into transglycosylases. J Mol Catal B Enzym 74:89–96
31. Monge EC, Tuveng TR, Vaaje-Kolstad G et al (2018) Systems analysis of the family glycoside hydrolase family 18 enzymes from *Cellvibrio japonicus* characterizes essential chitin degradation functions. J Biol Chem 293:3849–3859
32. Osada M, Miura C, Nakagawa YS et al (2012) Effect of sub- and supercritical water pretreatment on enzymatic degradation of chitin. Carbohydr Polym 88:308–312
33. Patil NS, Jadhav JP (2014) Enzymatic production of *N*-acetyl-D-glucosamine by solid state fermentation of chitinase by *Penicillium ochrochloron* MTCC 517 using agricultural residues. Int Biodeterior Biodegrad 91:9–17
34. Pichyangkura R, Kudan S, Kuttiyawong K et al (2002) Quantitative production of 2-acetamido-2-deoxy-D-glucose from crystalline chitin by bacterial chitinase. Carbohydr Res 337:557–559
35. Qin Z, Xiao YB, Yang XB et al (2015) A unique GCN5-related glucosamine *N*-acetyltransferase region exist in the fungal multidomain glycoside hydrolase family 3 β-*N*-acetylglucosaminidase. Sci Rep 5:18292
36. Rauvolfová J, Weignerová L, Kuzma M et al (2004) Enzymatic synthesis of *N*-acetylglucosaminobioses by reverse hydrolysis: characterisation and application of the library of fungal β-*N*-acetylhexosaminidases. J Mol Catal B Enzym 29:259–264
37. Sashiwa H, Fujishima S, Yamano N et al (2003) Enzymatic production of *N*-acetyl-D-glucosamine from chitin. Degradation study of *N*-acetylchitooligosaccharide and the effect of mixing of crude enzymes. Carbohydr Polym 51:391–395
38. Sashiwa H, Fujishima S, Yamano N et al (2002) Production of *N*-acetyl-D-glucosamine from α-chitin by crude enzymes from *Aeromonas hydrophila* H-2330. Carbohydr Res 337:761–763
39. Sinha S, Chand S, Tripathi P (2016) Enzymatic production of glucosamine and chitooligosaccharides using newly isolated exo-β-D-glucosaminidase having transglycosylation activity. Biotechnology 6:13
40. Slámová K, Krejzová J, Marhol P et al (2015) Synthesis of derivatized chitooligomers using transglycosidases engineered from the fungal GH20 β-*N*-Acetylhexosaminidase. Adv Synth Catal 357:1941–1950
41. Sørbotten A, Horn SJ, Eijsink VG et al (2005) Degradation of chitosans with chitinase B from *Serratia marcescens*. Production of chito-oligosaccharides and insight into enzyme processivity. FEBS J 272:538–549
42. Suginta W, Vongsuwan A, Songsiriritthigul C et al (2010) Enzymatic properties of wild-type and active site mutants of chitinase A from *Vibrio carchariae*, as revealed by HPLC-MS. FEBS J 272:3376–3386
43. Sumida T, Fujimoto K, Ito M (2011) Molecular cloning and catalytic mechanism of a novel glycosphingolipid-degrading beta-*N*-acetylgalactosaminidase from *Paenibacillus* sp. TS12. J Biol Chem 286:14065–14072
44. Suresh PV, Kumar PKA (2012) Enhanced degradation of α-chitin materials prepared from shrimp processing byproduct and production of *N*-acetyl-D-glucosamine by thermoactive chitinases from soil mesophilic fungi. Biodegradation 23:597–607
45. Tabasso S, Carnaroglio D, Calcio GE et al (2015) Microwave, ultrasound and ball mill procedures for bio-waste valorisation. Green Chem 17:684–693
46. Takiguchi Y, Shimahara K (1988) *N,N'*-Diacetylchitobiose production from chitin by *Vibrio anguillarum* strain E-383a. Lett Appl Microbiol 6:129–131

47. Tanaka T, Fukui T, Atomi H et al (2003) Characterization of an exo-β-D-glucosaminidase involved in a novel chitinolytic pathway from the hyperthermophilic archaeon *Thermococcus kodakaraensis* KOD1. J Bacteriol 185:5175–5181
48. Thamthiankul S, Suan-Ngay S, Tantimavanich S et al (2001) Chitinase from *Bacillus thuringiensis* subsp *pakistani*. Appl Microbiol Biotechnol 56:395–401
49. Udaya Prakash NA, Jayanthi M, Sabarinathan R et al (2010) Evolution, homology conservation, and identification of unique sequence signatures in GH19 family chitinases. J Mol Evol 70:466–478
50. Ueda M, Ohata K, Konishi T et al (2009) A novel goose-type lysozyme gene with chitinolytic activity from the moderately thermophilic bacterium *Ralstonia* sp. A-471: cloning, sequencing, and expression. Appl Microbiol Biotechnol 81:1077–1085
51. Uehara A, Takahashi N, Moriyama M et al (2018) Synthesis of chitin oligosaccharides using dried *Stenotrophomonas maltophilia* cells containing a transglycosylation reaction-catalyzing β-*N*-acetylhexosaminidase as a whole-cell catalyst. Appl Biochem Biotechnol 184:673–684
52. Umekawa M, Huang W, Li B et al (2008) Mutants of mucor hiemalis endo-β-*N*-acetylglucosaminidase show enhanced transglycosylation and glycosynthase-like activities. J Biol Chem 283:4469–4479
53. Umemoto N, Ohnuma T, Osawa T et al (2016) Modulation of the transglycosylation activity of plant family GH 18 chitinase by removing or introducing a tryptophan side chain. FEBS Lett 589:2327–2333
54. Vaikuntapu PR, Mallakuntla MK, Das SN et al (2018) Applicability of endochitinase of *Flavobacterium johnsoniae* with transglycosylation activity in generating long-chain chitooligosaccharides. Int J Biol Macromol 117:62–71
55. Vaikuntapu PR, Rambabu S, Madhuprakash J et al (2016) A new chitinase-D from a plant growth promoting *Serratia marcescens* GPS5 for enzymatic conversion of chitin. Bioresour Technol 220:200–207
56. van Munster JM, Sanders P, Geralt A et al (2015) Kinetic characterization of *Aspergillus niger* chitinase CfcI using a HPAEC-PAD method for native chitin oligosaccharides. Carbohydr Res 407:73–78
57. Wakita S, Kimura M, Kato N et al (2017) Improved fluorescent labeling of chitin oligomers: Chitinolytic properties of acidic mammalian chitinase under somatic tissue pH conditions. Carbohydr Polym 164:145–153
58. Wang Z, Zheng LH, Yang SL et al (2007) *N*-acetylchitooligosaccharide is a potent angiogenic inhibitor both *in vivo* and *in vitro*. Biochem Biophys Res Commun 357:26–31
59. Wang SL, Chang WT (1997) Purification and characterization of two bifunctional chitinases/lysozymes extracellularly produced by *pseudomonas aeruginosa* K-187 in a shrimp and crab shell powder medium. Appl Environ Microbiol 63:380–386
60. Wei G, Zhang AL, Chen K et al (2017) Enzymatic production of *N*-acetyl-D-glucosamine from crayfish shell wastes pretreated via high pressure homogenization. Carbohydr Polym 171:236–241
61. Wendland J, Schaub Y, Walther A (2009) *N*-acetylglucosamine utilization by *Saccharomyces cerevisiae* based on expression of *Candida albicans NAG* genes. Appl Environ Microbiol 75:5840–5845
62. Wu X, Wang J, Shi YQ et al (2017) *N*-Acetyl-chitobiose ameliorates metabolism dysfunction through Erk/p38 MAPK and histone H3 phosphorylation in type 2 diabetes mice. J Funct Foods 28:96–105
63. Yang SQ, Fu X, Yan QJ et al (2016a) Biochemical characterization of a novel acidic exochitinase from *Rhizomucor miehei* with antifungal activity. J Agric Food Chem 64:461–469
64. Yang SQ, Fu X, Yang QJ et al (2016b) Cloning, expression, purification and application of a novel chitinase from a thermophilic marine bacterium *Paenibacillus barengoltzii*. Food Chem 192:1041–1048

65. Yang SQ, Song S, Yang QJ et al (2014) Biochemical characterization of the first fungal glycoside hydrolyase family 3 β-*N*-Acetylglucosaminidase from *Rhizomucor miehei*. J Agric Food Chem 62:5181–5190
66. Yin H, Du YG, Dong ZM (2016) Chitin oligosaccharide and chitosan oligosaccharide: two similar but different plant elicitors. Front Plant Sci 7:522
67. Zakariassen H, Klemetsen L, Sakuda S et al (2010) Effect of enzyme processivity on the efficacy of a competitive chitinase inhibitor. Carbohydr Polym 82:779–785
68. Zhang AL, Wei G, Mo XF et al (2018) Enzymatic hydrolysis of chitin pretreated by bacterial fermentation to obtain pure *N*-acetyl-D-glucosamine. Green Chem 20:2320–2327
69. Zhang Y, Zhou XL, Ji LS et al (2017) Enzymatic single-step preparation and antioxidant activity of hetero-chitooligosaccharides using non-pretreated housefly larvae powder. Carbohydr Polym 172:113–119

Chitosanase

Huihui Sun

Chitosanase was initially identified as an enzyme with the capability to actively hydrolyze chitosan, as reported by Monaghan et al. in 1973 [67]. In 2004, the Enzyme Nomenclature Committee revised the definition of chitosanases (EC 3.2.1.132) to describe enzymes catalyzing the endohydrolysis of the β-1,4-glycosidic linkage within chitosan, resulting in the production of chitooligosaccharides (COSs) and D-glucosamine (GlcN) (Fig. 1a) (http://www.chem.qmul.ac.uk/iubmb/enzyme/). Furthermore, exochitosanases, recognized for their ability to cleave chitosan and/or COSs from the non-reducing end, liberating successive GlcN, were designated as exo-β-D-glucosaminidase (EC 3.2.1.165) in 2008 (Fig. 1b) [101].

1 Classification and Distribution of Chitosanases

1.1 Classification of Chitosanases

Chitosanases are categorized into six glycoside hydrolase (GH) families based on their amino acid sequences: GH5, GH7, GH8, GH46, GH75, and GH80 [8]. A phylogenetic tree was constructed to illustrate the relationships among chitosanases from these six families (Fig. 2). Notably, family GH46 encompasses bacterial chitosanases, with extensive studies conducted on their catalytic properties and mechanisms, particularly those derived from *Bacillus* and *Streptomyces* species [63, 84]. Another significant chitosanase family is GH75, primarily comprising fungal chitosanase [124]. Generally, chitosanases in families GH46, GH75, and GH80 exhibit substrate specificity primarily toward chitosan. In contrast, GH5, GH7, and GH8

H. Sun (✉)
Yellow Sea Fisheries Research Institute, Qingdao, China
e-mail: sunhh@ysfri.ac.cn

X. Mao (ed.), *Degrading Enzymes for Marine Polysaccharides and Their Applications*, https://doi.org/10.1007/978-981-96-9517-1_5

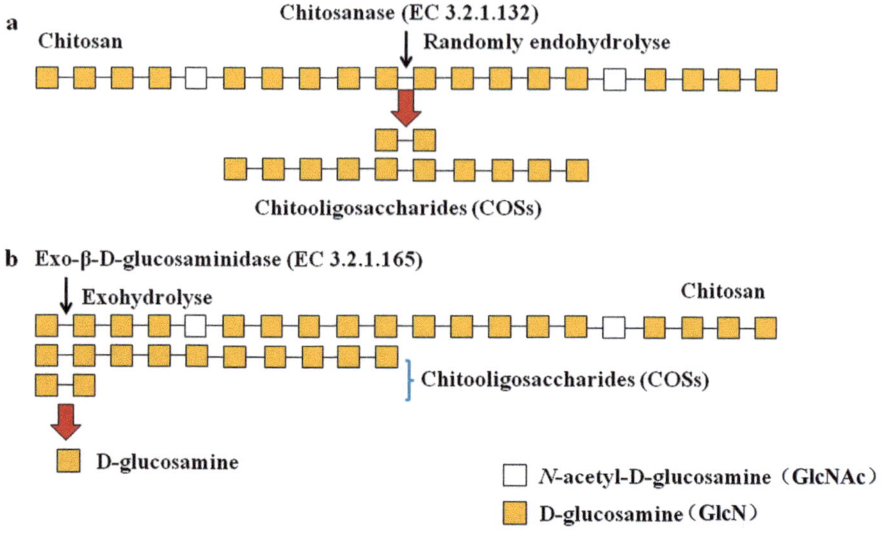

Fig. 1 Cleavage modes of chitosanase (**a**) and exo-β-D-glucosaminidase (**b**)

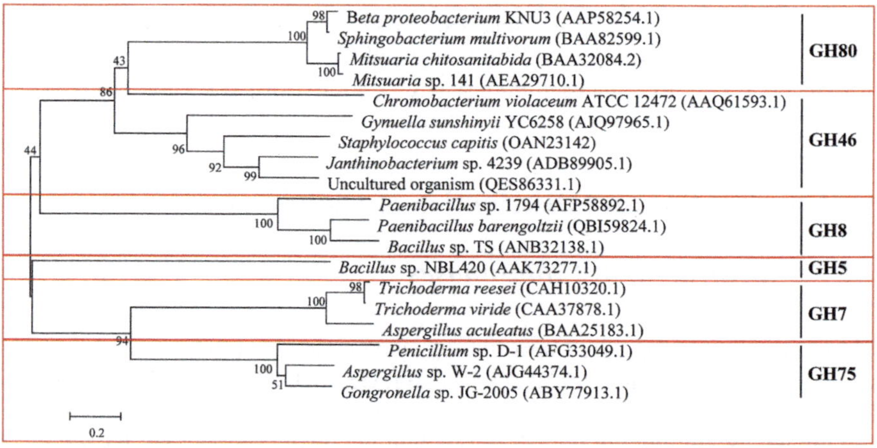

Fig. 2 The phylogenetic tree was constructed based on the amino acid sequence alignment of reported chitosanases belonging to GH5, GH7, GH8, GH46, GH75, and GH80. Phylogenetic analysis was carried out using MEGA 6.0. Groups with different GH families are in red framed

also demonstrated additional enzymatic activities, such as celluase and transglycosylation activities [39, 75, 100].

Based on the specificity of cleavage in chitosan, chitosanases can be categorized into classes I–IV (Fig. 3a) [109]. Class I chitosanases can cleave both GlcNAc-GlcN and GlcN-GlcN bonds, exemplified by the chitosanase from *Streptomyces* sp. N174 [22, 24, 25, 34, 35]. Class II chitosanases exclusively split only GlcN-GlcN

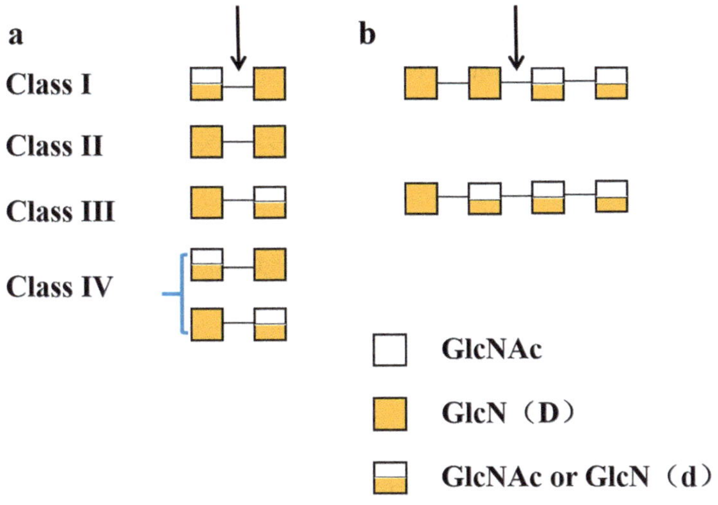

Fig. 3 Schematic illustrations showing the cleavage mode for chitosanases

linkages, as observed in the chitosanase from *Bacillus* sp. No. 7-M [40]. Class III chitosanases hydrolyze both GlcN-GlcN and GlcN-GlcNAc bonds, as demonstrated by the chitosanase from *B. circulans* MH-K1 [65, 66, 83]. Chitosanases from *Pseudomonas* sp. A-0126 and *Amycolatopsis* sp. CsO-227, capable of cleaving all three bond types, are classified into class IV [35]. While it is generally believed that chitosanases cannot hydrolyze GlcNAc-GlcNAc bonds in chitin, the chitosanase from *S. coelicolor* A3(2) can cleave all the mentioned bonds including GlcNAc-GlcNAc [46]. In this unique case, no specific class exists. Weikert et al. proposed a novel chitosanase classification system based on the specificities and preferences of subsites (−2) to (+2) [109].

Using a highly sensitive method for quantitatively sequencing partially acetylated COSs, Weikert et al. concluded that all currently reported chitosanases could be classified into two categories based on specificities and preferences of subsites (−2) to (+2) (Fig. 3b, [109]). These categories are as follows: (1) DD-dd type, where one end of the cleaved glycosidic bond is occupied by GlcN, and the other end might be GlcN or GlcNAc; (2) Dd-dd type, where the (−2) subsite has absolute specificity for GlcN, and the remaining subsites, including (−1), (+1), and (+2), have a certain preference for GlcN, which may also be occupied by GlcNAc. This study challenges the traditional classification of chitosanases based on amino acid sequence similarity and the type of cleaved glycosidic bonds. The analogy with chitinase and homogalacturonan-degrading enzymes provides a more accurate differentiation between various chitosanases. Chitosanases of different types will vary in the details of their subsite preferences, leading to variations in the exact composition of their products. Before this classification can be universally accepted, more chitosanases need to be thoroughly analyzed to encompass their full diversity. It is also likely that more classes and/or subclasses will be defined over time as additional chitosanases are investigated [109].

Based on the differences in enzymatic cleavage mechanism, chitosanases can be broadly categorized into endo-type and exo-type. Endo-type chitosanases act randomly along the interior of the chitosan chain, resulting in the cleavage of internal bonds. This process releases COSs with degrees of polymerization (DP) of two or three, typically lacking GlcN residues. In contrast, exo-type chitosanases act specifically on the non-reducing end of chitosan, sequentially removing monosaccharide residues and generating GlcN as their end products. There are reports of exo-type chitosanases that can release $(GlcN)_2$ from the non-reducing end [47]. While the majority of chitosanases adhere to endo-type mechanism of action, as per their definition, fewer reports exist on exo-type chitosanases. Exo-type chitosanases are more commonly found in fungi, such as *Aspergillus* sp. [115], *Xylella* sp. [71] and *Penicillium* sp. [2].

Another classification parameter for chitosanases is based on their production method, leading to the division into inducible and constitutive enzymes. When screening for chitosanase-producing microorganisms, chitosan is often employed as the sole carbon source to induce the expression of chitosanase. Consequently, the majority of chitosanases derived from microorganisms are inducible enzymes, whereas constitutive chitosanases are rare [86].

1.2 Distribution of Chitosanases

Since the first chitosanase was described in different soil microorganisms in 1973, numerous chitosanases from various sources have been identified, primarily originating from bacteria and fungi [67]. Additionally, chitosanases have been found in cyanobacteria, viruses, and certain plant tissues [95,101,102] (Table 1).

1.2.1 Bacterial Chitosanase

Extracellular chitosanases have been documented and studied in several bacteria, including *Bacillus* [26, 27], *Streptomyces* [41], *Serratia* [106], *Janthinobacterium* [44], *Paenibacillus* [125], and *Acinetobacter* [107]. Among them, *Bacillus* and *Streptomyces* species have been extensively investigated, reaching molecular and structural levels. Additionally, metagenomics studies have reported chitosanases from uncultured bacteria, such as CHIS5 and CHIS1754 from marine metagenomes [29, 32], CsnA from a soil sample near shrimp and crab shell storage areas [54], and CSN4 from marine mud metagenomes [96].

Table 1 Sources and biochemical properties of chitosanases

Microorganism	GH family	Molecular mass (kDa)	Optimal pH	Optimal temperature (°C)	Activity (U/mg)	Stability: Temperature (°C)	Stability: pH	References
Aquabacterium sp. A7-Y	5	50.7	5	40	18	40–50	6–11	Wang et al. [104]
Paenibacillus sp. 1794	8	40	4.8	–	–	70 ($t_{1/2}$ > 20 h)	–	Zitouni et al. [125]
Bacillus cereus D-11	8	41	6	60	347.8	20–50	6–10	Gao et al. [26]
Paenibacillus barengoltzii	8	57.7	5.5	70	360	55	5–11	Jiang et al. [42]
Bacillus sp. TS	8	47	5	60	555.3	30–40	–	Zhou et al. [123]
Janthinobacterium sp. strain 4239	46	33	5–7	45	1500	–	–	Johnsen et al. [44]
Marine metagenome (CHIS5)	46	29.24	5.5	40	–	–	–	Guan et al. [29]
Marine metagenome (CHIS1754)	46	30.2	5.5	55	–	60	4–6	Han et al. [32]
Marine metagenome (CSN4)	46	26	7	30	–	–	–	Sun et al. [96]
Soil metagenome (CsnA)	46	31	6	55	2373	50–55	4.5–6.5	Li et al. [54]
Staphylococcus capitis	46	35	7	30	89.2	20–50	5–7	Sun et al. [93]
Chromobacterium violaceum	46	36	6.11	50	10,000	30–40	2–12	Azevedo et al. [4]
Mitsuaria sp. C4	46	34	7.2	40	14.52	40	7.2 (30 min)	Chen et al. [9]
Bacterium Serratia sp. QD07	46	27.1	5.8	60	412.6	40	3.4–8.8	Zheng et al. [118]
Gynuella sunshinyii	46	29.7	5.5	30	260.87	15–35	4–8.5	Qin et al. [79]
Penicillium sp. D-1	75	24.6	4	48	26.4	0–40	3–8	Zhu et al. [124]
Aspergillus sp. W-2	75	28	6	55	34	30–70	3–10	Zhang et al. [114]
Gongronella sp. JG (Csn2)	75	28	5.6	55–60	–	50 ($t_{1/2}$ = 30 min)	–	Wang et al. [108]

(continued)

Table 1 (continued)

Microorganism	GH family	Molecular mass (kDa)	Optimal pH	Optimal temperature (°C)	Activity (U/mg)	Stability		References
						Temperature (°C)	pH	
Aspergillus sp. CJ22–326 (ChiA)	2	109	4	50	6.46	30–60	3.5–7.5	Chen et al. [11]
Sheaths of bamboo shoots	–	24.5	3–4	70	0.753	30–80	–	Hsu et al. [36]
Bacillus cereus GU-02	–	16	9	37	–	50	–	Goo and Park [27]
Alternaria alternate	–	17	4	60–70	180 nkat/mg	60	–	Kohlhoff et al. [48]
Streptomyces roseolus	–	41	5	50	217	30–60	5–7	Jiang et al. [41]
Serratia marcescens TKU011	–	21	5	50	0.1	<50	4–8	Wang et al. [106]
Acinetobacter calcoaceticus TKU024 (CHSA1)	–	27	6	50	5.2	<90	4–10	Wang et al. [107]
Gongronella sp. JG	–	90	4.6–4.8	50	82	50 ($t_{1/2}$ = 65 min)	–	Zhou et al. [122]
Paenibacillus sp. TKU042	–	70	–	–	3.09	–	–	Doan et al. [20]
Bacillus cereus HMRSC30	–	41	5.5	55	81.09	30–50	4–6	Cahyaningtyas et al. [6]
Paenibacillus sp. TKU047	–	23	7	60	6.45	<40	6–9	Doan et al. [19]

1.2.2 Fungal Chitosanase

Compared to bacterial chitosanases, fewer chitosanases have been isolated from fungi, primarily from *Aspergillus* sp. [116], *Gongronella* sp. [108, 122] and *Trichoderma* sp. [15].

1.2.3 Cyanobacterial Chitosanase

Cyanobacteria, encompassing various photosynthetic prokaryotes with diverse morphological and physiological characteristics, have been found to contain chitosanase. For instance, *Anabaena fertilissima* RPAN1 exhibits chitosanase activity, inhibiting fungal growth [31].

1.2.4 Plant Tissues Chitosanase

Chitosanase is rarely obtained from plant tissues. Active chitosanase substances have been identified in onion roots and certain plant leaves infected by pathogens. Different chitosanase activities can be detected in various parts of healthy plant tissues. Both chitosan and COSs can induce plants to produce defense responses, including the induction of chitosanase and 1,3-glucanase isoforms. However, the defense mechanism involving endogenous chitosanase in plants remains unclear [36]. Notably, Hsu et al. recently discovered two thermostable chitosanases from the outer sheath of bamboo shoots [36].

2 Measurement of Chitosanase Activity

The direct action of most chitosanases on solid or crystalline chitosan is challenging. While few chitosanases have been reported to degrade chitosan powder [93], their applications are limited due to low activity. Colloidal chitosan or glycol chitosan is commonly used as substrate for chitosanase activity detection [91]. There are two main methods for detecting chitosanase activity, involving the measurement of substrate decrease and the product increase. The first method measures the decrease in viscosity of chitosan, providing insights into whether the chitosanase is endo-type or exo-type. Endo-type chitosanases typically rapidly decrease chitosan viscosity in the initial stage, while exo-type chitosanases exhibit a linear decrease. However, this method is infrequently used due to operational challenges and relative insensitivity [7, 91]. The most widely used methods for determining chitosanase activity involve measuring the content of reducing sugars generated [38, 69]. The 3,5-Dinitrosalicylic acid colorimetric (DNS method) is common, but it may have a large margin of error when the reducing sugar content is relative low, and it is not suitable for long-term hydrolysis [50]. Other reported methods include the

Somogyi–Nelson method, Park–Johnson ferricyanide submicromethod [13], 2,2′-bicinchoninate (BCA) method [82], 3-methyl-2-benzothiazolinone (MBTH) method [117], spectrophotometric method based on Calcofluor white dye binding (a new method), fluorometry [73], radiometric analysis [22], and enzyme spectrum method [28]. However, these methods may have limitations related to sensitivity, measurement range, toxicity, etc. High-performance liquid chromatography (HPLC) is another commonly used method [23], enabling the simultaneous separation and identification of products. Additionally, electrospray ionization–mass spectrometry (ESI-MS) [17], thin-layer chromatography (TLC) [81], and paper chromatography [74] are alternative methods for separating and identifying chitosan products.

3 Basic Biochemical Properties of Chitosanases

Numerous chitosanases from diverse microorganisms have undergone comprehensive characterization, unveiling fundamental biochemical properties closely tied to their respective sources. Table 1 provides an overview of the basic properties reported for some chitosanases.

3.1 Molecular Mass

Chitosanases typically exhibit molecular masses within the 20–70 kDa range. For instance, chitosanases from *Mitsuaria* sp. C4 and *Paenibacillus barengoltzii* have molecular masses of 34 kDa and 58 kDa, respectively [9, 42]. In contrast, exo-β-D-glucosaminidases fall within the 90–140 kDa range [11]. Notably, reported chitosanases span a molecular mass spectrum, with examples ranging from the highest at 90 kDa in *Gongronella* sp. JG to the lowest at 16 kDa in *B. cereus* GU-02 [27, 122]. The molecular characteristics of chitosanases can provide insights into their properties. For instance, the molecular masses of chitosanases can vary based on their GH families. Chitosanases from GH8 typically have higher molecular masses than those from GH46 and GH75, usually exceeding 40 kDa [20, 27, 95].

3.2 Effects of pH/Temperature on Chitosanase Activities and Stabilities

The optimal pH range for most chitosanases typically falls between 4 and 8 (Table 1). For example, chitosanases from *B. cereus* HMRSC30 and *P. barengoltzii* exhibit acidic pH optima of 5.5, with the latter maintaining stability in a broad pH range of 5–11 [6, 42]. In contrast, the chitosanase from *Paenibacillus* sp. TKU047 displays its highest activity at a neutral pH of 7 and stability in the pH range from 6 to 9 [19]. However, some chitosanases deviate from this optimal range. For instance, two acidic chitosanases from bamboo shoots sheaths show optimal pH at 3–4, retaining 70–100% relative activity between pH 3 and 6 [36]. Alkaline chitosanases, such as the one from *B. cereus* GU-02, display optimal activity at pH 9 [27]. Notably, chitosanase CvCsn46 exhibits dual optimum pH values for catalysis (6 and 11) and stability over a wide pH range, recovering >85% relative activity when incubated at pH between 2 and 12 [4].

The optimum temperature for various chitosanases varies depending on their sources and ranges between 40 and 60 °C (Table 1). Recently reported chitosanases from *Aquabacterium* sp. A7-Y, *Chromobacterium violaceum* ATCC 12472, and a deep-sea bacterium *Serratia* sp. QD07 exhibited their highest activities at 40 °C, 50 °C, and 60 °C, respectively [4, 104, 118]. Due to the high viscosity and low solubility of chitosan at low temperatures, chitosanases with high thermostabilities offer distinct advantages in industrial applications [125]. Moreover, a high-temperature environment can also accelerate reactions and decrease the risk of microbial contamination [101]. In this context, some thermally stable chitosanases have been reported and characterized. For example, two chitosanase isoforms A and B from the sheaths of chitosan-treated bamboo shoots showed their optimum reaction temperature at 70 and 60 °C, respectively. Isoforms A remained stable between 30 and 80 °C [36]. Another chitosanase with high thermostability was reported by Chen et al., with half-lives of 2.5 h, 1 h, and 32 min at 80 °C, 90 °C, and 100 °C, respectively [125]. Similarly, cold-adapted chitosanases showing high activities at low temperatures are also required in special instances, such as transfection and transformation experiments [44, 78]. A cold-active chitosanase from *Staphylococcus capitis* was reported, maintaining >90% of its initial activity after incubation at 20–50 °C for 1 h. However, the enzyme was rapidly inactivated and retained only 50% of its initial activity when incubated at 55 °C for 1 h [93]. Another cold-adapted chitosanase, GsCsn46A from *G. sunshinyii*, exhibited 70% of its maximum activity at 15 °C and was used for controlled conversion of chitosan under mild conditions to produce COSs [79].

3.3 *Effects of Metal Ions on Chitosanase Activities*

Metal ions play a crucial role in modifying the activities of chitosanases. The enzyme protein molecules may be activated by metal ions or metal complexes with an electrical charge, making certain metal ions activators for achieving normal reactions with the substrate. The effects of metal ions on chitosanase activities vary depending on the enzyme's source, and various metal ions exhibit diverse effects. Common metal ions, including Na^{+}, K^{+}, Ag^{+}, Mn^{2+}, Cu^{2+}, Zn^{2+}, Mg^{2+}, Ba^{2+}, Fe^{2+}, Ca^{2+}, Co^{2+}, Hg^{2+}, Sr^{2+}, Fe^{3+}, and Al^{3+}, have been studied to analyze their impacts on chitosanase activities. In most cases, Mn^{2+} served as an activator, enhancing chitosanase activities to varying degrees. For example, chitosanases from *Paenibacillus* sp. TKU047, *Bacillus* sp. BY01, and *A. fumigatus* CJ22–326 showed 1.45-, 2.57-, and 4-fold increased activities in the presence of Mn^{2+} [19, 112, 120]. However, there are also reports indicating inhibitory or negligible effects of Mn^{2+}, such as the chitosanase from *S. roseolus*, which exhibited decreased activity to 45% after Mn^2 addition [41]. Other metal ions, including Cu^{2+}, Co^{2+}, Zn^{2+}, and Fe^{2+}, act as either activators or inhibitors of chitosanases from different sources. For instance, Cu^{2+} and Co^{2+} increased the activity of CsnQ from marine *Bacillus* sp. Q1098, while they strongly inhibited BbCSN-1 from *Beauveria bassiana* [58, 62]. Ag^+, Hg^{2+}, and Fe^{3+} are generally considered inhibitors of the chitosanase group. This is attributed to the involvement of thiol groups in the active catalytic site, essential for maintaining the structure of the active protein [120]. The hypothesis was further supported by evaluating the effect of dithiothreitol (DTT), an inhibitor of thiol groups, on chitosanase activity. The results demonstrated a significant decrease in activity in the presence of DTT [120].

4 Hydrolysis Pattern of Chitosanase/Mode of Enzymatic Degradation

The hydrolysis pattern of chitosanase is deduced by analyzing the composition and sequence of COSs obtained after enzymatic digestion. A thorough understanding of the hydrolysis pattern, coupled with the control of the hydrolysis process, can enhance the enzymatic preparation technology of COSs for various applications [10]. Due to the complexity of chitosan's structure and the challenge of separating COSs with DP greater than 10. COSs substrates like chitotetraose, chitopentaose, and chitohexaose are often used in the analysis of chitosanase hydrolysis mechanism. For example, Fukamizo et al. conducted a hydrolysis reaction using chitohexaose as the substrate and analyzed the products of chitosanase from *Streptomyces* sp. N174 by HPLC. The results indicated that the primary products were chitotriose, with some chitobiose and chitotetraose also produced. The content of chitobiose was higher than that of chitotetraose, suggesting that chitohexaose could be hydrolyzed into two chitotrioses or to chitobiose and chitotetraose by chitosanase N174. Chitotetraose could further be hydrolyzed to two chitobioses, demonstrating that the smallest substrate in the binding pocket of this chitosanase was chitotrtraose [24].

While COSs are convenient substrates, understanding the complete chitosan hydrolysis process requires studying the initial reaction. Heggset et al. investigated the hydrolysis process of chitosanase ScCsn46A toward chitosan with different degrees of deacetylation. They found that the long-chain substrate was rapidly degraded during hydrolysis to produce small-molecule products, indicating a non-processive hydrolysis pattern [33]. Kinetic analysis revealed an initial fast followed by a slow hydrolytic rate toward highly deacetylated chitosan, with the hydrolytic rate related to the length of the substrates chain and the acetyl group content. The study demonstrated that the degradation degree of chitosan decreased with the decrease of chitosan deacetylation, with fully deacetylated chitosan primarily producing dimers (DD or DA, D stands for GlcN, A stands for GlcNAc) [33]. Lyu et al. investigated the degradation pattern of another chitosanase, OU01, using chitosan with 50% deacetylation as a substrate. The main products were DD and DA, with small amounts of DDA and DAA. Docking data of the enzyme-substrate indicated that the (−2) subsite of the enzyme has an absolute specificity for D, while it is acceptable for A at all other subsites [59].

Weikert et al. employed a high sensitive method for the quantitative sequencing of partially deacetylated chitosan, reassessing the substrate specificities of well-characterized class I–III chitosanases [109]. Selected model chitosanases, including CSN-174 from *Streptomyces* sp. N174 (representing class I), CSN-7 M from *Bacillus* sp. No. 7-M (representing class II), CSN-MHKI from *B. circulans* MH-K1 (representing class III), and CSN-MN, an unclassified chitosanase belonging to GH8 from *Bacillus* sp. MN (with 97% sequence identity to CSN-7 M), were studied. Hydrolysis results revealed distinct preferences for D and A at different subsites among these chitosanases. The (−2) subsite exhibited an absolute specificity for D in all four chitosanases, and the (−1) subsite showed this specificity in some chitosanases. While many chitosanases share absolute specificity for D at the (−2) subsite, it is not universally observed, as seen in the chitosanase form *B. circulans* WL-12, which can accept two units at this subsite [66]. The subsites (+1) and (+2) were more flexible, with a preference for the D unit at both subsites, especially at the glycosidic bond to be cleaved (the new non-reducing end). Both subsites efficiently bound with the A unit, yet the flexibility of these sites to bind with A varied among different chitosanases. This study suggested that the current chitosanase classification system, based on the enzyme's ability to cleave a specific sequence in the substrate, overlooks the quantitative efficiency of the cleavage pattern by focusing solely on the specificities of subsites (−1) and (+1).

The ability to cleave the D-D bond is often considered the fundamental difference between chitinase and chitosanase. However, a GH46 family chitosanase from *Streptomyces coelicolor* A3(2) was reported to cleave all four kinds of bonds (D-D, D-A, A-D, and A-A), even the chitin hexamer, albeit showing a clear preference for D-D [33]. Therefore, a potential criterion for distinguishing between chitinase and chitosanase is that chitosanases typically prefer D units, leading to decreased activity with increasing acetylation of substrates, while the opposite is true for chitinases [109].

5 Structure and Catalytic Mechanism of Chitosanases

Despite extensive research on the genetic resources and basic biochemical properties of various chitosanases, structural studies have been limited. To date, only eight crystal structures of chitosanases have been successfully resolved, with six belonging to the GH46 family, one to GH8, and one to GH80 (Fig. 4) [55, 60, 63, 84, 99, 105, 113, 1]. Unfortunately, the structural information for GH75 family

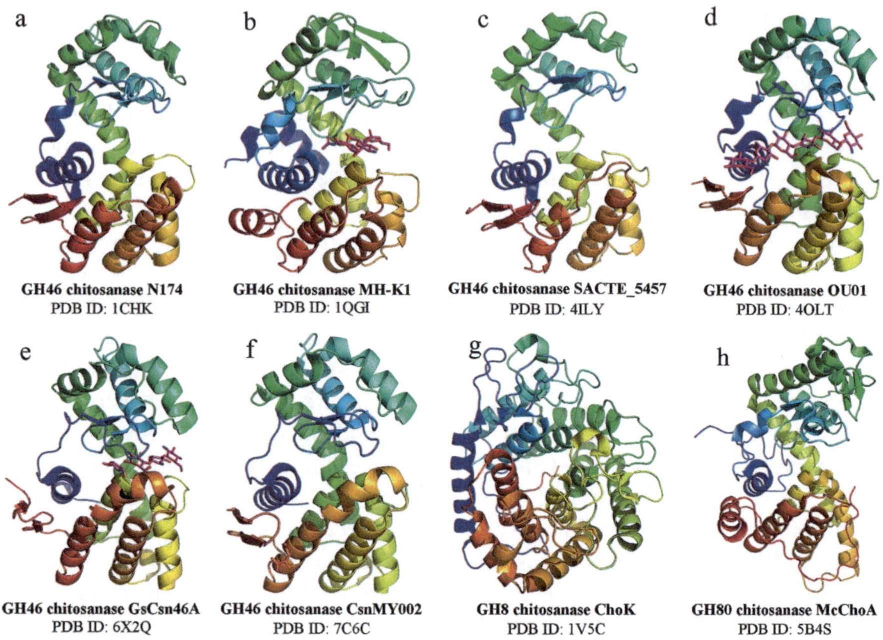

Fig. 4 Three-dimensional structures of chitosanases. (**a**) chitosanase (PDB ID: 1CHK) of family GH46 from *Streptomyces* sp. N174; (**b**) chitosanase (PDB ID: 1QGI) of family GH46 from *B. circulans* MH-K1; (**c**) chitosanase SACTE_5457 (PDB ID: 4ILY) of family GH46 from *Streptomyces* sp. SirexAA-E; (**d**) chitosanase OU01 mutant E25A (PDB ID: 4OLT) of family GH46 from *Microbacterium* sp.; (**e**) chitosanase GsCsn46A mutant E19A (PDB ID: 6X2Q) of family GH46 from *G. sunshinyii*; (**f**) chitosanase CsnMY002 (PDB ID: 7C6C) of family GH46 from *B. subtilis* MY002; (**g**) chitosanase ChoK (PDB ID: 1V5D) of family GH8 from *Bacillus* sp. K17; (**h**) chitosanase McChoA (PDB ID: 5B4S) of family GH80 from *M. chitosanitabida*

chitosanases, primarily found in fungi, has not been reported due to difficulties in obtaining their crystal structures.

Chitosanases are categorized based on the configuration of the reducing-end anomer of enzymatic product, leading to two mechanisms: retaining (the configuration of the reducing-end anomer of enzymatic product is consistent with the substrate, Fig. 5a), and inverting (the configuration of the reducing-end anomer of enzymatic product is reversed with the substrate, Fig. 5b) [32]. Retaining glycosidases, including chitosanases from GH5 and GH7 families, operate through a two-step, double-displacement reaction. In this mechanism, one essential amino acid acts as a nucleophile, and the other functions as a general acid/base [101]. On the other hand, inverting glycoside hydrolases undergo a single substitution reaction during hydrolysis, requiring the assistance of a general base and a general acid. The general base polarizes a water molecule to form a stronger nucleophile, while the general acid protonates the oxygen atom to accelerate the reaction. Experimental verification of this inverting mechanism has been reported for GH8, GH46, GH75, and GH80 family chitosanases [1, 60, 63, 99, 101, 103, 113].

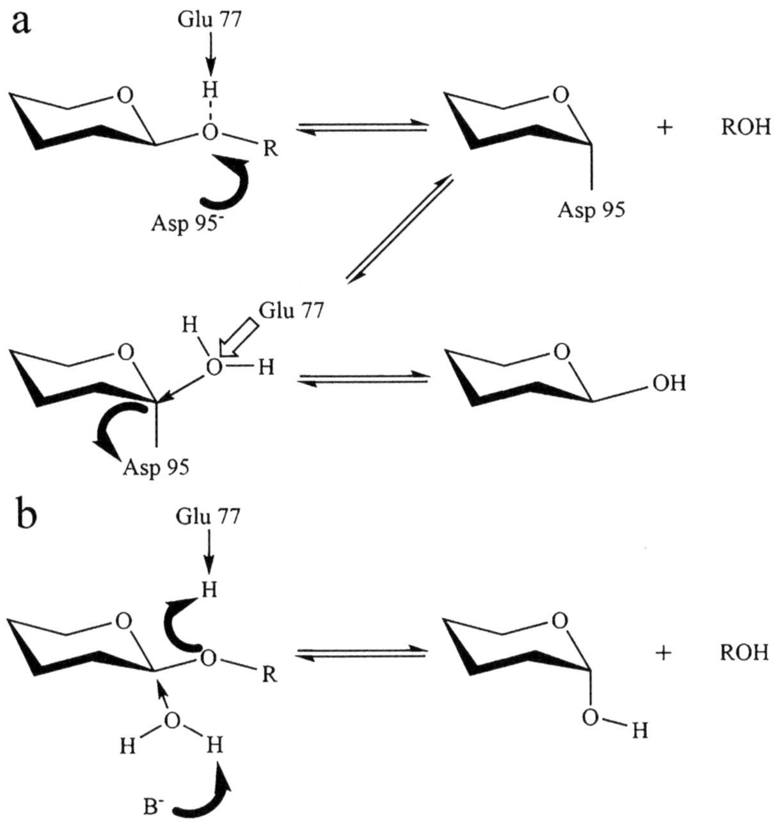

Fig. 5 Two catalytic mechanisms of chitosanase. (**a**) retaining mechanism; (**b**) inverting mechanism. (Csn-PD from *P. dendritiformis* as an example [94])

5.1 GH46 Family Chitosanases

The first crystal structure of a GH46 family chitosanase was reported for the chitosanase from *Streptomyces* sp. N174 by Marcotte et al. in 1996 (PDB ID: 1CHK) [63]. The structure revealed two spherical regions connected by a curved helix, forming a catalytic cleft in the middle groove for substrate binding. The key catalytic residues were hypothesized to be Glu22 and Asp40 (Fig. 6). The secondary structure of N174 primarily consists of ten α-helixes and three β-strands, with α-helixes accounting for approximately 64% of the entire structure. The surface of the catalytic groove is negatively charged, adding in the attraction of the positive charge of the amino group on the substrate (Fig. 7).

The second crystal structure in the GH46 family was reported for the chitosanase from *B. circulans* MH-K1 by Saito et al. in 1999 (PDB ID: 1QGI, Fig. 8). [84]. The crystal structure provided insights into the substrate recognition mechanism.

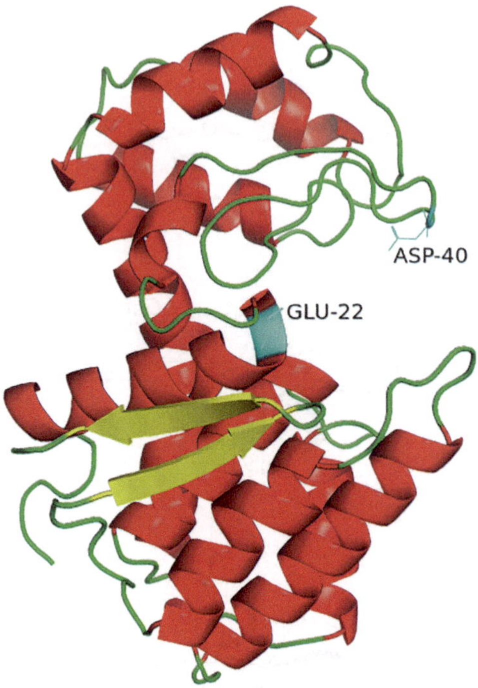

Fig. 6 The X-ray structure of chitosanase from *Streptomyces* sp. N174. The putative catalytic acids Glu22 and Asp40 are shown in cyan bonds

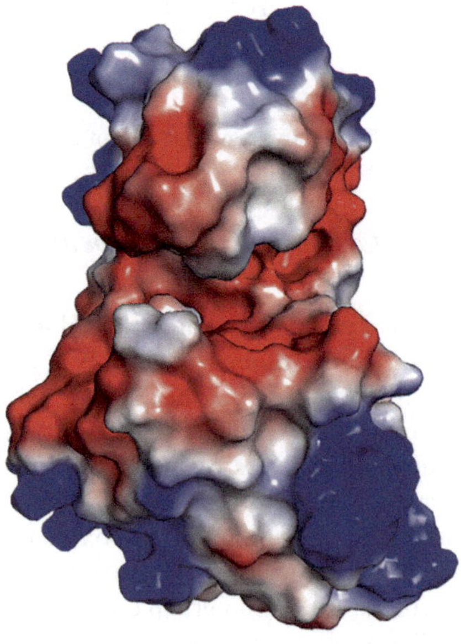

Fig. 7 Stereo diagram of molecular surface renderings of chitosanase. Surfaces are colored according to their electrostatic potentials. The surface potential is displayed as a color gradient from red (electronegative) to blue (electropositive)

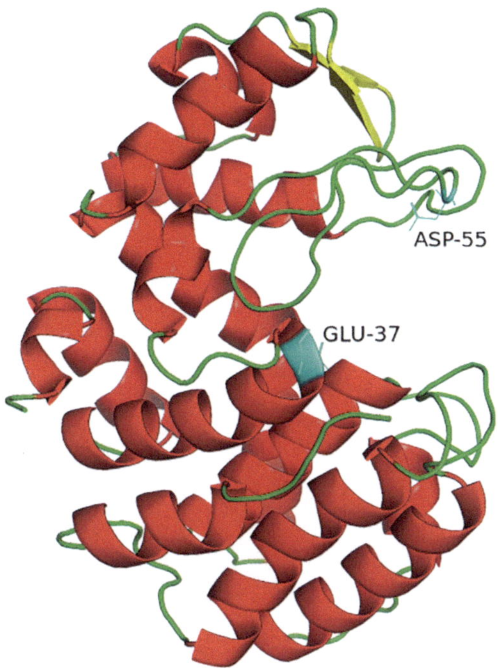

Fig. 8 The X-ray structure of chitosanase from *B. circulans* MH-K1. The putative catalytic acids Glu37 and Asp55 are shown in cyan bonds

The crystal structures of the GH46 family chitosanases from *Streptomyces* sp. N174 and *B. circulans* MH-K1 exhibited significant differences in amino acid sequences, yet only three apparent structural distinctions were observed. Notably, a disulfide bonds was formed between two cysteines, connecting the β1 chain and the α7 helix in MH-K1. Additionally, MH-K1 demonstrated the ability to hydrolyze the GlcN-GlcN bond and GlcN-GlcNAc bond but not the GlcNAc-GlcN bond, a feature distinct from N174. The crystal structures of these two chitosanases have served as highly representative models, prompting researchers to conduct site-directed/multi-point mutagenesis near the catalytic groove based on their structure information. These studies have provided a more detailed understanding of the substrate recognition and catalytic mechanisms of chitosanases [98].

In 2014, Takasuka et al. analyzed the crystal structure of the GH46 family chitosanase SACTE_5457 from *Streptomyces* sp. SirexAA-E (PDB ID: 4ILY, Fig. 9) [99]. The protein structure of SACTE_5457 closely resembles that of the previously mentioned chitosanases, particularly N174. Through structural analysis and examination of product distributions, it was evident that the catalytic groove of SACTE_5457 preferentially binds substrates in the −2, −1, +1, and +2 subsites, rendering it incapable of hydrolyzing COS with a DP less than 4. Additionally, SACTE_5457 cannot accommodate GlcNAc at the +2 subsite due to the steric overlap of the acetyl group. The study also proposed that motions of two α-helical structural domains lead to the opening and closing conformation of the active site cleft during the reaction. This interdomain motion is facilitated by the α-helix

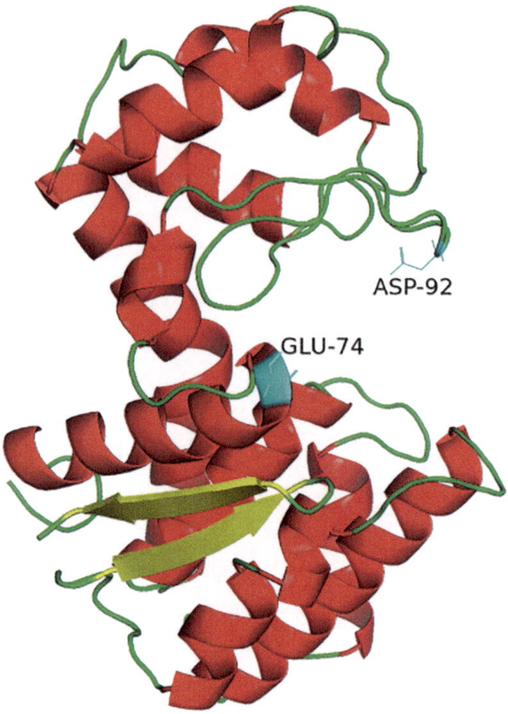

Fig. 9 The X-ray structure of chitosanase of SAXTE_5457 from *Streptomyces* sp. SirexAA-E. The putative catalytic acids Glu74 and Asp92 are shown in cyan bonds

(Arg170-Arg172) connecting the two structural domains and a loop (Glu78-Lys81) bridging between the two α-helix domains.

In 2015, Lyu et al. expressed the chitosanase OU01 from *Microbacterium* sp. and obtained complex structures of inactivating mutants (E25A) with chitohexaose (PDB ID: 4OLT) [60]. The structure of chitosanase OU01 is divided into upper and lower parts (Fig. 10). The upper lobe consists of six α-helixes and two β-strands, while the lower lobe is composed of five α-helixes and four β-strands. The substrate is located at the substrate-binding cleft, formed between the upper and lower lobes. This structure is similar to previously reported GH46 family chitosanases and closely resembles the overall folding trend of the peptide segment in T4 lysozyme [68].

In 2020, Wang et al. published the complex structures of the E19A mutant of GH46 family chitosanase GsCsn46A from *G. sunshinyii* with $(GlcN)_5$ (PDB ID: 6LWU) and $(GlcN)_4$ (PDB ID: 6X2Q), respectively (Fig. 11, [105]). The overall structure contains nine α-helixes and two β-strands, exhibiting a division into upper and lower parts, similar to other GH46 and GH80 family chitosanases. Notably, a segment of α-helical H10, composed of three amino acid residues, is observable in the chitotetraose-bound complex. The upper lobe consists of five α-helixes (H1-H5), while the lower lobe consists of α-helixes H6-H9 and β-strands S1-S2. The longest among all α-helixes is H5, spanning the upper and lower parts and serving as the backbone connecting the two lobes. This structural pattern shares similarities

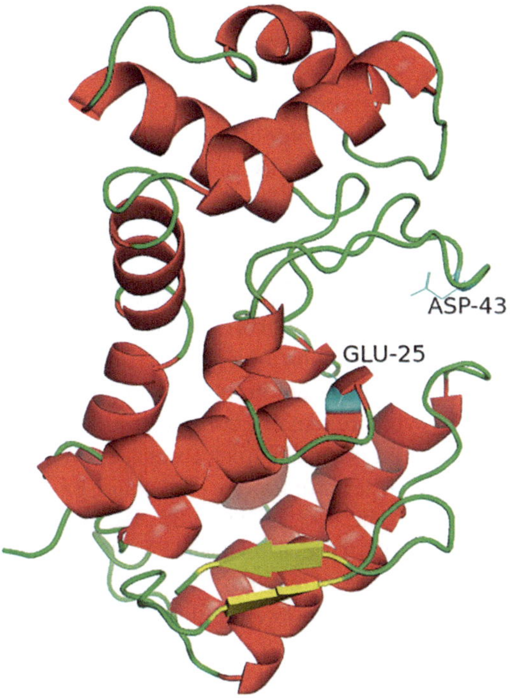

Fig. 10 The X-ray structure of chitosanase OU01 from *Microbacterium* sp. The putative catalytic acids Glu25 and Asp43 are shown in cyan bonds

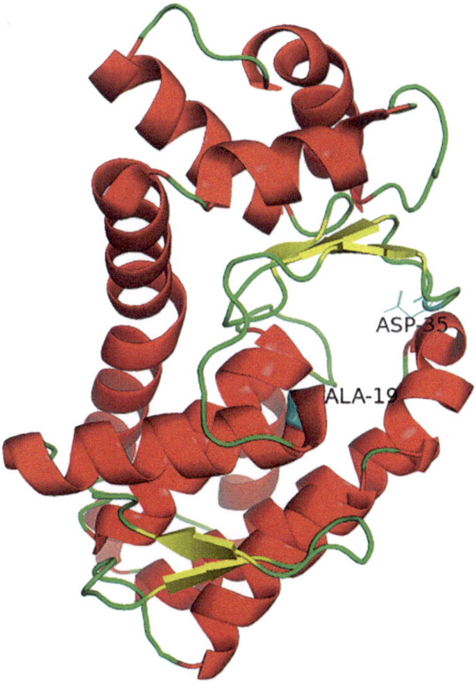

Fig. 11 The X-ray structure of chitosanase GsCsn46A (E19A) from *G. sunshinyii*. The putative catalytic acids Glu25 and Asp43 are shown in cyan bonds

with the GH80 family of chitosanases [113]. The substrate resides within the catalytic cleft formed between the two structural domains, and this cleft is closed in both complex structures. Different from other reported GH46 family chitosanases, GsCsn46A exhibits a flexible loop region near the −1 and +1 subsites. The distinctive binding positions of $(GlcN)_4$ and $(GlcN)_5$ uncover the existence of two new negative subsites in GsCsn46A, suggesting the presence of at least eight glycosyl residues (−5/−4 to +3) in the catalytic cleft of GH46 family chitosanase. Furthermore, alternations in interactions near the +3 subsite contribute to the generation of COS with a high DP.

In 2021, Li et al. published the crystal structure of chitosanase CsnMY002 from *B. subtilis* MY002 (PDB ID: 7C6C, [55]), presenting a tertiary structure comprising an N-terminal minor lobe and a C-terminal major lobe. These two lobes are connected by a long bent helix (α5) and a hinge segment (Fig. 12). While the overall protein structure remains consistent, CsnMY002 exhibits specific structural features distinct from other GH46 chitosanases. Within the CsnMY002 structure, the minor lobe is in closer proximity to the major lobe, forming two hydrogen bonds between Arg37 and Thr200. The catalytic cleft is found in a closed state, mirroring the configuration observed in GsCsn46A. The distance between the backbone carbonyl groups of catalytic residues Asp35 and Glu19 is approximately 14.6 Å, significantly shorter than distances found in CsnN174 (17.6 Å), CsnMH-K1 (16.0 Å) and CsnSACTE_5457 (16.1 Å). Despite displaying a tunnel-like substrate binding site, CsnMY002 operates as a nonprocessive enzyme with a "4 + 2" splitting mode,

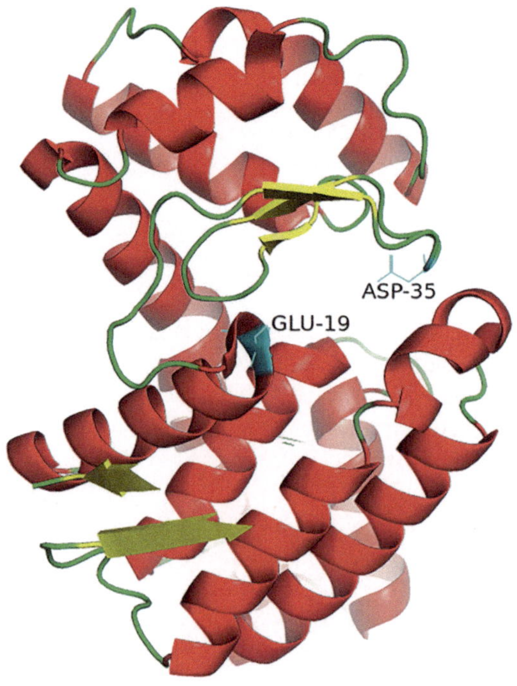

Fig. 12 The X-ray structure of chitosanase CsnMY002 from *B. subtilis* MY002. The putative catalytic acids Glu19 and Asp35 are shown in cyan bonds

primarily converting $(GlcN)_6$ to $(GlcN)_2$ and $(GlcN)_3$. Li et al. also confirmed that the mutant enzyme maintained the "4 + 2" splitting mode even after introducing mutations disrupting the binding of the +3 subsite. Consequently, $(GlcN)_2$ became the predominate product [55].

5.2 GH8 Family Chitosanases

In 2004, Adachi et al. reported two crystal structures of chitosanase ChoK from *Bacillus* sp. K17, belonging to the GH8 family (PDB IDs 1V5D and 1V5C), at pH 6.4 and 3.7, respectively [1]. ChoK's three-dimensional structure significantly differs from that of GH46 family chitosanases. The overall structure consists of six α-helixes forming a central barrel, surrounded by six additional α-helixes (Fig. 13). This forms a typical α6/α6 double barrel structure, and the substrate-binding pocket is located in the long cleft at the top of the double barrel, similar to cellulase and xylanase structures in the GH8 family [3, 30, 77, 85]. The GH8 family can be further divided into three subfamilies, with ChoK belonging to the GH-8b subfamily. Unlike GH46 family chitosanases that use Glu/Asp as the catalytic acid (proton donor)/base (proton acceptor), ChoK employs Glu122 as the catalytic acid to donate a proton, with the dissociated Glu309 serving as a proton acceptor. The distance

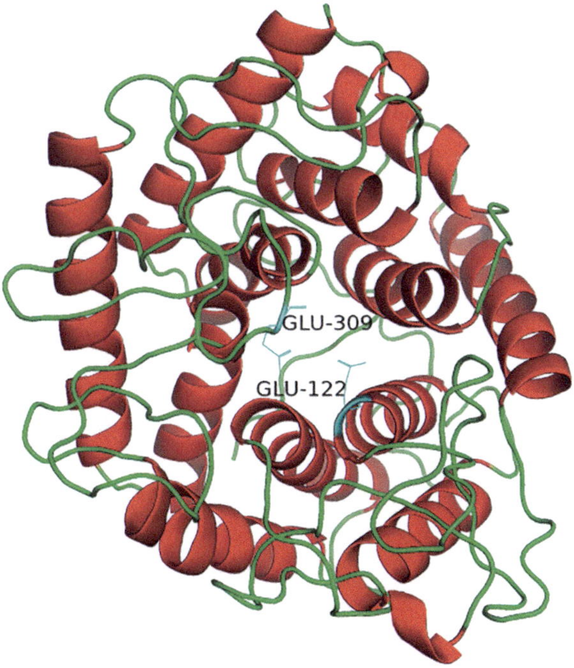

Fig. 13 The X-ray structure of chitosanase ChoK from *Bacillus* sp. K17. The putative catalytic acids Glu122 and Asp309 are shown in cyan bonds

(9.7 Å) between the two carboxyl groups of these two catalytic residues falls within the range (~10 Å) proposed for the inverting catalytic reaction [64,103,110].

5.3 GH80 Family Chitosanases

In 2017, Yorinaga et al. reported the crystal structure of chitosanase McChoA from *M. chitosanitabida*, belonging to the GH80 family, at a resolution of 1.75 Å (PDB ID: 5B4S) [113]. Despite the sequence identity between GH80 and GH46 chitosanases being less than 20%, they share a similar two-domain arrangement, with only some topological differences. McChoA contains 12 α-helixes and 5 β-strands that fold into upper and lower domains of comparable size, employing an inverting hydrolysis reaction (Fig. 14). Notably, McChoA differs from other chitosanases in having six cysteines that form three pairs of disulfide bonds (Cys56-Cys151, Cys227-Cys235, and Cys260-Cys297). Mutational analysis has shown that these six cysteines are essential for stabilizing the proper folding structure, and any substitution results in the loss of catalytic activity [88].

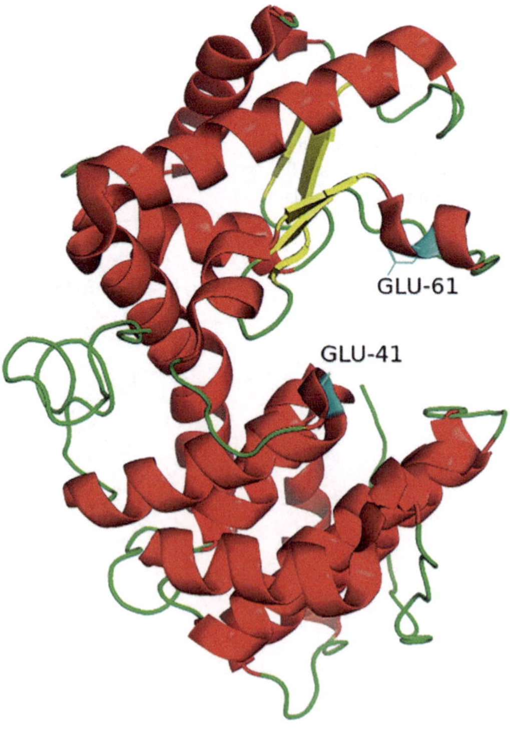

Fig. 14 The X-ray structure of chitosanase MoChoA from *M. chitosanitabida*. The putative catalytic acids Glu41 and Asp61 are shown in cyan bonds

6 Modification and Design of Chitosanases

6.1 Heterologous Expression and Genetic Improvement

Numerous strains producing chitosanase have been screened; however, the yield from wild microorganisms is generally insufficient to meet the industrial demands. Consequently, researchers have focused on heterologous expression and genetic enhancement of chitosanase genes from diverse sources using genetic engineering techniques. For the expression of chitosanase genes, the primary hosts are *E. coli* and yeast. In 2009, Liu et al. Successfully expressed a GH8 family from a *Bacillus* strain in *E. coli* in a soluble form, achieving an expression level as high as 500 mg/L. The enzyme activity reached 140 U/mL, effectively converting chitosan to COSs, with 1 g of the enzyme capable of degrading about 100 kg of chitosan [57]. Zhou et al. achieved high enzyme activity (186 U/mL) by expressing the chitosanase from *Bacillus* sp. TS in *E. coli* through high cell-density cultivation in a 7 L fermenter [123]. Chen et al. secretorily expressed the endo-chitosanase gene from *A. fumigatus* in *Pichia pastoris* GS115, reaching a CSN-P yield of 3 mg/mL with high-density fermentation in a 14 L fermenter and an enzyme activity was 25,000 U/mL [12]. Huang et al. optimized the fermentation process for expressing this enzyme in *E. coli*, achieving an extracellular chitosanase activity of 14,000 U/mL after 2 day by a fed-batch fermentation strategy in a 5 L fermenter [37]. Recently, Jiang et al. heterologously expressed the GH8 family chitosanase PbCsn8 from *B. thuringiensis* JAM-GG01 in *B. subtilis* as an extracellular protein. The highest chitosanase yield of 1108 U/mL was achieved through high-cell density fermentation in a 5 L fermenter, and the enzyme activity toward chitosan was 360 U/mg. It was further utilized for COSs production, reaching a maximum yield of 79.3% (with COSs content reaching 39.7 mg/mL) after 4 h of incubation [42]. Additionally, researchers have successfully expressed chitosanase in plants. The chitosanase Cho1 from *B. circulans* MH-K1, known for its antifungal activity against the rice blast fungus *Magnaporthe oryzae*, was introduced into rice cells by Kouzai et al. The expression in the apoplast demonstrated increased resistance to *M. royzae* [49].

6.2 Molecular Modification of Enzymes

While heterologous expression and genetic improvement have mitigated the issue of low original chitosanase yield to some extent, challenges persist in the development of chitosanase. Issues such as poor stability and low efficiency mean relying solely on the natural enzyme may not meet the requirements of industrial production. Consequently, the modification of chitosanase through protein engineering to enhance its properties has garnered increasing attention from researchers. Li et al., based on the protein structure of chitosanase CsnMY002 (PDB ID: 7C6C), successfully designed a mutant for chitobiose production by interfering with substrate binding in the +3 subsite. The mutant CsnMY002_G21K exhibits the ability to produce chitobiose (with a relative content >87%) from chitosan with a low degree of acetylation. Notably, 50.65 mg chitobiose (with a purity >98%) was prepared from 100 mg of chitosan [55]. In the case of chitosanase ScCsn46 from *Streptomyces* sp. N174, structural analysis revealed protrusions of residues 198A, 199A, 200H, and 201D in the catalytic cleft. Ding et al. removed these four residues from SsCsn46, resulting in the mutant m-SsCsn46. Catalytic results demonstrated a significant reduction in the proportion of chitotriose and chitotetraose, with the main product being chitopentaose [18].

The industrial applicability of chitosanase relies on its thermostability, as a more stable enzyme enhances the biotransformation of chitosan. Zhou et al. focused on mutating serine residues in chitosanase derived from *Bacillus* sp. TS. The mutants, S265G, S276A, and S347G, exhibited increased half-lives at 60 °C extending from 5.32 min to 34.57 min, 36.79 min, and 7.2 min, respectively. Through combined mutations, the thermostabilities of all double-point and triple-point mutants were improved compared to single mutants and the wild type [121]. Han et al. obtained the chitosanase CHIS1754 from the marine metagenome. To enhance catalytic efficiency and thermostability, a multi-point mutant, CHIS1754T, was designed and constructed. It displayed a significant increase in thermostability, with the T_m showing an increase of 7.63 °C. The residual enzyme activity after 1 h at 60 °C remained at 80%, and the k_{cat}/K_m of CHIS1754T increased by around 4.8 times compared to CHIS1754 [32]. In another approach, Sheng et al. inserted an artificial disulfide bond by mutating a site in the flexible region to a cysteine, resulting in a sixfold increase in the half-life of chitosanase at 50 °C [87].

6.3 *Immobilization of Chitosanase*

Immobilized enzyme technology has become integral, offering enhance stability, enzyme regeneration, and reusability. This approach ensures continuity in the degradation process and facilitates the utilization of chitosanases under bioreactor conditions [89]. Yamasaki et al. pioneered the immobilization of chitosanase G-1 from *Enterobacter* sp. on various carriers, including anion exchange resin, chitosan pellets, and calcium alginate. Among these, chitosanase immobilized on the strong-base anion exchanger FE-3901 exhibited the highest enzyme-specific activity, while chitosan particles provided optimal stability, retaining over half of its initial activity after 17 day of storage [111].

Since the early 2000s, researchers have explored the immobilization of bacterial-derived chitosanase on agar or agarose gel particles through multi-point covalent binding. This approach proved efficient in continuously degrading chitosan for a month in a bioreactor with a carrier layer [51]. After further optimization, 7 kg/m^3 of chitopentaose and chitohexaose were obtained from 20 kg/m^3 of chitosan [51]. Song et al. employed the cross-linking method to immobilize chitosanase on silica gel, demonstrating negligible loss of enzyme activity after several successive uses [92].

The viscosity of the chitosan solution significantly influences the degree of hydrolysis, particularly in systems with limited mass transfer patterns. Conventional porous carriers with millimeter-sized or submillimeter-sized particles may face challenges due to resistance to mass transfer. To address this, magnetic nanoparticles and nanofibers have been explored as carriers for chitosanase immobilization [52, 90]. Magnetic nanoparticles offer advantages such as good mass transfer properties, a large specific surface area, and ease to recycling. Kuroiwa et al. immobilized chitosanase on magnetic nanoparticles, resulting in 1.4–2 times more chitosanase immobilization compared to traditional physical adsorption methods, with a COS yield reaching 40% in one reaction cycle [52]. Polyacrylonitrile nanofibrous membranes have also been developed for chitosanase immobilization, exhibiting over 70% retained activity after 10 cycles of use and stability at 4 °C for up to 60 days [90].

Traditional enzyme immobilization often involves expensive and time-consuming protein purification steps. Researchers have explored methods to directly immobilize fusion enzymes from cell extracts using immobilization tags, such as His-Tag, HaloTag™, and Ig-G binding domain tags [5,21,76]. Oliveira et al. developed a one-step purification and immobilization method for enzymes from crude cell extracts using the unique property of the carbohydrate-binding modules to specifically bind carbohydrates [72]. Qin et al. established a one-step immobilization-purification strategy for chitosanase by CBM56-Tag fusion, achieving continuous production of COSs in a biocatalytic packed-bed reactor [80]. Zhou et al. successfully expressed the recombinant chitosanase Csn75-CBM fused with carbohydrates (CBM) in *P. aeruginosa*. The enzyme was then purified and immobilized in a gel polysaccharide-filled bed reactor in one step, demonstrating excellent stability and reusability, significantly improving COSs production efficiency [119].

7 Preparation Methods (Isolation and Purification, In Vitro Recombination)

7.1 Fermentation Production of Chitosanase

The majority of reported chitosanases are inducible enzymes, and their production through fermentation involves either liquid or solid-state fermentation, often requiring the addition of specific inducers to the culture medium. Various substances, such as colloidal chitosan, chitosan powder, chitosan, squid pen powder, and shrimp shell powder, have been identified as effective inducers for chitosanase production [41, 106, 107, 122, 125].

To scale up chitosanase production to meet commercial demands, optimization of parameters such as media components and fermentation conditions is crucial. Zhou et al. optimized bacterial growth parameters, including pH, temperature, carbon, and nitrogen sources, achieving an enzyme production of 800 U/L at 72 h under the optimized conditions [122]. However, traditional single-factor experiments have limitations in addressing the complexity of multiple factors in culture media optimization. Consequently, researchers have turned to statistical methods like orthogonal experiments and response surface experiments. Zhang et al. used Box–Behnken design and response surface analysis to optimize culture conditions for chitosanase production by fermentation of *Aspergillus* sp. QD-2, resulting in an increased chitosanase yield from 26.5 to 85.8 U/mL [116]. Ma et al. improved the yield of chitosanase CsnA from the marine bacterium *Renibacterium* sp. QD1 by statistically optimizing the fermentation medium, achieving an optimized enzyme yield of 608.11 U/mL, 52.03% higher than the original yield [61]. Cahyaningtyas et al. isolated a strain from *B. cereus* HMRSC30 in Terasi, capable of producing chitosanase from shrimp shell waste. Statistical optimization analysis determined optimal conditions, resulting in a simultaneously produce proteases, indicating its potential for the dual production of chitosanase and chitin in a single fermentation [6].

7.2 *Isolation and Purification of Chitosanase*

While the use of crude enzyme solutions suffices for industrial COSs production, further biochemical and structural studies necessitate the isolation and purification of chitosanase. Typically derived from microbial sources, chitosanase purification employs classical protein isolation methods. The process involves obtaining a crude enzyme solution through centrifugation, followed by selective precipitation/concentration using ammonium sulfate or polyethylene glycol. The concentrated chitosanases undergo additional purification through various chromatographic techniques, including gel filtration, ion exchange, and affinity methods. The purity of the target protein is assessed using SDS-PAGE. Numerous studies have reported the purification and characterization of chitosanase from different bacteria and fungi. For example, Zitouni et al. isolated and purified a heat-resistant chitosanase, Csn1794 (identified as a novel *Paenibacillus* member), from shrimp shells compost. Purification steps included ice-cold acetone protein precipitation, Q-SepHarose FF ion exchange chromatography, mini-column Hi-Trap Q enzyme concentration, hydroxyapatite chromatography, and dialysis. The specific activity of the purified chitosanase was 122 U/mg [125]. Nguyen et al. induced chitosanase production in *P. janthinellum* D4 using squid pen powder and purified the enzyme for COSs preparation through ammonium sulfate precipitation and combined column chromatography [70]. Although these traditional purification methods involve multiple steps and are time-consuming, the expanded bed adsorption technique offers a streamlined approach for separation, concentration, and target protein capture in a single step. de Araújo et al. utilized a Streamline DEAE ionic exchanger for expanded bed adsorption to purify and recover chitosanase produced by *B. cereus* C-01, significantly reducing purification steps and associated cost [16]. The group achieved a 5.4-fold increase in the purification rate of this chitosanase (from 1.75 to 9.5) at a lower salt concentration, using it for COSs production [14].

8 Application of Chitosanase

Chitosanases derived from microorganisms exhibit diverse biological functions, and their extensive applications across various fields have garnered significant attention. In the industrial sector, one of the pivotal applications of chitosanase is the degradation of chitosan to produce COSs with specific DPs [18, 19, 55, 120]. COSs, known for their water solubility, easy absorption, antibacteria, antifungi, immune-regulating, and anticancer properties, offer substantial economic benefits. They find applications in functional foods, healthcare drugs, and various industries [43, 53]. Chitosanase also plays a crucial role in the bioconversion and value addition of marine biomaterials, as well as the production of bioactive molecules. The global seafood processing industry generates substantial bio-waste and by-products, contributing to environmental pollution. The application of chitosanase becomes instrumental in addressing these environmental challenges [97]. Additionally, as chitosan constitutes a primary structural component of fungal cell wall, chitosanases are widely used to degrade chitosan in the cell wall for protoplast preparation [45]. Fungal protoplasts find application in genetic hybridization experiments, particularly for zygomycetes with chitosan-based cell walls, playing a vital role in mycological studies and strain improvement for enhanced industrial applications. Chitosanase finds yet another application in chitosan-mediated gene delivery, enhancing cellular transformation efficiency. Studies have demonstrated that predelivery of chitosanase into cells before transfection accelerates chitosan degradation, promoting gene unwrapping and significantly increasing gene expression levels [56].

Moreover, owing to the prevalence of chitosan in most fungal cell walls, chitosanases are employed to degrade the cell walls of plant pathogenic fungi. Consequently, chitosanases serve as biocontrol agents, increasing plant resistance to phytopathogenic fungi. Some research groups have reported the in vitro antifungal activity of chitosanases, showing their potential for enhancing plant resistance to various phytopathogenic fungi [26, 49, 56, 101]. For instance, Cho1 from *B. circulans* MH-K1 was found to have antifungal activity against the rice blast fungus *M. oryzae*. When the gene was inserted into rice, the transgenic rice exhibited increased resistance to *A. oryzae* along with evaluated hydrogen peroxide production in infected epidermal cells [49].

References

1. Adachi W, Sakihama Y, Shimizu S, Sunami T, Fukazawa T, Suzuki M, Yatsunami R, Nakamura S, Takénaka A (2004) Crystal structure of family GH-8 chitosanase with subclass II specificity from *Bacillus* sp. K17. J Mol Biol 343:785–795
2. Aktuganov GE, Galimzianova NF, Gilvanova EA, Pudova EA, Kuzmina LY, Melentiev AI, Safina VR (2019) Purification and characterization of exo-β-1,4-glucosaminidase produced by chitosan-degrading fungus, *Penicillium* sp. IB-37-2A. World J Microbiol Biotechnol 35:18
3. Alzari PM, Souchon H, Dominguez R (1996) The crystal structure of endoglucanase CelA, a family 8 glycosyl hydrolase from *Clostridium thermocellum*. Structure 4:265–275
4. Azevedo MIG, Oliveira ST, Silva CFB, Carneiro RF, Nagano CS, Gadelha ACS, Torres DC, Monteiro-Júnior JE, Girão MS, Muniz CR, Freitas CDT, Grangeiro TB (2020) Secretory production in *Escherichia coli* of a GH46 chitosanase from *Chromobacterium violaceum*, suitable to generate antifungal chitooligosaccharides. Int J Biol Macromol 165:1482–1495
5. Barbosa O, Ortiz C, Berenguer-Murcia A, Torres R, Rodrigues RC, Fernandez-Lafuente R (2015) Strategies for the one-step immobilization-purification of enzymes as industrial biocatalysts. Biotechnol Adv 33:435–456
6. Cahyaningtyas HAA, Suyotha W, Cheirsilp B, Yano S (2021) Statistical optimization of halophilic chitosanase and protease production by *Bacillus cereus* HMRSC30 isolated from *Terasi* simultaneous with chitin extraction from shrimp shell waste. Biocatal Agric Biotechnol 31:101918
7. Cabib E (1987) The synthesis and degradation of chitin. Adv Enzymol 59:84–86
8. Cantarel BL, Coutinho PM, Rancurel C, Bernard T, Lombard V, Henrissat B (2009) The carbohydrate-active EnZymes database (CAZy): an expert resource for glycogenomics. Nucleic Acids Res 37:233–238
9. Chen D, Chen CC, Zheng XH, Chen JN, He WJ, Lin CT, Chen HB, Chen YQ, Xue T (2021) Chitosan oligosaccharide production potential of *Mitsuaria* sp. C4 and its whole-genome sequencing. Front Microbiol 12:695571
10. Chen XE, Fang XB, Yu H, Xia WS (2008) Hydrolytic model of chitosanases produced with *Aspergillus* CJ22-326. Chin J Process Eng 8:350–354
11. Chen XE, Xia WS, Yu XB (2005) Purification and characterzation of two types of chitosanase from *Aspergillus* sp. CJ22-326. Food Res Int 38:315–322
12. Chen X, Zhai C, Kang L, Li C, Yan H, Zhou Y, Yu X, Ma L (2012) High-level expression and characterization of a highly thermostable chitosanase from *Aspergillus fumigatus* in *Pichia pastoris*. Biotechnol Lett 34:689–694
13. Chen H, Zhou B, Huang G (2019) Content determination of reducing sugar in pharmaceutical maltodextrin by park-Johnson ferricyanide submicromethod. China Pharm 28:30–32
14. Dantas JMDM, de Araújo NK, da Silva NS, Torres-Rego M, Furtado AA, de Assis CF, Araújo RM, Teixeira JA, Ferreira LDS, Fernandes-Pedrosa MDF, dos Santos ES (2021) Purification of chitosanases produced by *Bacillus toyonensis* CCT 7899 and functional oligosaccharides production. Prep Biochem Biotech:1–9
15. da Silva LC, Honorato TL, Franco TT, Rodrigues S (2012) Optimization of chitosanase production by *Trichoderma koningii* sp. under solid-state fermentation. Food Bioprocess Tech 5:1564–1572
16. de Araújo NK, Pagnoncelli MGB, Pimentel VC, Xavier MLO, Padilha CEA, de Macedo GR, dos Santos ES (2015) Single-step purifification of chitosanases from *Bacillus cereus* using expanded bed chromatography. Int J Biol Macromol 82:291–298
17. Dennhart N, Fukamizo T, Brzezinski R, Lacombe-Harvey MÈ, Letzel T (2008) Oligosaccharide hydrolysis by chitosanase enzymes monitored by realtime electrospray ionization-mass spectrometry. J Biotechnol 134:253–260
18. Ding M, Zhang T, Sun C, Zhang H, Zhang Y (2020) A chitosanase mutant from *Streptomyces* sp. N174 prefers to produce functional chitopentasaccharide. Int J Biol Macromol 151:1091–1098

19. Doan C, Tran T, Nguyen V, Tran T, Nguyen A, Wang S (2020) Bioprocessing of squid pens waste into chitosanase by *Paenibacillus* sp. TKU047 and its application in low-molecular weight chitosan oligosaccharides production. Polymers (Basel) 12:1163
20. Doan C, Tran T, Nguyen V, Nguyen A, Wang S (2018) Conversion of squid pens to chitosanases and proteases via *Paenibacillus* sp. TKU042. Mar Drugs 16:83
21. Doebber J, Gerlach T, Offermann H, Rother D, Pohl M (2018) Closing the gap for efficient immobilization of biocatalysts in continuous processes: HaloTag$^{(TM)}$ fusion enzymes for a continuous enzymatic cascade towards a vicinal chiral diol. Green Chem 20:544–552
22. Fenton DM, Eveleigh DE (1981) Purification and mode of action of a chitosanase from *Penicillium islandicum*. J Gen Microbiol 126:151–165
23. Fukamizo T, Fleury A, Côté N, Mitsutomi M, Brzezinski R (2006) Exo-b-D-glucosaminidase from *Amycolatopsis orientalis*: catalytic residues, sugar recognition specificity, kinetics, and synergism. Glycobiology 16:1064–1072
24. Fukamizo T, Ohkawa T, Ikeda Y, Goto S (1994) Specificity of chitosanase from Bacillus pumilus. Biochim Biophys Acta 1205:183–188.
25. Fukamizo T, Honda Y, Goto S, Boucher I, Brzezinski R (1995) Reaction mechanism of chitosanase from *Streptomyces* sp. N174. Biochem J 311:377–383
26. Gao XA, Ju WT, Jung WJ, Park RD (2008) Purification and characterization of chitosanase from *Bacillus cereus* D-11. Carbohydr Polym 72:513–520
27. Goo BG, Park JK (2014) Characterization of an alkalophilic extracellular chitosanase from *Bacillus cereus* GU-02. J Biosci Bioeng 117:684–689
28. Grenier J, Asselin A (1990) Some pathogenesis related proteins are chitosanases with lytic activity against fungal spores. Mol Plant Microb Interact 3:401–407
29. Guan FF, Han YS, Yan K, Zhang Y, Zhang ZF, Wu NF, Tian J (2020) Highly efficient production of chitooligosaccharides by enzymes mined directly from the marine metagenome. Carbohydr Polym 234:115909
30. Guérin DMA, Lascombe MB, Costabel M, Souchon H, Lamzin V, Béguin P, Alzari PM (2002) Atomic (0.94 Å) resolution structure of an inverting glycosidase in complex with substrate. J Mol Biol 316:1061–1069
31. Gupta V, Prasanna R, Srivastava AK, Sharma J (2012) Purification and characterization of a novel antifungal endo-type chitosanase from Anabaena fertilissima. Ann Microbiol 62:1089–1098.
32. Han Y, Guan F, Sun J, Wu N, Tian J (2020) Identification of a chitosanase from the marine metagenome and its molecular improvement based on evolution data. Appl Microbiol Biotechnol 104:6647–6657
33. Heggset EB, Dybvik AI, Hoell IA, Norberg AL, Sørlie M, Eijsink VG, Vårum KM (2010) Degradation of chitosans with a family 46 chitosanase from *Streptomyces coelicolor* A3(2). Biomacromolecules 11:2487–2497
34. Heggset E B. et al. Mode of action of a family 75 chitosanase from Streptomyces avermitilis. Biomacromolecules 13:1733–1741 (2012).
35. Hirano K, Watanabe M, Seki K, Ando A, Saito A, Mitsutomi M (2012) Classification of chitosanases by hydrolytic specificity toward N^1,N^4-diacetylchitohexaose. Biosci Biotechnol Biochem 76:1932–1937
36. Hsu SK, Chung YC, Chang CT, Sung HY (2012) Purification and characterization of two chitosanase isoforms from the sheaths of bamboo shoots. J Agric Food Chem 60:649–657
37. Huang L, Wang Q, Jiang S, Zhou Y, Zhang G, Ma Y (2016) Improved extracellular expression and high-cell-density fed-batch fermentation of chitosanase from *Aspergillus Fumigatus* in *Escherichia coli*. Bioprocess Biosyst Eng 39:1–9
38. Imoto T, Yagishita K (1971) A simple activity measurement of lysozyme. Agric Biol Chem 35:1154–1160
39. Ike M, Ko Y, Yokoyama K, Sumitani JI, Kawaguchi T, Ogasawara W, Okada H, Morikawa Y (2007) Cellobiohydrolase I (Cel7A) from *Trichoderma reesei* has chitosanase activity. J Mol Catal B Enzym 47:159–163

40. Izume M, Nagae S, Kawagishi H, Mitsutomi M, Ohtakara A (1992) Action pattern of Bacillus sp. no. 7-M chitosanase on partially N-acetylated chitosan. Biosci Biotechnol Biochem 56:448–453.
41. Jiang X, Chen D, Chen L, Yang G, Zou S (2012) Purification, characterization and action mode of a chitosanase from *Streptomyces roseolus* induced by chitin. Carbohydr Res 355:40–44
42. Jiang Z, Ma S, Guan L, Yan Q, Yang S (2021) Biochemical characterization of a novel bifunctional chitosanase from *Paenibacillus barengoltzii* for chitooligosaccharide production. World J Microb Biot 37:1–13
43. Jitprasertwong P, Khamphio M, Petsrichuang P, Eijsink VGH, Poolsri W, Muanprasat C, Rangnoi K, Yamabhai M (2021) Anti-inflammatory activity of soluble chito-oligosaccharides (CHOS) on VitD3-induced human THP-1 monocytes. PLoS One 16:e0246381
44. Johnsen MG, Hansen OC, Stougaard P (2010) Isolation, characterization and heterologous expression of a novel chitosanase from *Janthinobacterium* sp. strain 4239. Microb Cell Factories 9:5
45. Kauss H, Jeblick W, Domard A (1989) The degree of polymerization and *N*-acetylation of chitosan determine its ability to elicit callose formation in suspension cells and protoplasts of *Catharanthus roseus*. Planta 178:385–392
46. Kiyohiko S, Yasue N, Masaru M (2018) Characterization of a novel exo-chitosanase, an exo-chitobiohydrolase, from *Gongronella butleri*. J Biosci Bioeng 127:425–429
47. Kim SK, Rajapakse N (2005) Enzymatic production and biological activities of chitosan oligosaccharides (COS): A review. Carbohydr Polym 62:357–368.
48. Kohlhoff M, Niehues A, Wattjes J, Bénéteau J, Cord-Landwehr S, El Gueddari NE, Bernard F, Rivera-Rodriguez GR, Moerschbacher BM (2017) Chitinosanase: a fungal chitosan hydrolyzing enzyme with a new and unusually specific cleavage pattern. Carbohyd Polym 174:1121–1128
49. Kouzai Y, Mochizuki S, Saito A, Ando A, Minami E, Nishizawa Y (2012) Expression of a bacterial chitosanase in rice plants improves disease resistance to the rice blast fungus *Magnaporthe oryzae*. Plant Cell Rep 31:629–636
50. Kramer DN, Gamson RM, Miller FM (1959) A study of the physical and chemical properties of the esters of indophenols I. preparation. J Org Chem 24:1742–1747
51. Kuroiwa T, Ichikawa S, Sato S, Mukataka S (2003) Improvement of the yield of physiologically active oligosaccharides in continuous hydrolysis of chitosan using immobilized chitosanases. Biotechnol Bioeng 84:121–127
52. Kuroiwa T, Noguchi Y, Nakajima M, Sato S, Mukataka S, Ichikawa S (2008) Production of chitosan oligosaccharides using chitosanase immobilized on amylose-coated magnetic nanoparticles. Process Biochem 43:62–69
53. Liaqat F, Eltem R (2018) Chitooligosaccharides and their biological activities: a comprehensive review. Carbohyd polym 184:243–259
54. Li H, Fei Z, Gong J, Yang T, Xu Z, Shi J (2015) Screening and characterization of a highly active chitosanase based on metagenomic technology. J Mol Catal B Enzym 111:29–35
55. Li Y, Gou Y, Liu Z, Xie T, Wang G (2021) Structure-based rational design of chitosanase CsnMY002 for high yields of chitobiose. Colloids Surf B 202:111692
56. Liang DC, Liu WG, Zuo AJ, Sun SJ, Cheng N, Guo G, Zhang JY, Yan KD (2006) Pre-deliver chitosanase to cells: a novel strategy to improve gene expression by endocellular degradation-induced vector unpacking. Int J Pharm 314:63–71
57. Liu Y, Jiang S, Ke Z, Wu H, Chi C, Guo Z (2009) Recombinant expression of a chitosanase and its application in chitosan oligosaccharide production. Carbohydr Res 344:815–819
58. Liu Y, Li Y, Tong S, Yuan M, Wang X, Wang J, Fan Y (2020) Expression of a *Beauveria bassiana* chitosanase (BbCSN-1) in *Pichia pastoris* and enzymatic analysis of the recombinant protein. Protein Expr Purif 166:105519
59. Lyu Q, Shi Y, Wang S, Yang Y, Han B, Liu W, Jones DNM, Liu W (2015) Structural and biochemical insights into the degradation mechanism of chitosan by chitosanase OU01. BBA-Gen Subjects 1852:1953–1961

60. Lyu Q, Wang S, Xu W, Han B, Liu W, Jones DN, Liu W (2014) Structural insights into the substrate-binding mechanism for a novel chitosanase. Biochem J 461:335–345
61. Ma J, Guo X, Liu H, Xing P, Yu W, Lu X (2014) Statistical optimization of fermentation medium for chitosanase production from marine bacteri *Renibacterium* sp. QDl. Sci Technol Food Ind 21:146–149
62. Ma C, Li X, Yang K, Li S (2020) Characterization of a new chitosanase from a marine *Bacillus* sp. and the anti-oxidant activity of its hydrolysate. Mar Drugs 18:126
63. Marcotte EM, Monzingo AF, Ernst SR, Brzezinski R, Robertus JD (1996) X-ray structure of an anti-fungal chitosanase from *Streptomyces* N174. Nat Struct Bio 3:155–162
64. McCarter JD, Withers SG (1994) Mechanisms of enzymatic glycoside hydrolysis. Curr Opin Struct Biol 4:885–892
65. Mitsutomi M, Ueda M, Arai M, Ando A, Watanabe T (1996) Action patterns of microbial chitinases and chitosanases on partially *N*-acetylated chitosan. Chitin Enzym 2:273–284
66. Mitsutomi M, Isono M, Uchiyama A, Nikaidou N, Ikegami T, Watanabe T (1998) Chitosanase activity of the enzyme previously reported as β-1,3-1,4-glucanase from *Bacillus circulans* WL-12. Biosci Biotechnol Biochem 62:2107–2114
67. Monaghan RL, Eveleigh DE, Tewari RP, Reese ET (1973) Chitosanase, a novel enzyme. Nat New Biol 245:78–80
68. Monzingo AF, Marcotte EM, Hart PJ, Robertus JD (1996) Chitinases, chitosanases, and lysozymes can be divided into procaryotic and eucaryotic families sharing a conserved core. Nat Struct Biol 3:133–140
69. Nelson W (1944) A photometric adaptation of the Somogyi method for the determination of glucose. J Biol Chem 153:375–380
70. Nguyen AD, Huang CC, Liang TW, Nguyen VB, Pan PS, Wang SL (2014) Production and purifification of a fungal chitosanase and chitooligomers from *Penicillium janthinellum* D4 and discovery of the enzyme activators. Carbohyd Polym 108:331–337
71. Nogawa M, Takahashi H, Kashiwagi A, Ohshima K, Okada H, Morikawa Y (1998) Purification and characterization of exo-β-d-glucosaminidase from a cellulolytic fungus, *Trichoderma reesei* PC-3-7. Appl Environ Microbiol 64:890–895
72. Oliveira C, Carvalho V, Domingues L, Gama FM (2015) Recombinant CBM fusion technology-applications overview. Biotechnol Adv 33:358–369
73. Osswald WF, McDonald RE, Niedz RP, Shapiro JP, Mayer RJ (1992) Quantitative fluorometric analysis of plant and microbial chitosanases. Anal Biochem 204:40–46
74. Patridge SM (1948) Filter paper partition chromatography of sugars. Biochem J 42:238.
75. Pelletier A, Sygusch J (1990) Purification and characterization of three chitosanase activities from *Bacillus megaterium* P1. Appl Environ Microbiol 56(4):844–848
76. Peschke T, Skoupi M, Burgahn T, Gallus S, Ahmed I, Rabe KS, Niemeyer CM (2017) Self-immobilizing fusion enzymes for compartmentalized biocatalysis. ACS Catal 7:7866–7872
77. Petegem FV, Collins T, Meuwis MA, Gerday C, Feller G, Beeumen JV (2003) The structure of a cold-adapted family 8 xylanase at 1.3 Å resolution: structural adaptations to cold and investigation of the active site. J Biol Chem 278:7531–7539
78. Prabaharan M, Mano JF (2005) Chitosan-based particles as controlled drug delivery systems. Drug Deliv 12:41–57
79. Qin Z, Chen QM, Lin S, Luo S, Qiu YJ, Zhao LM (2018) Expression and characterization of a novel cold-adapted chitosanase suitable for chitooligosaccharides controllable preparation. Food Chem 253:139–147
80. Qin Z, Lin S, Qiu Y, Chen Q, Zhang Y, Zhou J, Zhao L (2019) One-step immobilization-purification of enzymes by carbohydrate-binding module family 56 tag fusion. Food Chem 299:125037
81. Rondle CJM, Morgan WTJ (1955) The determination of glucosamine and galactosamine. Biochem J 61:586–589
82. Sabine W, Lothar J (1987) Assay of reducing sugars in the nanomole range with 2,2′-bicinchoninate. Anal Biochem 165:337–340

83. Sakai K, Katsumi R, Isobe A, Nanjo F (1991) Purifification and hydrolytic action of a chitosanase from *Nocardia orientalis*. Biochim Biophys Acta 1079:65–72
84. Saito J, Kita A, Higuchi Y, Nagata Y, Ando A, Miki K (1999) Crystal structure of chitosanase from *Bacillus circulans* MH-K1 at 1.6-Å resolution and its substrate recognition mechanism. J Biol Chem 274:30818–30825
85. Schmidt A, Gonzalez A, Morris RJ, Costabel M, Alzari PM, Lamzin VS (2002) Advantages of high-resolution phasing: MAD to atomic resolution. Acta Crystallog Sect D 58:1433–1441
86. Shan X, Zheng R, Zhang T, Wang Y (2007) Development and potential applicationn in study on chitosanase. J Chengdu Med Coll 2:192–198
87. Sheng J, Ji XF, Zheng Y, Wang Z, Sun M (2016) Improvement in the thermostability of chitosanase from *Bacillus ehimensis* by introducing artificial disulfide bonds. Biotechnol Lett 38:1809–1815
88. Shimono K, Shigeru K, Tsuchiya A, Itou N, Ohta Y, Tanaka K, Nakagawa T, Matsuda H, Kawamukai M (2002) Two glutamic acids in chitosanase A from *Matsuebacter chitosanotabidus* 3001 are the catalytically important residues. J Biochem 131:87–96
89. Sinha S, Chand S, Tripathi P (2016) Recent progress in chitosanase production of monomer-free chitooligosaccharides: bioprocess strategies and future applications. Appl Biochem Biotech 180:883–889
90. Sihna S, Dhakate SR, Kumar P, Mathur RB, Tripathi P, Chand S (2012) Electrospun polyacrylonitrile nanofibrous membranes for chitosanase immobilization and its application in selective production of chitooligosaccharides. Bioresour Technol 115:152–157
91. Somashekar D, Joseph R (1996) Chitosanases-properties and applications: a review. Bioresour Technol 55:35–45
92. Song JY, Alnaeeli M, Park JK (2014) Efficient digestion of chitosan using chitosanase immobilized on silica-gel for the production of multisize chitooligosaccharides. Process Biochem 49:2107–2113
93. Sun H, Cao R, Li L, Zhao L, Liu Q (2018a) Cloning, purification and characterization of a novel GH46 family chitosanase, Csn-CAP, from *Staphylococcus capitis*. Process Biochem 75:146–151
94. Sun HH, Mao XZ, Guo N, Zhao L, Cao R, Liu Q (2018b) Discovery and characterization of a novel chitosanase from *Paenibacillus dendritiformis* by phylogeny-based enzymatic product specificity prediction. J Agric Food Chem 66:1139–1145
95. Sun H, Gao L, Xue C, Mao X (2020a) Marine-polysaccharide degrading enzymes: status and prospects. Compr Rev Food Sci F 19:2762–2796
96. Sun H, Yang G, Cao R, Mao X, Liu Q (2020b) Expression and characterization of a novel glycoside hydrolase family 46 chitosanase identified from marine mud metagenome. Int J Biol Macromol 159:904–910
97. Suresh PV (2012) Biodegradation of shrimp processing bio-waste and concomitant production of chitinase enzyme and *N*-acetyl-D-glucosamine by marine bacteria: production and process optimization. World J Microb Biot 28:2945–2962
98. Tamo F, Satoko A, Kei Y, Tomoye Y, Tomomi K, Jun-ichi S, Michihiko S, Kunio M, Yoshiho N, Akikazu A (2005) *Bacillus circulans* MH-K1 chitosanase: amino acid residues responsible for substrate binding. J Biochem 138:563–569
99. Takasuka TE, Bianchetti CM, Tobimats Y, Bergeman LF, Ralph J, Fox BG (2014) Structure-guided analysis of catalytic specificity of the abundantly secreted chitosanase SACTE_5457 from *Streptomyces* sp. SirexAA-E. Proteins 82:1245–1257
100. Tanabe T, Morinaga K, Fukamizo T, Mitsutomi M (2003) Novel chitosanase from *Streptomyces griseus* HUT 6037 with transglycosylation activity. Biosci Biotechnol Biochem 67:354–364
101. Thadathil N, Velappan SP (2014) Recent developments in chitosanase research and its biotechnological applications: a review. Food Chem 150:392–399
102. Viens P, Lacombe-Harvey, Brzezinsk R (2015) Chitosanases from family 46 of glycoside hydrolases: from proteins to phenotypes. Mar Drugs 13:6566–6587

103. Wang Q, Graham R, Trimbur D, Warren RAJ, Withers S (1994) Changing enzymic reaction mechanisms by mutagenesis: conversion of a retaining glucosidase to an inverting enzyme. J Am Chem Soc 116:11594–11595
104. Wang YX, Li D, Liu MX, Xia CY, Fan QW, Li X, Lan ZJ, Shi GL, Dong WL, Li ZK, Cui ZL (2021) Preparation of active chitooligosaccharides with a novel chitosanase *Aq*CoA and their application in fungal disease protection. J Agric Food Chem 69:3351–3361
105. Wang YN, Qin Z, Fan LQ, Zhao LM (2020) Structure-function analysis of *Gynuella sunshinyii* chitosanase uncovers the mechanism of substrate binding in GH family 46 members. Int J Biol Macromol 165:2038–2048
106. Wang SL, Peng JH, Liang TW, Liu KC (2008a) Purification and characterization of a chitosanase from *Serratia marcescens* TKU011. Carbohydr Res 343:1316–1323
107. Wang SL, Tseng WN, Liang TW (2011) Biodegradation of shellfish wastes and production of chitosanases by a squid pen-assimilating bacterium, *Acinetobacter calcoaceticus* TKU024. Biodegradation 22:939–948
108. Wang J, Zhou W, Yuan H, Wang Y (2008b) Characterization of a novel fungal chitosanase Csn2 from *Gongronella* sp. JG. Carbohydr Res 343:2583–2588
109. Weikert T, Niehues A, Cordlandwehr S, Hellmann MJ, Moerschbacher BM (2017) Reassessment of chitosanase substrate specificities and classification. Nat Commun 8:1698
110. Withers SG (2001) Mechanisms of glycosyl transferases and hydrolases. Carbohydr Polym 44:325–337
111. Yamasaki Y, Fukumoto I, Kumagai N, Ohta Y, Nakagawa T, Kawamukai M, Matsuda H (1992) Continuous chitosan hydrolyzate production by immobilized chitosanolytic enzyme from *Enterobacter* sp. G-1. Biosci Biotechnol Biochem 56:1546–1551
112. Yang Y, Zheng Z, Xiao Y, Zhang J, Zhou Y, Li X, Li S, Yu H (2019) Cloning and characterization of a cold-adapted chitosanase from marine bacterium *Bacillus* sp. BY01. Molecules 24:3915
113. Yorinaga Y, Kumasaka T, Yamamoto M, Hamada K, Kawamukai M (2017) Crystal structure of a family 80 chitosanase from *Mitsuaria* chitosanitabida. FEBS Lett 591:540–547
114. Zhang J, Cao H, Li S, Zhao Y, Wang W, Xu Q, Du Y, Yin H (2015) Characterization of a new family 75 chitosanase from *Aspergillus* sp. W-2. Int J Biol Macromol 81:362–369
115. Zhang XY, Dai AL, Zhang XK, Kuroiwa K, Kodaira R, Shimosaka M, Okazaki M (2000) Purification and characterization of chitosanase and exo-β-D-glucosaminidase from a Koji Mold, *Aspergillus oryzae* IAM2660. Biosci Biotechnol Biochem 64:1896–1902
116. Zhang H, Sang Q, Zhang W (2012) Statistical optimization of chitosanase production by *Aspergillus* sp. QD-2 in submerged fermentation. Ann Microbiol 62:193–201
117. Zhang Y, Zhang J, Chang H, Zhang K, Liu Z, Luo C, Mou X (2013) A novel sensitive assay for chitosanolytic enzyme activity and its comparison with two others. Food Chem 34:277–281
118. Zheng Q, Meng X, Cheng M, Li Y, Liu Y, Chen X (2021) Cloning and characterization of a new chitosanase from a deep-sea *Bacterium Serratia* sp. QD07. Frontiers Microbiol 12:619731
119. Zhou J, Harindintwali JD, Yang W, Han M, Deng B, Luan H, Zhang W, Liu X, Yu X (2021) Engineering of a chitosanase fused to a carbohydrate-binding module for continuous production of desirable chitooligosaccharides. Carbohyd Polym 273:118609
120. Zhou J, Liu X, Yuan F, Deng B, Yu X (2020) Biocatalysis of heterogenously-expressed chitosanase for the preparation of desirable chitosan oligosaccharides applied against phytopathogenic fungi. ACS Sustain Chem Eng 8:4781–4791
121. Zhou Z, Wang X (2021) Improve thermostability of *Bacillus* sp. TS chitosanase through structure-based alignment. Sci Rep-UK 11:15846
122. Zhou W, Yuan H, Wang J, Yao J (2008) Production, purification and characterization of chitosanase produced by *Gongronella* sp. JG. Lett Appl Microbiol 46:49–54

123. Zhou Z, Zhao S, Wang S, Li X, Su L, Ma Y, Li J, Song J (2015) Extracellular overexpression of chitosanase from *Bacillus* sp. TS in *Escherichia coli*. Appl Biochem Biotechnol 175:3271–3286
124. Zhu XF, Tan HQ, Zhu C, Liao L, Zhang XQ, Wu M (2012) Cloning and overexpression of a new chitosanase gene from *Penicillium* sp. D-1. AMB Express 2:13
125. Zitouni M, Fortin M, Scheerle RK, Letzel T, Matteau D, Rodrigue S, Brzezinski R (2013) Biochemical and molecular characterization of a thermostable chitosanase produced by the strain *Paenibacillus* sp. 1794 newly isolated from compost. Appl Microbiol Biot 97:5801–5813

Alginate Lyases

Hang Su, Nghi-Van Phung, Yan-jun Gong, Xiao-Qing Ma, Ming Lu, and Fu-Li Li

1 Introduction

Algae, as a representative resource of the third-generation biofuel, are conducive to the release of monosaccharides (as they do not contain lignin). They exhibit many advantages such as rapid reproduction, low energy consumption, no arable land requirement, no freshwater consumption, and no food conflicts. At present, the total global production of macroalgae is 25 million tons/year, which is mainly contributed by Asia countries (96%). China is the largest producer accounting for 54% of the global share, followed by Indonesia, South Korea, Japan, and Malaysia. Brown algae and red algae are the two main types of macroalgae cultivated. The polysaccharides in brown algae include mainly alginate, mannitol, and a small portion of

H. Su · F.-L. Li (✉)
State Key Laboratory of Photoelectric Conversion and Utilization of Solar Energy, Shandong Provincial Key Laboratory of Synthetic Biology, Shandong C1 Refinery Engineering Research Center, Qingdao New Energy Shandong Laboratory, Qingdao Institute of Bioenergy and Bioprocess Technology, Chinese Academy of Sciences, Qingdao, China

Shandong Energy Institute, Qingdao, China

N.-V. Phung · X.-Q. Ma · M. Lu
State Key Laboratory of Photoelectric Conversion and Utilization of Solar Energy, Shandong Provincial Key Laboratory of Synthetic Biology, Shandong C1 Refinery Engineering Research Center, Qingdao New Energy Shandong Laboratory, Qingdao Institute of Bioenergy and Bioprocess Technology, Chinese Academy of Sciences, Qingdao, China
e-mail: lifl@qibebt.ac.cn

Y.-j. Gong
Weifang Key Laboratory of Marine Functional Oligosaccharides Green Biomanufacturing, Weifang Macroalga Biotechnology Co., Ltd, Weifang, China

X. Mao (ed.), *Degrading Enzymes for Marine Polysaccharides and Their Applications*, https://doi.org/10.1007/978-981-96-9517-1_6

kelp polysaccharides; and the former two occupy approximately 70% of total polysaccharides [1]. The products from brown algae, such as alginate and AOSs, have become high-value-added chemicals owning to their broad applications. Particularly, AOSs have attracted great attention due to their biological and chemical features, such as immunity regulation, antitumor activity, promotion of plant growth, blood sugar and lipids regulation. Additionally, alginate hydrogels can be used as a major biomaterial in engineering wound healing and drug delivery systems [2–7].

Therefore, selection of appropriate strategies for modifying alginate and AOSs chemical structures has been brought to the forefront. Compared with traditional chemical methods, the reactions catalyzed by microbial enzymes exhibit many advantages including ease of control, substrate specificity, high enzymatic activity, energy efficiency, and environmental-friendliness. Alginate lyase is a class of enzymes that can degrade alginate through cleavage reaction at their 1,4-glycosidic bond, resulting in the C4,5-double bond in uronic acid. In addition, as brown algae play a very important role in the ecosystem, alginate lyase is also involved in the process of the ocean carbon cycle. Thus, studies on alginate lyase are useful for application prospects and in providing new ideas for global carbon sink research and people's understanding of ecosystems. Despite their precious values, most of the reported alginate lyases are medium- and low-temperature enzymes, meaning that they are inappropriate for large-scale industrial applications. For these reasons, great attention and efforts have been made to discover more highly efficient and thermally stable alginate lyase and to improve the available enzymes' properties.

2 The Source, Structure, and Application of Alginate

Alginate is mainly found in brown algae and several bacteria. The alginate-producing brown algae include *Laminaria hyperborean*, *Macrocystis pyrifera*, *Laminria digitata*, *Ascophyllum nodosum*, *Laminaria japonica*, *Lessonia nigrescens*, and *Antarctic seaweed* (*Durvillea antarctica*) [4]. Besides, many pathogenic bacteria such as *Pseudomonas* and *Azotobacter* can also synthesize alginate. Interestingly, the strain *Pseudomonas aeruginosa*, a cystic fibrosis pathogen, can form a biofilm containing alginate to improve its defense against antibiotics [8, 9]. Similar to terrestrial plants, the main biological function of cellulose and alginate is to support the cell wall in macroalgae structurally.

Alginate is one of the three major polysaccharide components in brown algae, and its content varies according to seasons, maximally reaching 40% of the total algae dry weight. It is a linear polysaccharide made up of β-D-mannuronic acid (M) and α-L-guluronic acid (G), connected by 1,4-glycosidic bonds. The arrangement of these units forms poly-mannuromte (Poly-M), poly-guluronate (Poly-G), and mannuromte-guluronic acid hybrid fragment (MG blocks for abbreviation Poly-MG), which differ greatly in the spatial structure of aggregation. The four types of glycosidic bonds include diequatorial (in M), diaxial (in GG), equatorial–axial (in MG), and axial–equatorial (in GM) (Fig. 1). Among them, the flexible diaxial bond is the determining factor for the rigidity and stretchability of

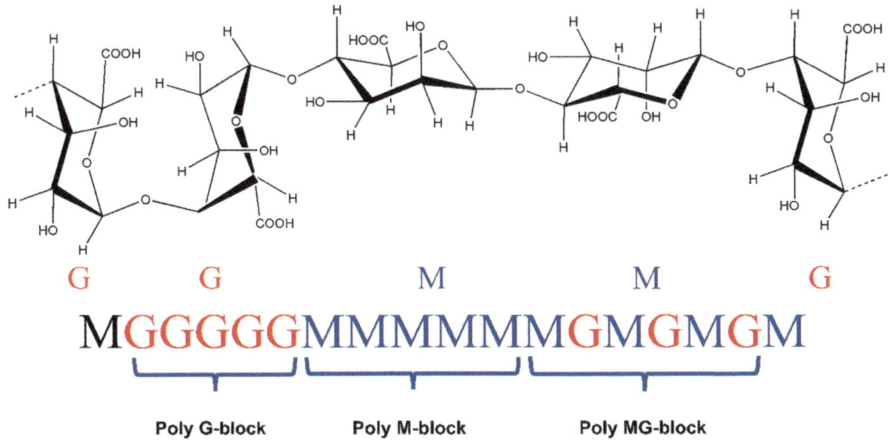

Fig. 1 Schematic representation of structure of alginate in algae

alginate [10]. Besides, as a long-chain macromolecule, alginate forms a complex with metal ions, resulting in polyanionic copolymers such as sodium alginate and calcium alginate.

Alginates in both the forms of polymer and oligosaccharide have attracted many attentions due to their potential applications across various areas. Alginate polymers possess unique physical and chemical properties such as hydrophobicity, gel formation, viscosity, and biocompatibility, allowing for their wide use in various fields such as food, medicine, and the chemical industry. For example, as a dietary fiber, it has functions in controlling blood sugar and blood lipids, delaying aging, antitumor, enhancing the immune function of the body, and other ecological effects and health care functions. Additionally, the good biocompatibility allows alginate to be used as the matrix material to wrap the drug for site-specific drug delivery. Furthermore, alginate can also be used as a biological scaffold material for medical purposes [11, 12]. The degradation of alginate results in AOSs, which have many valuable physiological properties such as immunomodulation, antioxidation, antihypertension, and antidiabetes. Nowadays, these compounds have attracted even more attention than their polymeric counterparts. Therefore, methods to degrade alginate have always been concerned, and degradation by alginate lyase via the enzymatic route is preferred due to its environmental-friendliness and energy efficiency [13].

3 Enzyme Classification and Source of Alginate Lyase

According to existing literature reports, alginate lyases are produced by a wide variety of organisms, including seaweed, marine molluscs, bacteria, and viruses, etc. In the past 10 years, more than 100 alginate lyases have been discovered [14], among which bacterial-derived alginate lyases are the most widely studied, mainly coming from the following species: *Zobellia galactanivorans* [15], *Agarivorans* sp. [16], *Agrobacterium* sp. [17], *Alteromonas* sp. [18], *Azotobacter* (*Azotobacter* sp.) [19], *Bacillus* sp. [20], *Enterobacter* [21], *Flavobacterium* sp. [22], *Klebsiella* sp. [23], *Pseudoalteromonas* sp. [24], *Pseudomonas* sp. [25], *Sphingomonas* sp. [26], and *Vibrio* sp. [15, 27, 28].

There are several methods of alginate lyase classification. According to the sequence identity of the catalytic domain of alginate lyases, most of them can be classified into the polysaccharide lyase families. More specifically, there are 42 polysaccharide lyase (PL1–42) families [29], which are defined based on the sequence and structure homology of polysaccharide lyases, according to the CAZy (Carbohydrate Active enZYme database, http://www.cazy.org/Home.html) database. Alginate lyases belong to PL-5, PL-6, PL-7, PL-15, PL-17, PL-18, PL-31, PL-32, PL-34, PL-36, PL-39, and PL-41 families. Another method is based on the substrate specificity for the degradation reactions, alginate lyases can be divided into three categories: polymannitol acid (Poly-M) lyase, polyguluronic acid (Poly-G) lyase, and bifunctional lyase that can cleave all substrates. Depending on their different modes of action, the enzymes can be divided into endo-type and exo-type. Most endo-type alginate lyases are derived from the PL-5 and PL-7 families. The endo-alginate lyase degrades alginate into di- and tri-saccharides, while the final product of exo-alginate lyase is a monosaccharide. Besides, on the basis of the size and weight of the enzyme molecules, it can be distributed into three types: large (>60 kDa), medium (40 kDa), and small (<30 kDa). Finally, based on protein's three-dimensional (3D) structure, they are classified into three categories: parallel β-helix, $(\alpha/\alpha)_6$ barrel, and β-jelly roll. Among them, PL-6 and PL-31 belong to parallel β-helix, PL-5 and PL-15 belong to (α/α) 6, PL-7, PL-14, PL-17, PL-18, and PL-36 belong to β-jelly roll. The structural types of PL-31, PL-34, and PL-41 have remained unclear (Tables 1 and 2).

4 Enzyme Activity Detection Method

4.1 DNS

The DNS method can be distributed into five steps. (1) Preparation of enzyme and substrate solutions: a certain amount (0.5 g) of the target substrate (e.g., sodium alginate) is dissolved in a 0.1 M phosphate buffer at their corresponding pH (50 mL); simultaneously, the enzyme of interest is diluted using the same buffer with the substrate. (2) Reaction: the prepared substrate and enzyme solutions are mixed and incubated at 40 °C for 30 min. (3) End of reaction process: The reaction is stopped by adding high salt concentration (e.g., 0.25 mL of 1 M Na_2CO_3). (4) Measurement of reducing sugar: The supernatant of the reaction is harvested by centrifugation and is used to determine reducing sugar concentration by employing 3,5-dinitrosalicylic

Table 1 Classification and source of alginate lyase

PL family	Enzyme	Source	Action mode	Structure
5	AlyA1-III	*Sphingomonas* sp. A1	Endolytic	$(a/a)_6$ barrel
6	AlyMG	*Stenotrophomas* maltophilia KJ-2	Endolytic	Parallel β-helix
7	PA1167	*Pseudomonas aeruginosa* PAO1	Endolytic	β-jelly roll
14	vAL-1	*Chlorella virus* ATCV-1	Endolytic	β-jelly roll
15	Atu3025	*Agrobacterium tumefaciens*	Exolytic	$(a/a)_6$ barrel
17	Alg17C	*Saccharophagus degradans*	Exolytic	β-jelly roll
18	Aly-SJ02	*Pseudoalteromonas* sp. SM0524	Endolytic	β-jelly roll
31	PsAly	*Paenibacillus* sp. FPU-7	Endolytic	Parallel β-helix
32	HMPREF1071_02769	*Bacteroides salyersiae* CL02T12C01	Endolytic	–
34	OPIT5_18480	*Opitutaceae bacterium* TAV5	Endolytic	–
36	CK934_20815	*Chitinophaga* sp. MD30	Endolytic	β-jelly roll
39	Dp0100	*Defluviitalea phaphyphila* Alg1	Endolytic and exolytic	$(\alpha/\alpha)_6$ toroid + anti-parallel β-sheet
41	AlY	*Saccharina japonica*	–	–

acid method [30]. Each unit equals the enzyme amount needed for the generation of 1 μmol reducing sugar within 1 min. Gluconic acid is used as the standard [31].

4.2 OD235

Another widely used method for the detection of enzyme activity is OD235. This method is based on the absorbance at the wavelength 235 nm of the products generated during the reaction (e.g., unsaturated unronates). Firstly, the reaction solution is prepared with a proper amount of substrate (e.g., 0.2% alginate, 2 mL), buffer (e.g., 100 mM of acetate-sodium acetate buffer at pH 5.8), and the target protein (10 μL). Next, the solution is applied in a quartz cuvette; the temperature and duration of reaction vary on different enzymes, substrates, and buffer composition (e.g., the appropriate time and temperature for Dp0100 alginate lyase are 65 °C in 3 min). Every increase of 0.1 in the A_{235} of /min equals one unit of enzyme activity. Finally, the amount of reducing sugar in the reaction solution is determined by DNS method [30]. The concentration of protein is measured by the Bradford assay [32].

5 Physical and Chemical Properties of Alginate Lyases

As previously mentioned, the CAZy database (Carbohydrate Active enZYme database, http://www.cazy.org/Polysaccharide-Lyases.html) classifies alginate lyases into 13 families based on analyzing hydrophobic clusters present in their primary structures, which include PL5, PL6, PL7, PL14, PL15, PL17, PL18, PL31, PL32, PL34, PL36, PL39, and PL41 [3]. Generally, PL5, PL6, PL14, and PL17 are polyM-specific, while PL7, PL15, and PL18 enzymes can degrade polyM and polyG blocks [33]. However, exceptions always happen; for example, Patl3640 and Pedsa0631 from PL6 family perform strict substrate specificity toward polyG [34, 35]. Although many alginate lyases are active at different levels at a wide range of pH (5–12) and temperature (20–80 °C), their maximal enzymatic activities have usually been observed from neutral to medium alkaline pH conditions, ranging around 7–8.5 [36–38], and at low temperature (mostly below 50 °C) [38–40]. However, several enzymes A1-II from *Sphingomonas* sp. A1 (PL5), and NitAly from *Nitratiruptor* sp. SB155–2 (PL7), exhibit the best activity under harsh conditions at 70 °C; interestingly, all of them have optimum acidic pH of around 6.0 [26, 41, 42]. The activities of alginate lyases are also influenced by ionic strength and species. For instance, the enzyme KJ-2 was inactivated when incubated in a buffer containing 2 mM EDTA; its activity can be recovered by adding a high concentration of $CaCl_2$ [40].

Alginate lyase molecules are composed of multiple modules and arranged in various 3D structures. The addition of other domains (e.g., S-layer homology domains) or modules (e.g., carbohydrate esterase module and carbohydrate-binding module) allows the enhancement of various properties of the enzymes, such as cellular binding capability and enzymatic activity [33, 43–45]. The 3D structures of alginate lyases vary among different families. PL5 and PL15 fold in $(\alpha/\alpha)_6$-barrel structure; PL6 adopts a parallel β-helix; PL7, PL14, and PL18 assume a jelly roll [43]. Changing the 3D structures of enzymes can affect the structure of active sites, leading to alterations or inactivation of enzyme function [46]. Many alginate lyases from marine sources exhibit great enhancement in their activities with the presence of a certain amount of NaCl. From the perspective of environmental adaptation, ocean water has a high NaCl content; therefore, marine alginate lyase exhibits an adaptive mechanism to high salt concentration where they live. Another possibility is that NaCl provides an charge effect, which reduces the interactions between water and alginate molecules, clearing the way for alginate to combine with enzymes [47] (Table 2).

6 Enzyme Degradation Mode

Polysaccharide lyases (PL, EC 4.2.2.-) are a large group of enzymes that breakdown uronic acid–containing polysaccharide chains through a β-elimination mechanism to produce an unsaturated hexenuronic acid residue and a new reducing end. The

Table 2 Physical and chemical properties of alginate lyase

Family	Enzyme Name	Organism	Substrate specificity	Optimum pH	Optimum temp	Thermo-stability	Source	Length (amino acids)	Reference
PL5	algL	*Azotobacter chroococcum* mcd 1	Alginate	7.5	30 °C	ND		372	[48]
	AlgA	*Pseudomonas* sp. E03	polyG, polyM, polyMG, alginate	8.0	30 °C	ND		364	[49]
	aly1	*Pseudomonas* sp. KS-408	polyM	8.0	20–30 °C	40 °C 30 min		373	[39]
	A1-I	Sphingomonas sp. A1	polyM, polyG	8.0	45 °C	ND		641	[26]
	A1-II	Sphingomonas sp. A1	polyG	8.0	70 °C	ND		641	[26]
PL6	*Bcel*PL6	*Bacteroides cellulosilyticus*	polyM, alginate	7.5	30 °C		Human fecal samples	890	[34]
	AlyGC	*Glaciecola chathamensis* S18K6	Alginate sodium	7.0	30 °C	ND	Marine bacterium	726	[36]
	aly	*Pseudomonas* sp. OS-ALG-9	D-mannuronan					398	[50]
	AlyF	*Vibrio splendidus* OU02	PolyG	7.5	30 °C	ND		538	[51]
	KJ-2 alyMG	*Stenotrophomas maltophilia* KJ-2	polyMG	8.0	40 °C	ND	Salted guts from hairtails	475	[40]
	Sven0074	*Streptomyces venezuelae*	polyMG	ND	ND	ND		486	[35]
	oalS6	*Shewanella* sp. Kz7	polyG, alginate, polyM	7.2	40 °C		Marine bacterium	770	[52]

(continued)

Table 2 (continued)

Family	Enzyme Name	Organism	Substrate specificity	Optimum pH	Optimum temp	Thermo-stability	Source	Length (amino acids)	Reference
PL7_1	PA1167	*Pseudomonas aeruginosa* PAO1	PolyMG	8.5	40 °C	42 °C 10 min	Wound	221	[53]
PL7_3	alyPG	*Corynebacterium* sp. ALY-1	PolyG	7.0	55 °C	40 °C 1 h	Sewage	224	[3, 4]
	Alg	*Streptomyces* sp. ALG5	PolyG	7.0	30 °C	ND	Seaweeds	223	[41]
	AlyA1	*Zobellia galactanivorans* DsiJT	PolyG	7.0	30 °C	ND	Red alga	248	[15]
PL7_5	A1m	*Agarivorans* sp. JAM-A1m	PolyMG, PolyG	10	30 °C	30 °C 15 min	Deep-sea sediment	289	[16]
	AlyA	*Klebsiella pneumoniae* sub sp.	PolyG	7.6	25 °C	ND	Enterobacteria	285	[54]
	AlgMsp	*Microbulbifer* sp. 6532A	PolyM	8.0	30 °C	ND	Drifting macroalgae	265	[55]
	AlyPI	*Pseudoalteromonas* sp. CY24	PolyM, PolyG	7.0	40 °C	40 °C 1 h	Seawater	354	[56]
	Alg2A	*Flavobacterium* sp. S20	PolyG	8.5	45 °C	50 °C 1 h	Seaweed factory	288	[57]
	AlyDW11	Uncultured bacterium	PolyM	7.0	45 °C	55 °C 2 h	Abalone gut microflora	323	[58]
	AlyL2	*Agarivorans* sp. L11	PolyM	8.6	40 °C	35 °C 1 h	Coastal zone	273	[59]
	AlyA5	*Zobellia galactanivorans* DsiJT	PolyMG	7.0	30 °C	ND	Red alga	314	[60]

PL7	FlAlyA	*Flavobacterium* sp. UMI-01	PolyM	7.7	55 °C	47 °C 30 min	Seashore	288	[22]
	NitAly	*Nitratiruptor* sp. SB155–2	PolyM, PolyG	6.0	70 °C	50 °C 6 h	Deep-sea vents	219	[42]
	A1-II'	*Sphingomonas* sp. A1	PolyG	7.5	40 °C	45 °C 5 min	Soil	237	[61]
	A9mT	*Vibrio* sp. A9m	PolyM	9.0	30 °C	45 °C 15 min	Deep-sea bacteria	263	[62]
	Alyl1	*Agarivorans* sp. L11	PolyM	8.5	45 °C	55 °C 1 h	Coastal zone	349	[59]
	AlyVI	*Vibrio* sp. QY101	PolyM, PolyG	7.5	40 °C	45 °C 10 min	Decaying *Laminaria*	273	[63]
	AlgB	*Vibrio* sp. W13	PolyM, PolyG	8.0	30 °C	35 °C 30 min	Sea water	287	[64]
PL14	vAL-1	*Chlorella virus* CVK2	PolyG	10	37 °C	ND	Natural waters	244	[65]
	AkAly30	*Aplysia kurodai*	PolyM	6.0	55 °C	45 °C 20 min	Marine gastropod	295	[66]
PL15	Atu3025	*Agrobacterium tumefaciens* C58	PolyMG, PolyM, PolyG	7.3	30 °C	ND		776	[38]
	A1-IV'	*Sphingomonas* sp. A1	PolyM, PolyMG, PolyG	8.0–8.5	50 °C	50 °C for 10 min in 50 mM Tris–HCl, pH 7.5		812	[37]
	A1-IV	*Sphingomonas* sp. A1	Trisaccharide of A1-I, A1-II reaction, Tetrasaccharide of A1-II-reaction	7.5–8.5	37 °C	45 °C for 10 min in 20 mM Tris–HCl (pH 7.5)		761	[67]

(continued)

Table 2 (continued)

Family	Enzyme Name	Organism	Substrate specificity	Optimum pH	Optimum temp	Thermo-stability	Source	Length (amino acids)	Reference
	OalA	*Vibrio splendidus* 12B01	PolyM, PolyG	6.5	16 °C	35 °C 13 min	Coastal waters, intestines crustaceans or mammals	692	[68]
	OalB	*Vibrio splendidus* 12B01	PolyMG	7.0	30 °C		Coastal waters, intestines crustaceans or mammals	735	[68]
	OalC	*Vibrio splendidus* 12B01	PolyM	7.5	35 °C		Coastal waters, intestines crustaceans or mammals	717	[68]
PL17	AlgSH17	*Microbulbifer* sp. SH-1	Sodium alginate, polyM and polyG	7.0	30 °C		Coastal soil	747	[69]
	alyII	*Pseudomonas* sp. OS-ALG-9						713	[70]
	Alg17C	*Saccharophagus degradans* 2–40	Sodium alginate, oligoalginates, PolyM, PolyG	6.0	40 °C			733	[71]
	oalS17	*Shewanella* sp. Kz7	Alginate, PolyM, PolyG	6.2	50 °C			763	[72]
	MJ-3	*Sphingomonas* sp. MJ-3	Alginate, AOSs, PolyM, PolyG, PolyMG	6.5	50 °C			731	[73]
PL18	aly-SJ02	*Pseudoalteromonas* sp.SM0524	PolyM, PolyG	8.5	50 °C	50 °C 20 min	Marine kelp	227	[27]
	alyA	*Pseudoalteromonas atlantica* AR06	Alginate, PolyM, PolyG	7.4	40 °C		Coastal water	391	[74]

	Aly	*Alteromonas* sp. H-4	PolyM, PolyG, PolyMG PolyMG, Alg-Na	7.5	30 °C		Laminaria fronds	398	[18]
	Aly-SJ02	*Pseudoalteromonas* sp. SM0524	PolyG, PolyM	8.5	50 °C	41 min 40 °C, 20 min 50 °C	Marine rotten kelp	400	[27]
PL31	PsAly	*Paenibacillus* sp. FPU-7	Alg-Na, PolyM	7.0–7.5	65 °C	47 °C 15 min, 57 °C 3 min, 67 °C 40 s	Air, rhizosphere, soil, and marine environments	337	[75]
PL36	Aly36B	*Chitinophaga* sp. MD30	Alg-Na, PolyM, PolyG	9	35 °C			882	[76]
PL39	Dp0100	*Defluviitalea phaphyphila*	Alginate	5.8	65 °C	65 °C 45 min		1825	[77]

CAZy database (http://www.cazy.org/Polysaccharide-Lyases.html) presents a classification of PL enzymes in families and subfamilies based on amino acid sequence similarities, as well as their structural features [43]. Among them, alginate lyases can degrade the glycoside linkages in alginate via a β-elimination mechanism [14, 78]. Endolytic alginate lyases can cleave alginate into the unsaturated alginate oligosaccharides (uAOSs), which contain double bonds at the C-4 = C-5 positions of the non-reducing ends [42, 79]. The unsaturated sugar unit is often indicated with the symbol Δ, such as ΔGG and ΔMG, and the degree of polymerization (DP) of uAOSs is usually 2 ~ 5. Exolytic alginate lyases can produce monosaccharide products, such as G, M, and mainly Δ, from sugar chains [42, 80]. Besides, alginate lyases can be categorized into G-specific, M-specific, and bifunctional subsets. Screening of bifunctional alginate lyases has focused on substrate specificity, activity, and product inhibition [14]. However, the substrate-degrading modes and the degradation patterns of uAOSs are poorly understood. A novel PL-39 alginate lyase, Dp0100, was recently isolated from the marine thermophile *Defluviitalea phaphyphila*. The enzyme displays broad specificity endotype activity; interestingly, its catalytic domain structure exhibits three regions that strongly resemble those found in the exotype lyase families PL15 and PL17 [77].

Many alginate lyases, particularly thermophilic ones, exhibit a multimodular structure, comprising catalytic modules, substrate-binding modules, and noncatalytic regions [1]. The structures of alginate lyases, including the new members of the PL-6 and PL-39 families, have been characterized, highlighting the roles of catalytic modules, substrate-binding residues, and active-site residues in the degradation of alginate [35, 47, 52]. The catalytic modules play a crucial role in recognizing, binding, and degrading processes through biochemical characterization. The substrate-binding modules primarily recognize specific substrate, while the noncatalytic regions are incapable of degrading alginate but exhibit various effects on enzymatic properties. In case of the guluronate lyase Aly5 from *Flammeovirga* sp. strain MY04 [47] and the bifunctional enzyme Aly2 from *Agarivorans* sp. strain L11 [59], the noncatalytic regions are able to enhance the efficiencies of binding and degradation on small-sized AOSs substrates. The removal of the noncatalytic region in Aly2 enzyme significantly increased the M-specific degradation efficiency [59]. However, unlike Aly2, Aly5 from *Flammeovirga* sp. strain MY04 did not show such a phenomenon [47]. Furthermore, the removal of either the substrate-binding module or the noncatalytic region can potentially impact the thermostability of alginate lyases, leading to a subsequent reduction in enzymatic activity under elevated temperatures. Consequently, both the substrate-binding module and the noncatalytic region are vital for the proper functioning and application of multimodular alginate lyases (Fig. 2).

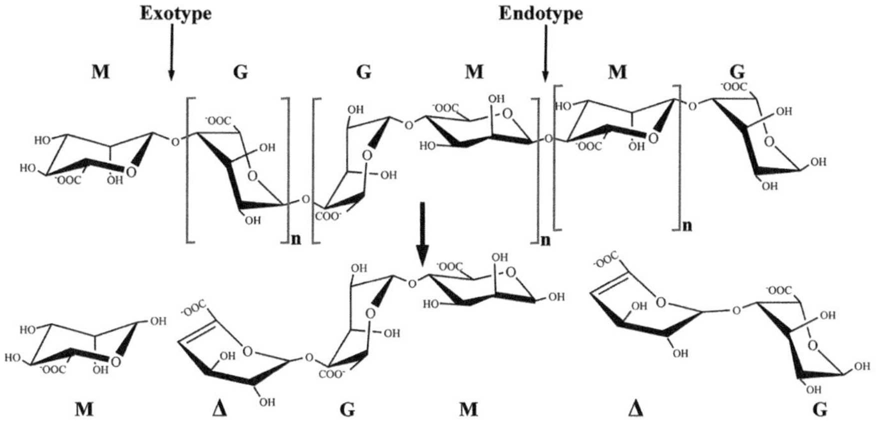

Fig. 2 Schematic representation of the structure of alginate and the action mode of alginate lyases

7 Structure and Catalytic Mechanism of Alginate Lyase

To date, a total of 35 3D structures of alginate lyases have been obtained, belonging to various families including PL5, PL6, PL7, PL14, PL15, PL17, PL18, PL31, PL32, PL34, PL36, and PL39. These structures can be categorized into four distinct types: the parallel β-helix fold, $(\alpha/\alpha)_n$ fold, β-jelly-roll fold, and the hybrid structure consisting of $(a/a)_n$ toroid plus β-jelly roll fold. Specifically, PL5 adopts the $(\alpha/\alpha)_n$ toroid fold, while PL6 and PL31 exhibit the parallel β-helix fold. On the other hand, PL7, PL14, PL18, and PL36 display the β-jelly-roll fold, while the hybrid structure of $(a/a)_n$ toroid plus β-jelly roll fold is observed in PL15, PL17, and PL39 families.

7.1 Alginate Lyases Belong to $(\alpha/\alpha)_n$ Toroid Fold Structure

The PL5 alginate lyases display $(\alpha/\alpha)_n$ circular structure, which consists of one or multiple antiparallel α-helices. Viewed from the top, the spiral toroids of α-helix toroid structure were arranged in a counterclockwise fashion [81]. The majority of the PL5 alginate lyases comprise a single domain with a $(\alpha/\alpha)_n$ ring fold; and the 3D structures of some of these enzymes have been reported (Fig. 3). Interestingly, certain members deviate from this $(\alpha/\alpha)_n$ rule. Specifically, the crystal structure of A1-III from *Sphingomonas* sp. A1 reveals a α6/α5 barrel structure, providing insights into four key catalytic amino acids: Asn191, His192, Arg239, and Tyr246 [81]. Another member, PA3547, shares a similar structure to A1-III; however, its properties have remained unknown. Other exception comes from the Smlt1473 enzyme from *Stenotrophomonas maltophilia* K279a, which is composed of 13 α-helical segments (α1 to α13) and two β-strands (β-sheet1 and β-sheet 2); this conformation is organized into an $(\alpha/\alpha)_n$ incomplete toroidal fold [82].

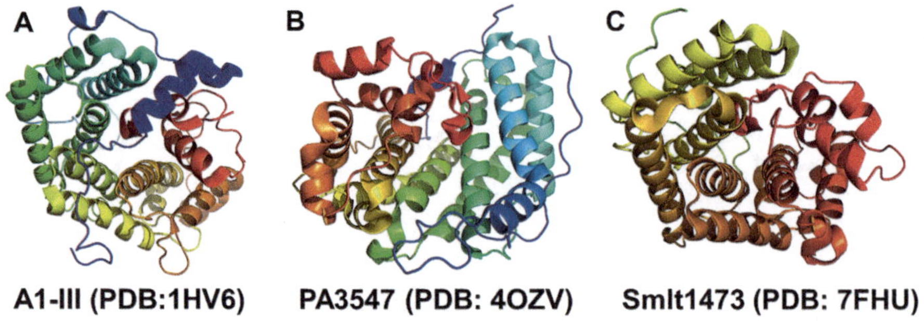

Fig. 3 The structure of alginate lyases that adopt $(a/a)_n$ toroid fold. (**a**) A1-III (PDB:1HV6) from *Sphingomonas* sp. A1. (**b**) PA3547 (PDB: 4OZV) from *Pseudomonas aeruginosa* PAO1. (**c**) Smlt1473(PDB: 7FHU) from *Stenotrophomonas maltophilia* K279a (https://www.rcsb.org/)

7.2 Alginate Lyases Belong to Parallel β-Helix Fold Structure

The main structure of the parallel β-helix fold consists of multiple β-sheets and turns (T); every single repetitive unit coil arrangement is described as β-sheets 1-T1-β-sheets 2-T2-β-sheets 3-T3 [31]. Figure 4 depicts the parallel β-helix fold structure of six alginate lyases, including five members from the PL6 and one member from the PL31 family. AlyGC, isolated from the marine bacterium *Glaciecola Chathamensis* S18K6T, presents the first crystal structure of the PL6 family of brown alginate lyases. The enzyme mainly consists of β-helix and catalyzes the reactions via arginine and lysine acid-base pairs [36]. Bcel PL6 (*Bacteroides cellulosilyticus*, Bcel PL6) was excavated from the genome of the gut bacterium *Bacteroides cellulosilyticus*. The structure of recombinant Bcel PL6 was resolved to 1.3 Å resolution by X-ray crystallography, revealing a single-domain, monomeric parallel β-helix characteristic of an alginate-converting parallel β-helix. Its conserved catalytic site residues include Lys-249, Arg-270, and His-271, etc. In addition, the enzyme's specific substrate is PolyM, which is determined by the presence of Asp153, Tyr155 and Trp184; and unlike other PL6 alginate lyases, Bcel PL6 was inactive on polyG [34]. Another enzyme, the BcAlyPL6 came from the human gut microbe *Bacteroides clarus*. It contains two domains, namely the N-terminal domain (NTD) and C-terminal domain (CTD), which share structural similarities with AlyGC; particularly, the domains of both enzymes exhibit a right-handed parallel β-helix fold. The substrate-binding function is mainly contributed by NTD, while the CTD is involved in substrate turnover and formation of a closed catalytic channel [83].

PL6 can be further divided into subgroups according to various characteristics. Based on most of action, the PL6 is categorized into three subfamilies PL6_1 to PL6_3, with PL6_1 is the most abundant one. The crystal structures of Patl3640 from *Pseudoalteromonas atlantica* T6c, a member of PL6_1, were investigated, revealing the conserved structure within its substrate-binding site, which is organized in the right-side parallel β-helix fold shared by all of thePL6_1 enzymes [84].

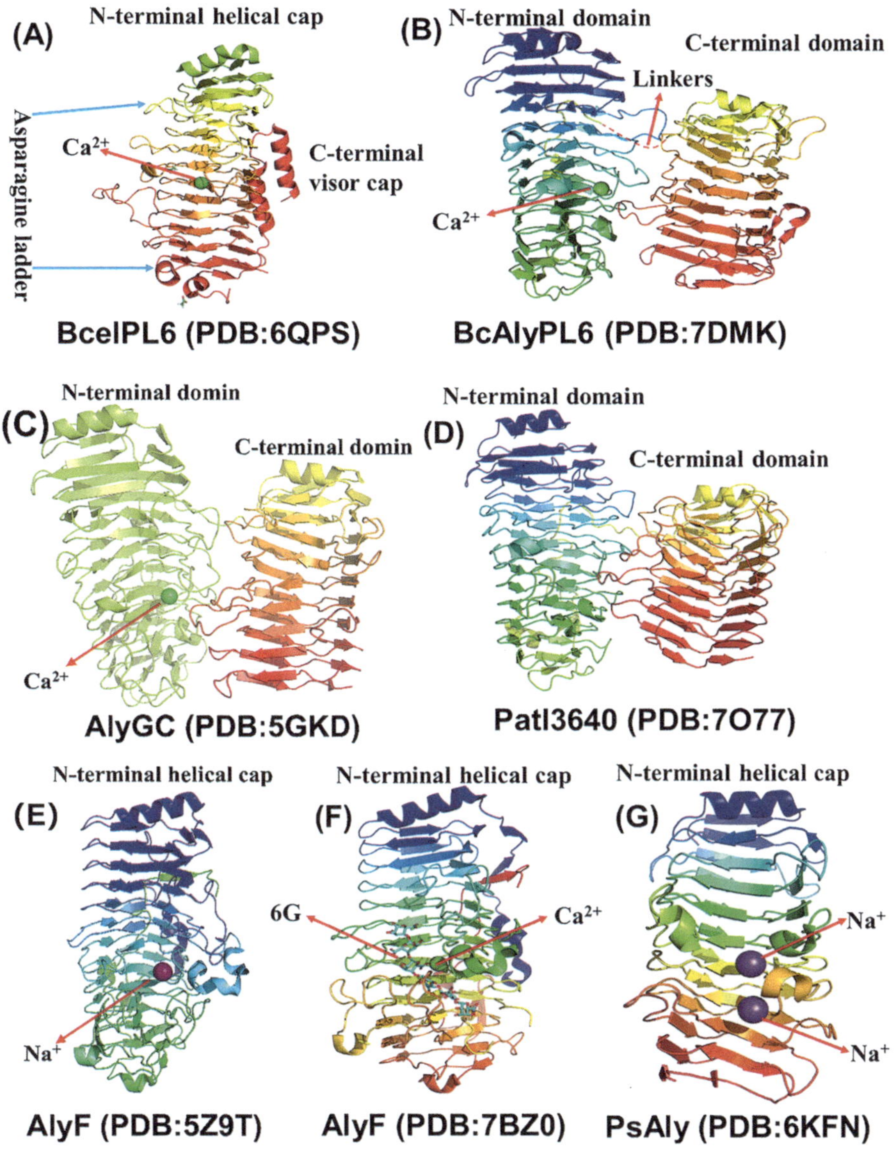

Fig. 4 The 3D structure of alginate lyases that adopt parallel β-helix fold. BcelPL6, BcelPL6, AlyGC, Patl3640, and AlyF are classified as PL6 family. PsAly is the first crystal structure of the PL31 family. (**a**) The structure of BcelPL6 (PDB:6QPS) from *Bacteroides cellulosilyticus* DSM 14838. (**b**) The structure of BcAlyPL6 (PDB:7DMK) from *Bacteroides clarus*. (**c**) The structure of AlyGC from *Vibrio* OU02. (**d**) The structure of Patl3640 from *Bacteroides cellulosilyticu*. (**e**) The structure of AlyF from *Paenibacillus* sp. str. FPU-7. (**f**) The complex structure of alginate lyase AlyF-OU02 with G6 from *Vibrio splendidus*. (**g**) The structure of PsAly from *Paenibacillus sp*. str. FPU-7 (https://www.rcsb.org/)

On the other hand, PL-6 alginate lyases can be distributed into two categories based on the number of domains: single-domain and dual- domain. AlyF possesses unique characteristics compared to other PL-6 enzymes, including: (1) a Ca^{2+}-independent catalytic mechanism, and (2) polyG-specific cleavage specificity with trisaccharides as the major products. The structures of AlyF and its complexes described herein reveal the structural basis of these unique features and substrate-binding mechanisms [51].

PsAly, belonging to the PL31 family, shares similarities with certain PL6 members as it also possesses a parallel β-helix fold. However, in contrast to AlyGC and BcelPL6, which have one metal ion binding site [36], PsAly contains 10 parts of the β-stranded coil [76] and two sodium-binding sites [76] within its structure.

7.3 Alginate Lyases Belong to β-Jelly-Roll Fold Structure

There are three PL families that belong to the β-jelly-roll fold, namely PL7, PL14, and PL18. Among these, PL7 is particularly well-studied and is considered one of the most extensively researched families among the 12 PL families. The superposition of the four PL7 alginate lyase structures showed that the catalytic region of the PL7 family is all conserved, while their loop regions are different (Fig. 5).

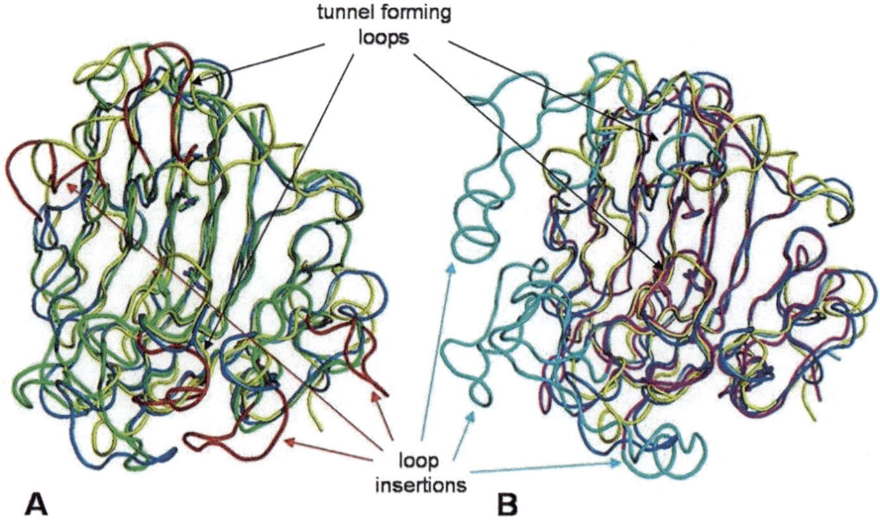

Fig. 5 Tube representation of the superimposition of the three-dimensional structure of PL7 alginate lyases. (**a**) superimposition of AlyA1, AlyPG, and A1-II′. (**b**) Superimposition of AlyA5 and AlyPG from *Corynebacterium* sp. ALY-1 (blue tube, PDB code 1UAI) and A1-II′ from *Sphingomonas* sp. A1 (yellow tube, PDB code 2ZAA). Loops that have different configurations or large loop insertions with respect to 1UAI and 2ZAA are highlighted by the arrows and colored in red for AlyA1PL7 and cyan for AlyA5 (https://www.rcsb.org/)

Site-directed mutagenesis reveals that the loop region undergoes distinct conformation changes during both the catalytic reaction and the entry and exit of the substrate. This flexibility of the loop region is critical for enabling substrate access to the catalytic center of alginate lyase. If the flexibility of the two loops is restricted, enzyme's activity will be lost, and it may also impact the endo- and exo-cutting mechanisms [60]. It was found that the amino acids responsible for the specific degradation of this family are all distributed on the lid loop around the catalytic center, rather than the generally considered conservative key amino acids [60]. Most of these enzymes consist of single catalytic domain. On the other hand, some other members of PL7 are multidomain enzymes, which consist of an N-terminal carbohydrate-binding module (CBM) and a C-terminal PL7 family alginate lyase module. AlyB from *Agarivorans* SP. L11 contains of the CBM13, while AlyQ from *Persicobacter* sp. CCB-QB2 consists of CBM32. In AlyQ, there is not only a CBM but also a CTD (catalytic domain). The CBM32 domain did not affect AlyQ's activity; however, it was able to bind and cleave alginate, indicating that the CBM32 domain recognizes substrate. Interestingly, CBM32 and the linker did not significantly enhance the catalytic activity but acted as a "fulcrum" during trisaccharide release (Figs. 6, 9 and 10).

AkAly30 in the PL14 family is a PolyM-specific enzyme whose substrate specificity is mainly controlled by the binding mode of the amino acid at subpoint +1. The hydroxyl group of Tyr140 is located near the C5 atom of the mannuronic acid residue, while the hydrogen atom at the C5 position is far away from the hydroxyl group of Tyr140, suggesting that the β-elimination of the G residue at subsite +1 cannot proceed efficiently. Therefore, it cannot catalyze the reaction of PolyG [66]. In the PL18 family, aly-SJ02 is a bifunctional enzyme with high activity toward PolyM and PolyG. Substrate recognition in aly-SJ02 occurs primarily at subsites +1 and +2, but some secondary interactions occur at subsites −1 and +3 [85]. Structural analysis revealed that aly-SJ02 and Aly272 (Fig. 7) contain a metal ion Ca^{2+} in their structures, yet it is located on the back of the catalytic center. From the perspective of spatial distance, as Ca^{2+} is far from the catalytic center, it is speculated that Ca^{2+} does not directly participate in the catalytic process. Actually, there are many alginate lyase reactions that require the presence of metal ions; however, the degree of their involvement in these reactions varies. Unlike the indirect role of Ca^{2+} in aly-SJ02 and Aly272, [36]; the structural analysis of AlyGC of the PL6 family found that each monomer chelates a Ca^{2+} in the catalytic center. Therefore, it can be explained that in addition to being chelated on the protein by coordination bonds, Ca^{2+} may also bind to the carboxyl group of the substrate through positive and negative charge interactions. Ca^{2+} participates in the activation of Cα-H at the +1 position and in the catalytic reaction process.

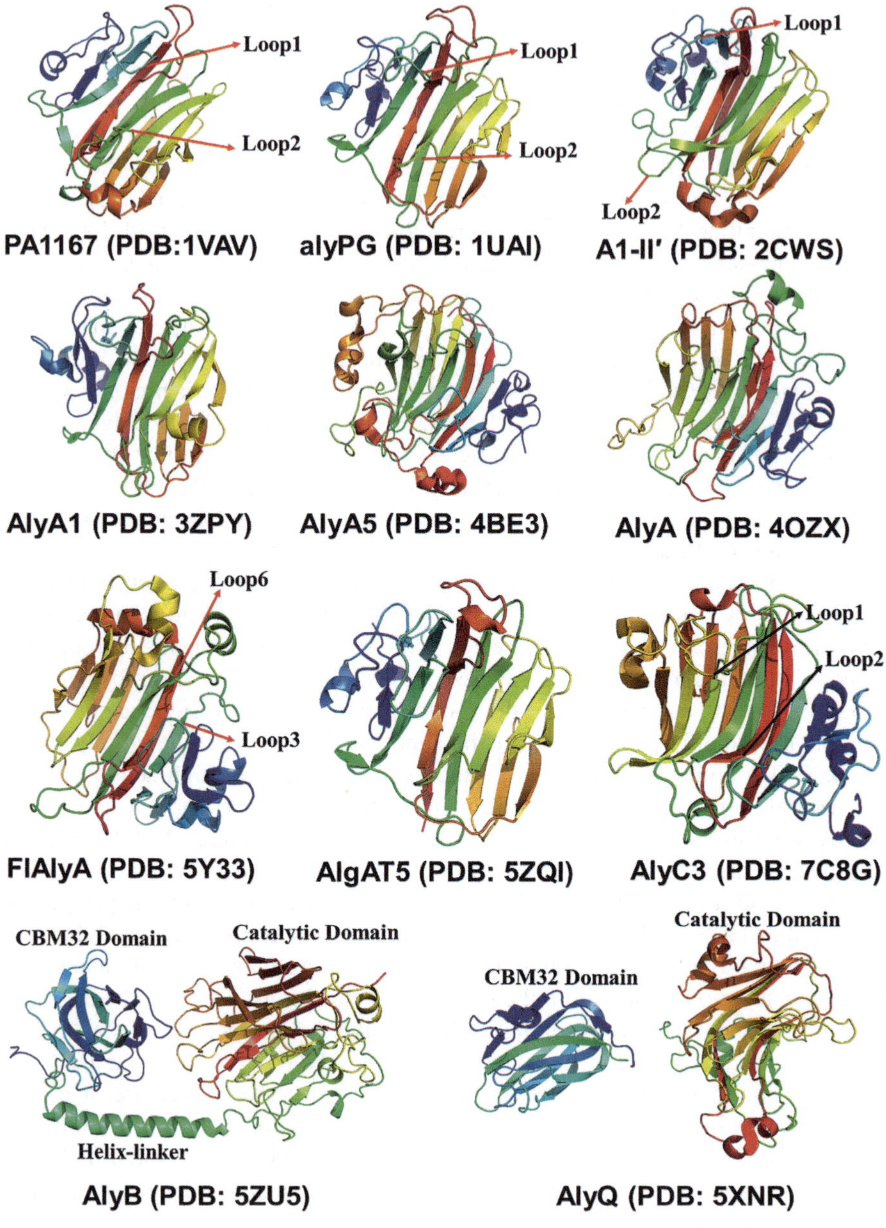

Fig. 6 The three-dimensional structure of PL7 alginate lyases (https://www.rcsb.org/)

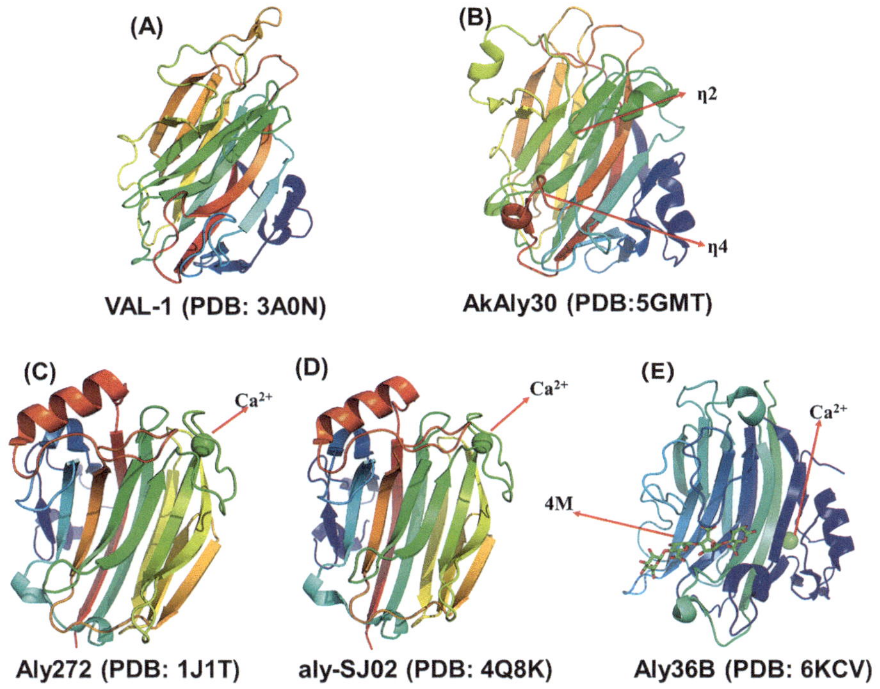

Fig. 7 The three-dimensional structure of PL14, PL18 and PL36 alginate lyases (https://www.rcsb.org/)

7.4 *Alginate Lyases Belong to the Hybrid Structure $(a/a)_n$ Toroid Plus β-Jelly Roll Fold*

As aforementioned, enzymes belonging to PL15, PL17, and PL39 families exhibit a hybrid structure $(a/a)_n$ toroid plus a β-jelly roll fold. As shown in Fig. 8a, Atu3025 from *Agrobacterium fabrum* str. C58 contains three domains, including the NTD, the center domain, and the CTD, which adopt a β-jelly roll fold, an α barrel fold, and a β-jelly roll fold, respectively. Alg17c, an enzyme derived from *Saccharophagus degradans* 2–40, contains two domains [86]. The NTD adopts a $(a/a)_6$-loop fold, while the CTD shares a β-jelly roll fold. Alya3 from marine flavobacteria *Zobellia galactanivorans* DsijT is composed of two domains. The NTD adopts an $(a/a)_6$-loop fold, whereas the CTD has a β-jelly roll fold [86]. Each domain contains metal ions—two magnesium ions in the NTD and two calcium ions in the CTD [86]. Another enzyme, the Dp0100 [77] from the marine thermophile *Defluviitalea phaphyphila*, was identified as the first PL39 alginate lyases. Its NTD folds in an α barrel model, while both the center domain and the CTD exhibit a β-jelly roll fold. Besides, there is one magnesium ion in its NTD, whereas its center β-jelly roll domain contains two calcium and one manganese ions.

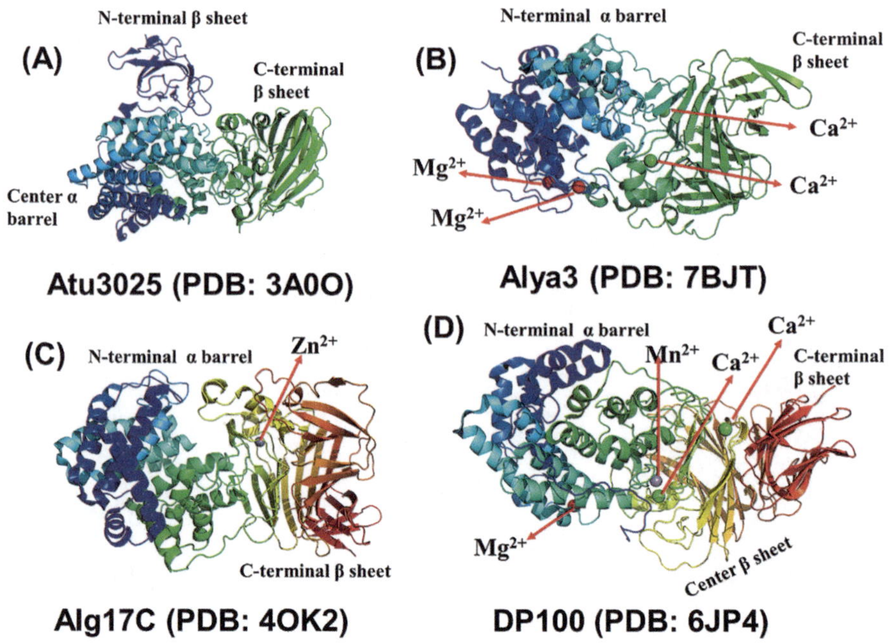

Fig. 8 The three-dimensional structure of PL15, PL17, and PL39 alginate lyases (https://www.rcsb.org/)

7.5 Catalytic Mechanism of Alginate Lyase

Alginate lyase can break the alginate polysaccharide chain via β-elimination reaction, and this catalytic mechanism can be explained by Gacesa theory: ① The neutralizing group neutralizes the carboxyl negative charge; ② The C5 proton transfer requires amino acid residues as proton acceptor and donor; ③ The transfer of electrons on the carboxyl group leads to the formation of a double bond between C4 and C5 and the rupture of the glycoside bond [87]. According to different residues neutralizing the C5 carboxyl and to the catalytic bases and acids, there are six groups of catalytic reacions of alginate lyases including metal-assisted β elimination, His/Tyr β-elimination, Tyr'/Tyr β-elimination, Tyr/Tyr β-elimination, Lys/lys-dependent β-elimination, and H_2O assisted β-elimination; in addition, another hybrid catalytic group describes enzymes that concurrently adopt at least two of six mentioned elimination mechanisms [88].

Metal-assisted β-elimination: Both alginate lyase families PL6 [23] and PL31 [66] adopt metal residues to neutralize the C5 carboxyl. A typical example of this mechanism is AlyGC from *Glaciecola chathamensis* S18K6T [36]. The complex structure of its inactive mutant R241A with tetramannuronic acid (M4), whose structure is mainly composed of β-helix and contains arginine and lysine as catalytic acid–base pair, contains Ca^{2+}. Ca^{2+} neutralizes the carboxylic group of its substrates for easier removal of the C-5 prone. The abstraction of a prone follows this from C5 by catalytic base Lys220, and the proton donation of Arg241 catalytic acid to create a C4–5 double bond, and an unsaturated monomer in the non-reducing end of the oligomer is generated as a result of glycosidic bond elimination [36, 88]. Besides, although the dimeric R241A-M4 contains larger α and smaller β active centers, conformational analysis of its structure indicated that only one substrate molecule can be accommodated in one of the two active centers at a time [36].

His/Tyr β-elimination: Enzymes of PL7 [89, 90], PL15 [17], and PL39 [77] families employ his/tyr β elimination for their catalytic action. In this group, positively charged, conserved Tyr and His residues act as neutralizers, catalytic acid, and base. The accurate activity of enzymes on their substrates is conducted via tightly joint operation of key residues. In the AlyC3 mutant-dimannuronate complex, the accurate position of substrates is obtained via R82 and Y190, while R78 and Q125 neutralize the negative charge of acidic groups of the substrate at þ1 site; abstraction and transfer of proton are then conducted by H127 and Y224, respectively [88, 90].

Tyr'/Tyr β-elimination: Alg17c from *Saccharophagus degradans* of PL17 family, which has an imperfect α6/α6 barrel at the N-terminus and a β-sheet at the C-terminus, uses this mode of action for catalysis. Four residues N201 and H202, Y450, and Y258, act as neutralizers, catalytic bases, and acids [86, 88].

Tyr/Tyr β-elimination: This mechanism refers to an alginate lyase that uses only one type of residue as a catalytic base and acid. For instance, A1-III-DMMM of the PL5 [81] and Aly-SJ02 of the PL18 [85] family use Y246 and Y353 as both catalytic acid and base for β-elimination [88].

H_2O-assisted catalytic mechanism: A member of PL6 family, AlyF from *Vibrio* OU02 uses a H_2O molecule to neutralize the C5 carboxylic group. Then, the catalytic base Arg293 abstracts the proton to generate a carboxylate dianion intermediate, followed by the proton donation of general acid Lys727 to the glycosidic oxygen [77, 88].

Lys/Lys-dependent catalytic mechanism: The mechanism is characterized by Lys143 residue acting as both general base and acid. So far, only one enzyme, Aly36B from *Chitinophaga* sp. MD30 belonging to PL36 family, employs Lys-dependent catalysis. Arg169, Tyr185, and Tyr187 conduct the neutralization of the negative charge. Dong, et al. suggested that as these four amino acid residues are conserved in the sequence of PL36 family, other members of this group might also adopt this mode of action [76, 88].

Hybrid catalytic mechanism: Unlike the afore-described groups, the β-elimination mechanism of FlAlyA (*Flavobacterium* sp. UMI-01) varies toward different substrates. Firstly, Arg74 and Gln122 play the role of neutralizers for negative charge stabilization. Next, when acting on the G unit of alginate, His/Tyr β elimination is employed, where His 124 and Tyr 239 act as a catalytic base and acid, respectively. In contrast, when the M unit of alginate is at þ1 subsite, Tyr239 acts as both a catalytic base and acid for the transfer of proton to oxygen, according to Tyr/Tyr β elimination mechanism [88, 91] (Figs. 9 and 10).

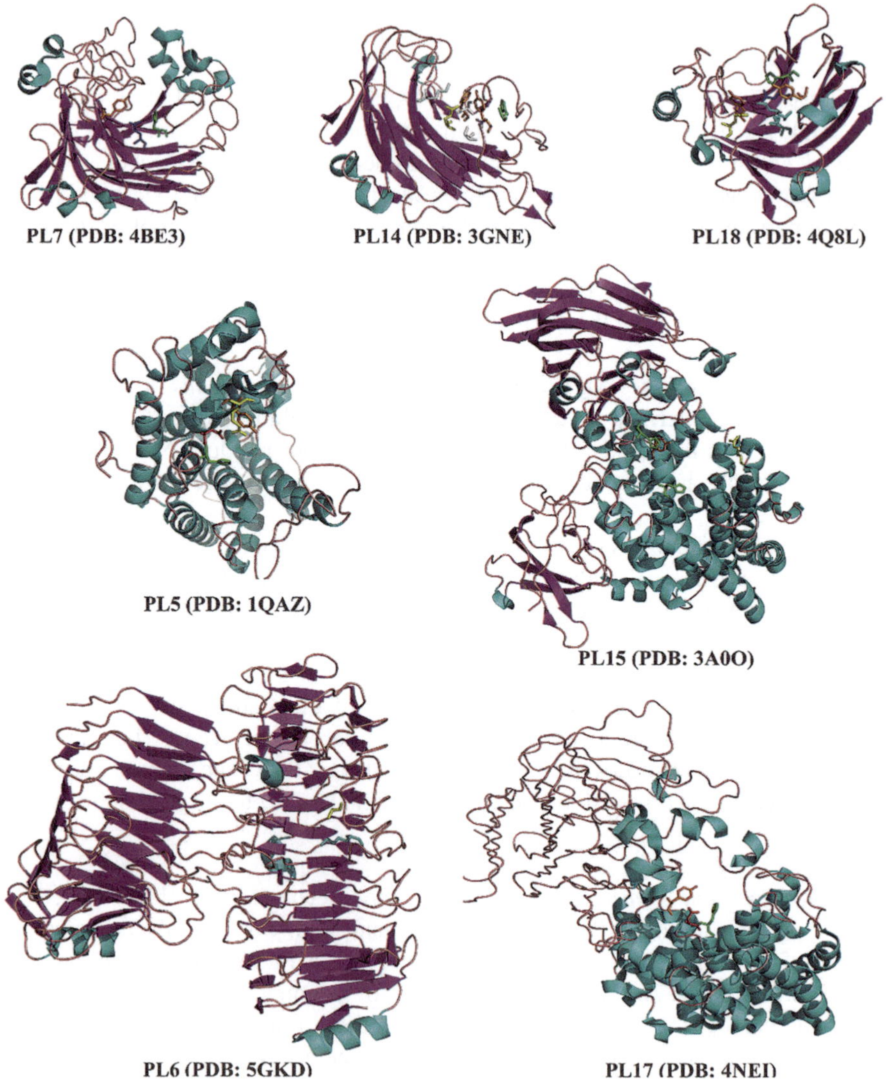

Fig. 9 Overall structures and active sites of alginate lyases (https://www.rcsb.org/)

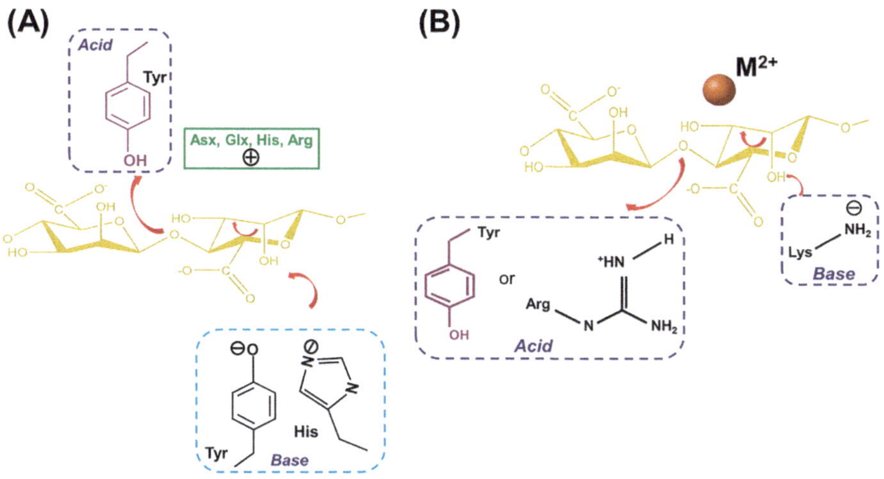

Fig. 10 Schematic diagram of the His/Tyr catalytic mechanisms of alginate lyases. (**a**) Catalytic mechanisms of alginate lyases except for PL6. Asx (Asp or Asn), Glx (Glu or Gln), His, or Arg neutralizes the acidic carboxyl group; Tyr functions as a Brønsted acid, and Tyr or His functions as a Brønsted base. (**b**) Divalent metal ion–participated β-elimination catalytic mechanisms

8 Enzyme Structural Modification and Molecular Modification

To enhance the industrial applicability of alginate lyases, a few strategies for enzyme modification, such as point (or random) mutation, conserved region recombination, and module deletion, have been applied. Many types of research focus on the thermostability of alginate lyases. For multi-modular enzymes, deletion of noncatalytic modules could increase alginate lyases' thermostability, since catalytic modules have a more compact structure than the full-length protein, which resists the effects of thermal degeneration [92]. However, in some cases, deletion of noncatalytic module decreases the thermostability of enzymes. For instance, an alginate lyase Aly7B shows a poor thermal stability at 35 °C [92]. Deletion of the noncatalytic module (amino acids R2-Y181) can maintain the activity of Aly7B-CDII at 35 °C [92]. It is very important to verify whether catalytic modules have intramolecular interaction with noncatalytic modules. If not, the noncatalytic module has a low influence on enzyme activities and thermostability. Thus, analyzing the interaction between the catalytic module and noncatalytic modules is necessary.

Disulfide bonds can increase thermostability by increasing the structural rigidity and reducing the entropy of the protein unfolding process [93]. Thus, adding one or several disulfide bonds into alginate lyases is a way to improve the thermostability and enzyme activity at elevated temperatures. The position of the disulfide bond is a key factor for alginate lyase thermostability. A comprehensive analysis, including protein homology modeling comparing the structural information of target protein with thermostable alginate lyases, is the first step. In the case of PyAly from *Pyropia yezoensis*, introducing two cysteines into PyAly based on homology modeling using *Pseudomonas aeruginosa* alginate lyase PA1167 as the template enhanced its heat stability [42].

Also, the thermostability of thermophilic alginate lyases could be improved by adding metal ions [94]. A new thermophilic alginate lyase AlgAT0 from *Defluviitalea phaphyphila* [94]. AlgAT0 has higher optimal temperature and thermal stability than other enzymes in the presence of Ca^{2+} ions. Based on a sequence alignment with other PL6 family alginate lyase, they found that AlgAT0 has a higher affinity with Ca^{2+} ions than other alginate lyase. Among the four conserved calcium-binding sites, one substitution of aspartate for glutamate in AlgAT0 stimulated Ca^{2+} affinity [94].

Enzyme immobilization technology can be defined as the restriction of the enzyme to a solid support or matrix that is different from the presence of the substrate and product. There have been many immobilization methods, including adsorption, covalent binding, embedding, and cross-linking. The performance of the carrier greatly affects immobilization. Good support needs to include many characteristics, such as being cheap, easy to obtain, and good mechanical strength. Based on the chemical composition, the carrier can be divided into inorganic, organic (divided into the natural source and synthetic carrier), and without a carrier.

Immobilized enzymes, unlike free enzymes, can be reused and are generally more stable, gaining more and more popularity. There are several studies on the immobilization of alginate lyase (Table 3). Li et al. (2019) constructed an immobilized alginate lyase using chitosan nanoparticles [95]. Although the optimum pH and temperature of the enzyme did not change after immobilization, the pH and thermal stability were greatly improved. Cross-linked enzyme aggregate is a carrier-free enzyme immobilization technology. The alginate lyase obtained better thermal stability [96].

9 Extraction and Purification Methods

Like all enzymes, purifying alginate lyases is critical to learn how they work and to manipulate their catalytic activity [3]. Cells secret the mixture of many different metabolites and enzymes that may directly or indirectly interact with alginate lyase and its substrates and products. Only when alginate lyase is purified to a grade where the presence of other enzymes cannot be detected and no other molecules interfere with the reactions it catalyzes, can we learn about its functional and regulatory aspects [101]. Furthermore, enzyme concentration plays an important role in all enzymatic reactions; thus, only when the enzyme of interest is of high purity, can the manipulation of its amount and its combination with cofactors be easy and accurate [102]. To satisfy aforementioned research requirements, the alginate lyase products should reach certain criteria, including high purity degree, high enzyme recovery activity, and reproducibility [103].

The separation process design is based on size/mass, physicochemical properties, binding affinity, and biological activity of alginate lyases. It generally

Table 3 Immobilization of alginate lyase

Source of enzyme	Immobilized support	Optimal pH of free / immobilized enzyme	Optimal temperature (°C) of free / immobilized enzyme	pH stability	Temperature stability	Reference
Aly08 from *Vibrio* sp. SY01	Chitosan nanoparticles	8.0/8.0	45/45	The free enzyme lost 60% of its initial activity after exposure at pH 8.0 for 40 min, while the immobilized one retained 76.8% even after 1 h	The thermal stability of immobilized Aly08 is better than that of free enzyme.	[97]
From *flavo bactierium*	Cross-linked enzymes aggregate of alginate lyase	6.3/6.3	70/70	Not reported.	The thermal stability of immobilized enzyme was higher than the native one.	[98]
AlgL17 from *Microbulbifer* sp. ALW1	Fe_3O_4 nanoparticles oxide	8.0/9.5	35/35	The free and immobilized AlgL 17 retained over 40% and 18% residual activity after exposure at pH 9.5 for 1 h, respectively.	The immobilized enzyme retained 97.8% residual activity after incubation at 30 °C for 1 h and the free one maintained only 16% residual activity after 30 min.	[99]
AlyPL6 from *Pedobacter hainanensis* NJ-02.	Mesoporous titanium oxide particles	10.0/Not reported	45/Not reported	Not reported	The free and immobilized AlyPL6 maintained 9.6% and 83.7% initial activity after incubation at 45 °C for 1 h, respectively.	[100]
Alg L from *Sphingomonas* sp.	Cellulose acetate ultrafiltration membrane	Not reported	Not reported	Not reported	The immobilized and free Alg L retained 80% and 20% residual activity after heating at 37 °C for 21 d.	[100]

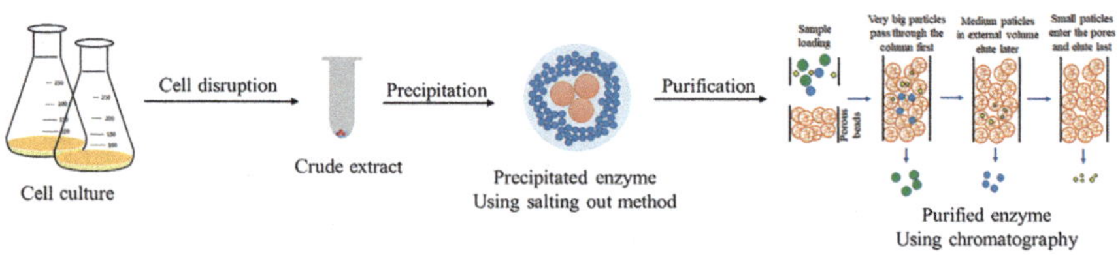

Fig. 11 Main steps of enzyme separation procedure

comprises three stages: (1) crude extraction, (2) precipitation, and (3) purification [104] (Fig. 11). The first stage aims to harvest alginate lyases containing the mixture. It can be as simple as a centrifugation step for extracellular alginate lyases extraction or more complicated by using different cell interruption methods to obtain intracellular enzymes. Then, the crude extracts undergo salting-out process to collect different fractions and increase the concentration of precipitate alginate lyases.

Last but not least, one or more sequential chromatography steps are then followed to improve enzyme purity [104–107]. Numerous alternative methods can be employed in each stage. It is recommended to use simple and effective methods in the early steps when the volume of the mixture is large. At the same time, more expensive and sophisticated techniques are applied when the handled volume gets smaller [103]. In this part, we will discuss methods frequently used for alginate lyase extraction and purification.

9.1 *Crude Extraction*

The extraction procedures for extra- and intracellular alginate lyases are different; however, centrifugation is usually the first step for both types of enzymes to separate cells and their culture medium. The supernatant of the former will be collected and optionally filtered before going to the precipitation stage [28, 106, 108]. In contrast, the pellet of the latter will be collected for the subsequent homogenization process to release intracellular components. Many studies have combined chemical methods (using lysis buffer) mechanical methods (such as the French press) or physical methods (sonication or temperature shock) to increase lysis efficiency. Besides, a second centrifugation step is needed to remove cell debris [104, 105, 107, 109, 110].

9.1.1 Lysis Buffer

A lysis buffer is used to disrupt the cell membrane and to maintain a stable environment for enzymes of interest. When selecting a buffer, it is important to consider the enzyme's optimal pH, the enzyme's optimal pH, the enzyme's optimal pH, the possibility of specific buffer effects influencing the enzyme, and possible interactions with substrates [111]. Since alginate lyases generally have a wide range of pH stability from 6 to 10, and their optimal activities have been mostly observed at 7.0–8.0, their buffers should have pKa close to these points. The lysis buffer of alginate lyases often contains buffering salts (e.g., Tris, pKa 8.06; Tris-HCl, pKa 8.0; and potassium sulfate, pK 7.2) and ionic salts (e.g., NaCl, $MgCl_2$, and $CaCl_2$). Previous studies have reported that the optimal concentration for buffering salts is 50 mM, while it is varying widely for different ionic salts [41, 109, 111, 112]. Moreover, these buffers and salts have nonspecific buffer effects on alginate lyases and substrates, such as alginate [111].

9.1.2 Sonication

Sonication is one of the most popular physical cell disruption methods for alginate lyases extraction experiments. During this process, bubbles or cavities form and explode, creating a shockwave of enormous mechanical energy spreading through the solution. The resulting energy disrupts the cell membrane and releases intracellular particles [110]. However, energy release will lead to heating up of the solution; therefore, the sample should be placed on ice throughout the sonication step. The sonication programs for alginate lyases extraction often use 30–60 cycles and at 60% power for 10 min [41, 109, 111, 112].

9.2 Precipitation—Salting-out

The precipitation step aims to concentrate and fractionate enzymes of interest from cellular components [113]. Various methods can be employed for enzyme fractionation; however, no doubt salting-out has been predominantly used for alginate lyases precipitation [28, 96, 106, 108, 114–116]. The most widely used salt for precipitation of enzymes generally and for alginate lyases particularly is ammonium sulfate $(NH_4)_2SO_4$ (AS) due to its high solubility, stabilization to enzyme structure, low cost, availability of pure material, and ease of removal by dialysis [117]. The underlying mechanism of salting-out is to reduce the solubility of alginate lyases in a solution by adding a very high ionic strength. The enzyme dissolves when exposed to polar and ionic groups to form hydrogen bonds with water molecules. The AS ions added will compete with these groups to associate with the solvent, leading to exposure of hydrophobic patches on the alginate lyase surface, resulting in hydrophobic interactions between alginate lyase molecules. These molecules then aggregate to form precipitation [113]. Usually, a high concentration of AS (70–80%) will be added for their aggregation, which can retain approximately 80% of the enzymes' yield and activity [33, 96, 114–116, 118]. Subsequently, an optional dialysis step against its corresponding buffer (e.g., phosphate, phosphate–citric acid buffer, and Tris-HCl) can be conducted to remove contaminated salts [114, 115, 118] (Fig. 12).

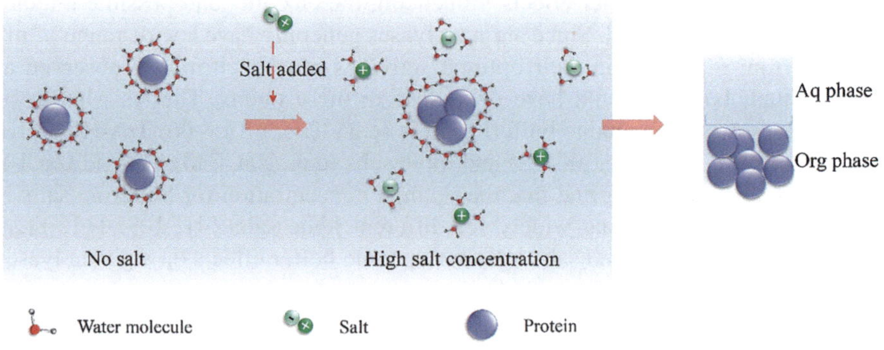

Fig. 12 Principles of salting-out effect

9.3 *Purification*

The precipitated enzyme will be further purified to reach the requirements of analytical experiments [119]. These purification methods via chromatography can be categorized into three main groups based on the properties of enzymes: (1) techniques based on ionic properties, (2) techniques based on adsorption ability, and (3) techniques based on molecular mass [120]. Depending on the degree of purity of enzymes required for downstream applications, one or more chromatography steps will be performed. There are numerous chromatographic strategies for alginate lyases; however, methods including ion-exchange chromatography—IEC [104, 112, 114, 115, 121], gel filtration or size exclusion chromatography—SEC [41, 104, 106, 114, 115, 121, 122], hydrophobic interaction chromatography—HIC [28, 118, 121], and the modern, powerful methods for recombinant alginate lyases, Immobilized Metal Affinity Chromatography—IMAC [105, 109], have been widely employed (Table 4). Besides, the purification efficiency is determined by various parameters such as the types of columns or resin, species, and concentration of salts.

9.3.1 Ion-Exchange Chromatography

The IEC separates enzymes based on their differences in ionic charges. The ion exchange resin (stationary phase) has the opposite charge of the target enzymes [120, 127]. For example, the anion exchange resin carries positive charges. It allows binding of enzyme molecules to its surface at pH values larger than their isoelectric point (pI), and vice versa. The IEC involves four main steps: (1) Equilibration: a low-salt buffer is passed through the column to prepare the opposite charged ions, allowing the binding of target molecules. (2) Sample application and adsorption: enzyme molecules with opposite charges bind to the matrix, and nonbinding molecules can be eluted first; it is worth noting that the protein loading buffer concentration should be low to avoid prevention of binding. (3) Elution of bound proteins:

Table 4 List of strategies commonly used in alginate lyase purification

Enzyme name	Mol. Mass (kDa)	Source/ Host	Purification procedure	Yield (%)	Activity (U/ mg)	Purification (fold)	Reference
Aly2 (His_6 tag)	60.58	*Flammeovirga* sp. MY04 / *E. coli* BL21	Sonication ➜ Ni-NTA (IMAC)	17.7	2025.8		[109]
NitAly (His_6 tag)	29	*Nitratiruptor* sp. SB155-2/ *E. coli* BL21	Sonication ➜ Ni-NTA agarose (IMAC)		1,620		[42]
ALG-5 (His_6 tag)	27.5	*Streptomyces* sp. ALG-5/ *Escherichia coli BL21*	Sonication ➜ Ni-Sepharose affinity chromatography (IMAC) ➜ HiTrap™ desalting colum	70	774		[41]
KS-408-1F (His_6 tag) KS-408-2F (His_6 tag)	44.5 42.4	*Pseudomonas* sp. strain KS-408/ *E. coli* BL21	Sonication ➜ Ni-Sepharose column (IMAC) ➜ HiTrap desalting column (SEC)	10.1 13.4	51.6 63.2		[39]
BcAlyPL6 (His_6 tag)	88	*Bacteroides clarus/ E. coli* BL21	Ultrasonication ➜ Ni-NTA-Sefinose column (IMAC) ➜ Reverse immobilized metal affinity chromatography (IMAC) ➜ Mono Q 5/50 GL [123] ➜ HiPrep 16/60 Sephacryl S200 HR column (SEC)				[83]
AlyA (His_6 tag)	25.32	*Isoptericola halotolerans NJ-05/ E. coli* BL21	Ni-NTA column➜ dialysis	72	7984.82	4.97	[123]

(continued)

Table 4 (continued)

Enzyme name	Mol. Mass (kDa)	Source/ Host	Purification procedure	Yield (%)	Activity (U/ mg)	Purification (fold)	Reference
PsAly (His_6 tag)	33	*Paenibacillus* sp. str. FPU-7/ *E. coli* BL21	Ultrasonication ➔ Ammonium sulfate precipitation ➔ HisTrap HP 5 ml column (IMAC) ➔ HiTrap Q HP 5 ml column [123]➔ dialysis		2200		[75]
ZH0-I ZH0-II ZH0-III ZH0-IV	91 52 67 113	*Sphingomonas* sp. ZH0/ *E. coli* BL21	Filtration ➔ GST-Sefinose Resin (AF)	22.7 31.9 41.6 21	53.2 103.9 13.7 12.3	6.7 7.2 34.3 13.7	[124]
–	40	*Pseudomonas stutzeri* MSEA04	70% ethanol precipitation ➔ DEAE- Cellulose A-52 ➔ Sephadex G-100 (SEC)	3.66	116	3.13	[125]
AlySY08	33	*Vibrio* sp. SY08	Ammonium sulfate precipitation ➔ Phenyl-Sepharose column [54]	43.6	1070.2	13.1	[52]
AlyV5	37	*Vibrio*sp. QY105	Ammonium sulfate precipitation ➔ Phenyl-Sepharose Fast Flow column [123] ➔ dialysis ➔ DEAE-Sepharosehigh performance column		2152		[108]

–	60	*Vibrio* sp. YKW-34	Polyethyleneimine precipitation ➔ DEAE-Sepharose FF column [123] ➔ Phenol Sepharose6 FF column [54] ➔ Sephacryl S-100 HR column (SEC)	7.6	55.93	25	[121]
A_1-II_{-1} A_2-I	60 60	Bacterial strain A-1 Bacterial strain A-2	Sonication ➔ DEAE-cellulose column [123]➔ Hydroxylapatite column ➔ Sephadex G-150 column (SEC) ➔ QAE-Sephadex A-25 column [123] ➔ Sephadex G-150 column (SEC) ➔ Butyl Toyopearl 650 M column [54] ➔ Bio-Gel P-60-alginate affinity colum (AF)	2.97 1.64	143 158	114 141	[126]

His_6 tag: recombinant enzymes fused with His_6 tag at N- or C-terminus.
IEC ion-exchange chromatography, *SEC* size-exclusion chromatography, *HIC* hydrophobic interaction chromatography, *IMAC* immobilized metal affinity chromatography, *AF* Affinity chromatography

Bound molecules are eluted by a gradient of salt concentration and pH values in the elution buffer, molecules with small differences in pI values can be eluted first. (4) Cleaning of the column: a high concentration of salts or alkaline is used to break the strong bonds between impurities and the matrix, followed by the final washing steps with water and ethanol [120] (Fig. 13).

9.3.2 Size-Exclusion Chromatography

In SEC, molecules are separated according to their sizes and shapes. The stationary column phase consists of porous beads, and the mobile phase containing target enzymes is continuously passed through the column. Very large molecules can pass through the column first, and molecules larger than the pores retain within the external volume of porous particles. In contrast, molecules small enough to enter the pores remain in the internal volume of the pores wholly or partly as very tiny molecules can easily leave the pore [120, 127]. Based on this principle, several parameters should be considered when designing the SEC experiment, including the bead diameter, the pore size, the molecule size and diameter, the flow rate of the mobile phase, and the buffer temperature and viscosity [120]. The alginate lyase separation often combines IEC and SEC to enhance the enzyme purity [28, 41, 105–107, 109, 112].

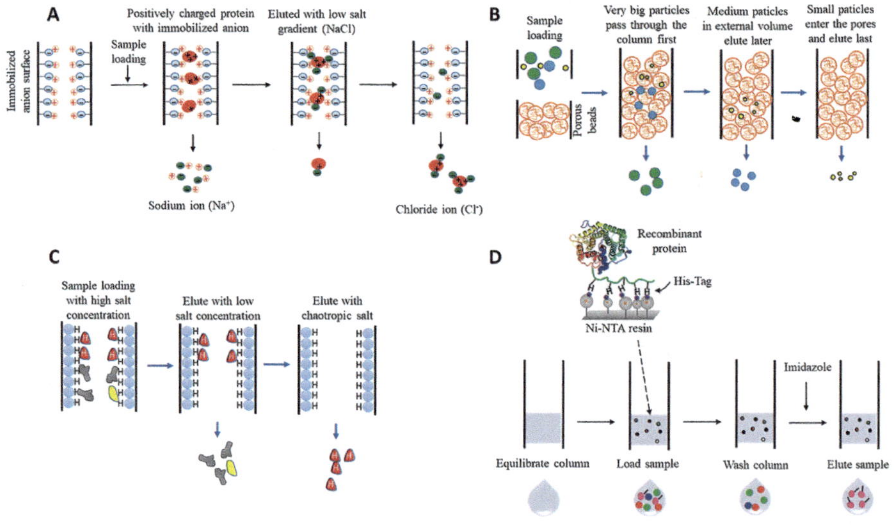

Fig. 13 Principles and procedures of chromatography methods are commonly used to purify alginate lyase. (**a**) Ion exchange chromatography—IEC. (**b**) Size-exclusion chromatography—SEC or gel filtration. (**c**) Hydrophobic interaction chromatography—HIC. (**d**) Immobilized Metal Affinity Chromatography—IMAC

9.3.3 Hydrophobic Interaction Chromatography

HIC separates proteins based on the degree of hydrophobicity on their surface. These proteins, therefore, can form hydrophobic interactions with the column matrix that carry hydrophobic ligands such as phenyl, octyl, and butyl [120]. However, these interactions are very weak in pure water; the solution is to apply high ionic strength (e.g., ammonium sulfate) to favor the interactions. The proteins are then eluted by either reducing ionic strength or applying chaotropic salts such as sodium chloride [105, 112, 120, 128].

9.3.4 Immobilized Metal Affinity Chromatography

Single-step purification of recombinant proteins can be achieved by employing IMAC [120]. The target enzyme, via genetic engineering, is tagged with polyhistidine (His-tag) consisting of 6–12 histidine residues at the N- or C-terminus of the molecule [122]. The principle of separation is based on the interaction between chelated metal ions (e.g., Co^{2+}, Ni^{2+}, Cu^{2+}, or Zn^{2+}), acting as electron acceptors, immobilized to the matrix and the imidazole nitrogen atoms of His-tags, as electron donors, at neutral pH. The elution of enzyme molecules takes place by altering parameters in the mobile phase such as increasing ionic strength (e.g., sodium chloride) or reducing the pH value [120, 122, 129].

10 Recombinant Alginate Lyase

Alginate lyases have been isolated from a variety of organisms such as algae, terrestrial and marine bacteria, fungi, marine invertebrates, and even viruses [1, 5, 14]. However, due to the increasing demand for alginate lyases, researchers have engineered the genetics of new or original organisms for higher production. Moreover, recombinant alginate lyases with affinity tags can facilitate the purification process for a higher enzyme recovery [120]. So far, alginate lyases have been heterologously synthesized by many other microorganisms, among which the two most popular hosts are *Escherichia coli* [54, 130–132] and *Yarrowia lipolytica* [133–135].

E. coli BL21 is predominantly used for heterologous production of alginate lyases due to its rapid growth, ease of transformation, and high expression of cloned genes (Table 4). In 2019, Sun et al. reported the successful enzyme synthesis of truncated Alg gene from *Vibro* sp. QY102 in *E. coli* with the enzyme activity and final yield of feed batch cultivation reached 1789 U/ml and 0.58 g/L, respectively [131]. Most recently, the novel enzyme Algpt from the marine bacteria *Paenibacillus* sp. was heterologously produced by *E. coli*; the recombinant enzymes can function efficiently and stably in a wide range of temperatures from 30 °C to 50 °C and pH 5.0–9.0 and its enzymatic activity was not affected by sodium chloride [130]. Numerous genes (e.g., Alg, FlAlyA, and alyA) encoding different alginate lyases from various sources have been successfully and efficiently cloned, expressed, and characterized using *E. coli*, which indicates a promising future for higher production of recombinant alginate lyases.

Another well-known host for recombinant alginate lyase biosynthesis is *Y. lipolytica*. Although its growth rate is not as fast as *E. coli*, *Y. lipolytica* possesses many advantages, including being safe and having a dominant ability for extracellular secretion [134, 136]. The enzyme Alyw20 originated from *Vibrio* sp. W2 was produced by *Y. lipolytica*, and it displayed many advanced properties, including thermostability at 25–40 °C, a wide pH range of 5.0–10.0, and salt tolerance [133]. A unique heat recovery property was performed by Alyw203, another recombinant enzyme also from *Vibrio* sp. The enzymatic activity of Alyw203 is above 80% at 40–55 °C and dramatically drops to 60% when the temperature exceeds 55 °C; however, after 3 h of incubation at its optimal temperature 45 °C, its activity can recover to more than 80% of its highest activity, which makes it a potential candidate for industrial application [134].

11 Application of Alginate Lyase

Pseudomonas aeruginosa is the main pathogen causing cystic fibrosis. Due to the secretion of alginate on their cell surfaces, they not only form fibrotic lesions in the host body but also greatly prevent antibiotics from entering their body. As a result, it is extremely difficult to be cleared by the host's immune system; they even produce antibiotic resistance, making it more difficult for treatment strategies. Luckily, alginate lyases can just degrade the biofilm of *P. aeruginosa*. Therefore, they exhibit an unparalleled advantage in curing the stubborn disease of cystic fibrosis. Combined use of alginate lyases and antibiotics greatly reduces the resistance of monospores and a half of the lethal dose. Alkawash [137] used alginate lyases from *Bacillus circulans* combined with gentamicin to treat the viscous *P. aeruginosa* biofilm and found that it can kill the bacteria in the biofilm, while gentamicin alone could not do this.

The brown algae cell wall contains a large amount of alginate, which prevents the penetration of foreign DNA, making it extremely difficult to establish an efficient genetic manipulation system. The use of alginate lyases to prepare brown algae protoplasts has greatly improved the efficiency and facilitated research on the physiological and biochemical, genetic, breeding, and molecular biology of brown algae [138].

Highly substrate-specific alginate lyases can also be used to prepare structure-specific alginate. For example, AlxMB, a polyM lyase from *Photobacterium* sp., can produce polyM and polyMG from *P. aeruginosa* alginate [139] by using a degradable polyG and polyM/G from *Flavobacterium multivolume*. The lyase can prepare polyM from sodium alginate [138].

Enzymatic preparation of oligosaccharides derived from brown algae has both pros and cons. First, advantages include mild reaction conditions, ease of control, and environmental-friendliness, and strong substrate specificity. Meanwhile, low catalytic efficiency and poor stability have become the bottleneck restricting the further promotion of the application. Therefore, research on new and highly efficient alginate lyases is of great significance for developing fucoidan oligosaccharides, improving the practicability of the enzyme, and expanding the scope of application. The high-efficiency and heat-stable alginate lyases not only improve the economics of preparing fucoidan oligosaccharides but also help to develop new types of oligosaccharides. Alginate lyases also play a key role in the saccharification of brown algae biomass. The presence of alginate is one of the main obstacles in the degradation of brown algae biomass [134]; thus, the investigation of highly efficient and heat-stable alginate lyases will facilitate the construction of a more efficient and higher-quality value-added brown algae bio-transformation system. Therefore, the research and development of novel, efficient, and stable alginate lyases have practical value [137–139].

Acknowledgments This study was supported by grants from the Key R&D Program of Shandong Province, China (2022SFGC0101), Key R&D Program of Shandong Province, China (2025CXGC011001), Weifang Science and Technology Development Program Project (2024 ZJ1074), Postdoctoral Innovation Program of Shandong Province (SDCX-ZG-202503099), and Qingdao Postdoctoral Program (QDBSH20250102165).

References

1. Zhu B, Yin H (2015) Alginate lyase: review of major sources and classification, properties, structure-function analysis and applications. Bioengineered 6:125–131. https://doi.org/10.1080/21655979.2015.1030543
2. Iwamoto M, Kurachi M, Nakashima T, Kim D, Yamaguchi K, Oda T et al (2005) Structure-activity relationship of alginate oligosaccharides in the induction of cytokine production from RAW264.7 cells. FEBS Lett 579:4423–4429. https://doi.org/10.1016/j.febslet.2005.07.007
3. Khan S, Tondervik A, Sletta H, Klinkenberg G, Emanuel C, Onsoyen E et al (2012) Overcoming drug resistance with alginate oligosaccharides able to potentiate the action of selected antibiotics. Antimicrob Agents Ch 56:5134–5141. https://doi.org/10.1128/Aac.00525-12
4. Lee KY, Mooney DJ (2012) Alginate: properties and biomedical applications. Prog Polym Sci 37:106–126. https://doi.org/10.1016/j.progpolymsci.2011.06.003
5. Wong TY, Preston LA, Schiller NL (2000) Alginate lyase: review of major sources and enzyme characteristics, structure-function analysis, biological roles, and applications. Ann Rev Microbiol 54:289–340. https://doi.org/10.1146/annurev.micro.54.1.289
6. Wang Y, Han F, Hu B, Li JB, Yu WG (2006) *In vivo* prebiotic properties of alginate oligosaccharides prepared through enzymatic hydrolysis of alginate. Nutr Res 26:597–603. https://doi.org/10.1016/j.nutres.2006.09.015
7. Iwamoto Y, Xu X, Tamura T, Oda T, Muramatsu T (2003) Enzymatically depolymerized alginate oligomers that cause cytotoxic cytokine production in human mononuclear cells. Biosci Biotech Bioch 67:258–263. https://doi.org/10.1271/bbb.67.258
8. Islan GA, Dini C, Bartel LC, Bolzan AD, Castro GR (2015) Characterization of smart auto-degradative hydrogel matrix containing alginate lyase to enhance levofloxacin delivery against bacterial biofilms. Int J Pharm 496:953–964. https://doi.org/10.1016/j.ijpharm.2015.10.050

9. Rehm BHA (2010) Bacterial polymers: biosynthesis, modifications and applications. Nat Rev Microbiol 8:578–592. https://doi.org/10.1038/nrmicro2354
10. Grasdalen H (1983) High-field, H-1-NMR spectroscopy of alginate – sequential structure and linkage conformations. Carbohydr Res 118:255–260. https://doi.org/10.1016/0008-6215(83)88053-7
11. Lopes M, Abrahim B, Veiga F, Seica R, Cabral LM, Arnaud P et al (2017) Preparation methods and applications behind alginate-based particles. Expert Opin Drug Del 14:769–782. https://doi.org/10.1080/17425247.2016.1214564
12. Fuenzalida JP, Nareddy PK, Moreno-Villoslada I, Moerschbacher BM, Swamy MJ, Pan S et al (2016) On the role of alginate structure in complexing with lysozyme and application for enzyme delivery. Food Hydrocolloid 53:239–248. https://doi.org/10.1016/j.foodhyd.2015.04.017
13. Cheng D, Jiang C, Xu J, Liu Z, Mao X (2020) Characteristics and applications of alginate lyases: a review. Int J Biol Macromol 164:1304–1320. https://doi.org/10.1016/j.ijbiomac.2020.07.199
14. Kim HS, Lee C-G, Lee EY (2011) Alginate lyase: structure, property, and application. Biotechnol Bioproc E 16:843–851. https://doi.org/10.1007/s12257-011-0352-8
15. Thomas F, Barbeyron T, Tonon T, Genicot S, Czjzek M, Michel G (2012) Characterization of the first alginolytic operons in a marine bacterium: from their emergence in marine *Flavobacteriia* to their independent transfers to marine *Proteobacteria* and human gut *Bacteroides*. Environ Microbiol 14:2379–2394. https://doi.org/10.1111/j.1462-2920.2012.02751.x
16. Kobayashi T, Uchimura K, Miyazaki M, Nogi Y, Horikoshi K (2009) A new high-alkaline alginate lyase from a deep-sea bacterium *Agarivorans* sp. Extremophiles 13:121–129. https://doi.org/10.1007/s00792-008-0201-7
17. Ochiai A, Yamasaki M, Mikami B, Hashimoto W, Murata K (2010) Crystal structure of exotype alginate lyase Atu3025 from *Agrobacterium tumefaciens*. J Biol Chem 285:24519–24528. https://doi.org/10.1074/jbc.M110.125450
18. Sawabe T, Ohtsuka M, Ezura Y (1997) Novel alginate lyases from marine bacterium *Alteromonas* sp. strain H-4. Carbohydr Res 304:69–76. https://doi.org/10.1016/s0008-6215(97)00194-8
19. Rehm BH, Ertesvag H, Valla S (1996) A new *Azotobacter vinelandii* mannuronan C-5-epimerase gene (algG) is part of an alg gene cluster physically organized in a manner similar to that in *Pseudomonas aeruginosa*. J Bacteriol 178:5884–5889. https://doi.org/10.1128/jb.178.20.5884-5889.1996
20. Chen P, Zhu Y, Men Y, Zeng Y, Sun Y (2018) Purification and characterization of a novel alginate lyase from the marine bacterium *Bacillus* sp. Alg07. Mar Drugs 16. https://doi.org/10.3390/md16030086
21. Shimokawa T, Yoshida S, Kusakabe I, Takeuchi T, Murata K, Kobayashi H (1997) Some properties and action mode of (1-->4)-alpha-L-guluronan lyase from *Enterobacter cloacae* M-1. Carbohydr Res 304:125–132. https://doi.org/10.1016/s0008-6215(97)00230-9
22. Inoue A, Takadono K, Nishiyama R, Tajima K, Kobayashi T, Ojima T (2014) Characterization of an alginate lyase, FlAlyA, from *Flavobacterium* sp. strain UMI-01 and its expression in *Escherichia coli*. Mar Drugs 12:4693–4712. https://doi.org/10.3390/md12084693
23. Ostgaard K, Knutsen SH, Dyrset N, Aasen IM (1993) Production and characterization of guluronate lyase from *Klebsiella pneumoniae* for applications in seaweed biotechnology. Enzym Microb Technol 15:756–763. https://doi.org/10.1016/0141-0229(93)90006-n
24. Iwamoto Y, Iriyama K, Osatomi K, Oda T, Muramatsu T (2002) Primary structure and chemical modification of some amino acid residues of bifunctional alginate lyase from a marine bacterium *Pseudoalteromonas* sp. strain no. 272. J Protein Chem 21:455–463. https://doi.org/10.1023/A:1021347019863
25. Linker A, Jones RS (1966) A new polysaccharide resembling alginic acid isolated from *Pseudomonads*. J Biol Chem 241:3845–3851. https://doi.org/10.1016/S0021-9258(18)99848-0

26. Yoon HJ, Hashimoto W, Miyake O, Okamoto M, Mikami B, Murata K (2000) Overexpression in *Escherichia coli*, purification, and characterization of *Sphingomonas* sp. A1 alginate lyases. Protein Expr Purif 19:84–90. https://doi.org/10.1006/prep.2000.1226
27. Li JW, Dong S, Song J, Li CB, Chen XL, Xie BB et al (2011) Purification and characterization of a bifunctional alginate lyase from *Pseudoalteromonas* sp. SM0524. Mar Drugs 9:109–123. https://doi.org/10.3390/md9010109
28. Li S, Wang L, Hao J, Xing M, Sun J, Sun M (2016) Purification and characterization of a new alginate lyase from marine bacterium *Vibrio* sp. SY08. Mar Drugs 15. https://doi.org/10.3390/md15010001
29. Keppley LJW, Walker SJ, Gademsey AN, Smith JP, Keller SR, Kester M et al (2020) Nervonic acid limits weight gain in a mouse model of diet-induced obesity. FASEB J 34:15314–15326. https://doi.org/10.1096/fj.202000525R
30. Miller GL (1959) Use of dinitrosalicylic acid reagent for determination of reducing sugar. Anal Chem 31:426–428. https://doi.org/10.1021/ac60147a030
31. Kurakake M, Kitagawa Y, Okazaki A, Shimizu K (2017) Enzymatic properties of alginate lyase from *Paenibacillus* sp. S29. Appl Biochem. Biotech 183:1455–1464. https://doi.org/10.1007/s12010-017-2513-5
32. Ji SQ, Wang B, Lu M, Li FL (2016) *Defluviitalea phaphyphila* sp. nov., a aovel thermophilic bacterium that degrades brown algae. Appl Environ Microb 82:868–877. https://doi.org/10.1128/Aem.03297-15
33. Dharani SR, Srinivasan R, Sarath R, Ramya M (2020) Recent progress on engineering microbial alginate lyases towards their versatile role in biotechnological applications. Folia Microbiol (Praha) 65:937–954. https://doi.org/10.1007/s12223-020-00802-8
34. Stender EGP, Dybdahl Andersen C, Fredslund F, Holck J, Solberg A, Teze D et al (2019) Structural and functional aspects of mannuronic acid-specific PL6 alginate lyase from the human gut microbe *Bacteroides cellulosilyticus*. J Biol Chem 294:17915–17930. https://doi.org/10.1074/jbc.RA119.010206
35. Mathieu S, Henrissat B, Labre F, Skjak-Braek G, Helbert W (2016) Functional exploration of the polysaccharide lyase family PL6. PLoS One 11:e0159415. https://doi.org/10.1371/journal.pone.0159415
36. Xu F, Dong F, Wang P, Cao HY, Li CY, Li PY et al (2017) Novel molecular insights into the catalytic mechanism of marine bacterial alginate lyase AlyGC from polysaccharide lyase family 6. J Biol Chem 292:4457–4468. https://doi.org/10.1074/jbc.M116.766030
37. Hashimoto W, Miyake O, Ochiai A, Murata K (2005) Molecular identification of *Sphingomonas* sp. A1 alginate lyase (A1-IV') as a member of novel polysaccharide lyase family 15 and implications in alginate lyase evolution. J Biosci Bioeng 99:48–54. https://doi.org/10.1263/jbb.99.48
38. Ochiai A, Hashimoto W, Murata K (2006) A biosystem for alginate metabolism in *Agrobacterium tumefaciens* strain C58: molecular identification of Atu3025 as an exotype family PL-15 alginate lyase. Res Microbiol 157:642–649. https://doi.org/10.1016/j.resmic.2006.02.006
39. Kam N, Park YJ, Lee EY, Kim HS (2011) Molecular identification of a polyM-specific alginate lyase from *Pseudomonas* sp. strain KS-408 for degradation of glycosidic linkages between two mannuronates or mannuronate and guluronate in alginate. Can J Microbiol 57:1032–1041. https://doi.org/10.1139/w11-106
40. Lee SI, Choi SH, Lee EY, Kim HS (2012) Molecular cloning, purification, and characterization of a novel polyMG-specific alginate lyase responsible for alginate MG block degradation in *Stenotrophomas maltophilia* KJ-2. Appl Microbiol Biotechnol 95:1643–1653. https://doi.org/10.1007/s00253-012-4266-y
41. Kim DE, Lee EY, Kim HS (2009) Cloning and characterization of alginate lyase from a marine bacterium *Streptomyces* sp. ALG-5. Mar Biotechnol (NY) 11:10–16. https://doi.org/10.1007/s10126-008-9114-9

42. Inoue A, Anraku M, Nakagawa S, Ojima T (2016) Discovery of a novel alginate lyase from *Nitratiruptor* sp. SB155-2 thriving at deep-sea hydrothermal vents and identification of the residues responsible for its heat stability. J Biol Chem 291:15551–15563. https://doi.org/10.1074/jbc.M115.713230
43. Lombard V, Bernard T, Rancurel C, Brumer H, Coutinho PM, Henrissat B (2010) A hierarchical classification of polysaccharide lyases for glycogenomics. Biochem J 432:437–444. https://doi.org/10.1042/Bj20101185
44. Fouet A, Mesnage S (2002) *Bacillus anthracis* cell envelope components. Curr Top Microbiol 271:87–113. https://doi.org/10.1007/978-3-662-05767-4_5
45. Fontes CMGA, Gilbert HJ (2010) Cellulosomes: highly efficient nanomachines designed to designed to deconstruct plant cell wall complex carbohydrates. Annu Rev Biochem 79:655–681. https://doi.org/10.1146/annurev-biochem-091208-085603
46. Gupta PK (2018) Environmental and ecotoxicology. In: Gupta PK (ed) Illustrated toxicology. Academic Press, Bareilly, pp 373–425
47. Han WJ, Gu JY, Cheng YY, Liu HH, Li YZ, Li FC (2016) Novel alginate lyase (Aly5) from a polysaccharide-degrading marine bacterium, *Flammeovirga* sp. strain MY04: effects of module truncation on biochemical characteristics, alginate degradation patterns, and oligosaccharide-yielding properties. Appl Environ Microb 82:364–374. https://doi.org/10.1128/Aem.03022-15
48. Pecina A, Pascual A, Paneque A (1999) Cloning and expression of the algL gene, encoding the *Azotobacter chroococcum* alginate lyase: purification and characterization of the enzyme. J Bacteriol 181:1409–1414. https://doi.org/10.1128/JB.181.5.1409-1414.1999
49. Zhu BW, Huang LSX, Tan HD, Qin YQ, Du YG, Yin H (2015) Characterization of a new endo-type polyM-specific alginate lyase from *Pseudomonas* sp. Biotechnol Lett 37:409–415. https://doi.org/10.1007/s10529-014-1685-0
50. Maki H, Mori A, Fujiyama K, Kinoshita S, Yoshida T (1993) Cloning, sequence-analysis and expression in *Escherichia coli* of a gene encoding an alginate lyase from *Pseudomonas* sp. Os-Alg-9. J Gen Microbiol 139:987–993. https://doi.org/10.1099/00221287-139-5-987
51. Lyu Q, Zhang K, Shi Y, Li W, Diao X, Liu W (2019) Structural insights into a novel Ca(2+)-independent PL-6 alginate lyase from Vibrio OU02 identify the possible subsites responsible for product distribution. BBA-Gen Subj 1863:1167–1176. https://doi.org/10.1016/j.bbagen.2019.04.013
52. Li S, Wang L, Han F, Gong Q, Yu W (2016) Cloning and characterization of the first polysaccharide lyase family 6 oligoalginate lyase from marine *Shewanella* sp. Kz7. J Biochem 159:77–86. https://doi.org/10.1093/jb/mvv076
53. Yamasaki M, Moriwaki S, Miyake O, Hashimoto W, Murata K, Mikami B (2004) Structure and function of a hypothetical *Pseudomonas aeruginosa* protein PA1167 classified into family PL-7 – a novel alginate lyase with a beta-sandwich fold. J Biol Chem 279:31863–31872. https://doi.org/10.1074/jbc.M402466200
54. Baron AJ, Wong TY, Hicks SJ, Gacesa P, Willcock D, McPherson MJ (1994) Alginate lyase from *Klebsiella pneumoniae*, subsp. *aerogenes*: gene cloning, sequence analysis and high-level production in *Escherichia coli*. Gene 143:61–66. https://doi.org/10.1016/0378-1119(94)90605-x
55. Wakabayashi M, Sakatoku A, Noda F, Noda M, Tanaka D, Nakamura S (2012) Isolation and characterization of *Microbulbifer* species 6532A degrading seaweed thalli to single cell detritus particles. Biodegradation 23:93–105. https://doi.org/10.1007/s10532-011-9489-6
56. Duan G, Han F, Yu W (2009) Cloning, sequence analysis, and expression of gene alyPI encoding an alginate lyase from marine bacterium *Pseudoalteromonas* sp. CY24. Can J Microbiol 55:1113–1118. https://doi.org/10.1139/w09-051
57. Huang LS, Zhou JG, Li X, Peng Q, Lu H, Du YG (2013) Characterization of a new alginate lyase from newly isolated *Flavobacterium* sp. S20. J Ind Microbiol Biot 40:113–122. https://doi.org/10.1007/s10295-012-1210-1

58. Sim SJ, Baik KS, Park SC, Choe HN, Seong CN, Shin TS et al (2012) Characterization of alginate lyase gene using a metagenomic library constructed from the gut microflora of abalone. J Ind Microbiol Biot 39:585–593. https://doi.org/10.1007/s10295-011-1054-0
59. Li SY, Yang XM, Bao MM, Wu Y, Yu WG, Han F (2015) Family 13 carbohydrate-binding module of alginate lyase from *Agarivorans* sp. L11 enhances its catalytic efficiency and thermostability, and alters its substrate preference and product distribution. FEMS Microbiol Lett 362. https://doi.org/10.1093/femsle/fnv054
60. Thomas F, Lundqvist LCE, Jam M, Jeudy A, Barbeyron T, Sandstrom C et al (2013) Comparative characterization of two marine alginate lyases from *Zobellia galactanivorans* reveals distinct modes of action and exquisite adaptation to their natural substrate. J Biol Chem 288:23021–23037. https://doi.org/10.1074/jbc.M113.467217
61. Miyake O, Ochiai A, Hashimoto W, Murata K (2004) Origin and diversity of alginate lyases of families PL-5 and-7 in *Sphingomonas* sp. strain A1. J Bacteriol 186:2891–2896. https://doi.org/10.1128/Jb.186.9.2891-2896.2004
62. Uchimura K, Miyazaki M, Nogi Y, Kobayashi T, Horikoshi K (2010) Cloning and sequencing of alginate lyase genes from deep-sea strains of *Vibrio* and *Agarivorans* and characterization of a new *Vibrio* enzyme. Mar Biotechnol 12:526–533. https://doi.org/10.1007/s10126-009-9237-7
63. Han F, Gong QH, Song K, Li JB, Yu WG (2004) Cloning, sequence analysis and expression of gene alyVI encoding alginate lyase from marine bacterium *Vibrio* sp. QY101. DNA Seq 15:344–350. https://doi.org/10.1080/10425170400019300
64. Zhu BW, Tan HD, Qin YQ, Xu QS, Du YG, Yin H (2015) Characterization of a new endo-type alginate lyase from *Vibrio* sp. W13. Int J Biol Macromol 75:330–337. https://doi.org/10.1016/j.ijbiomac.2015.01.053
65. Sugimoto I, Hiramatsu S, Murakami D, Fujie M, Usami S, Yamada T (2000) Algal-lytic activities encoded by *Chlorella* virus CVK2. Virology 277:119–126. https://doi.org/10.1006/viro.2000.0590
66. Qin HM, Miyakawa T, Inoue A, Nishiyama R, Nakamura A, Asano A et al (2017) Structure and polymannuronate specificity of a eukaryotic member of the polysaccharide lyase family 14. J Biol Chem 292:2182–2190. https://doi.org/10.1074/jbc.M116.749929
67. Miyake O, Hashimoto W, Murata K (2003) An exotype alginate lyase in *Sphingomonas* sp. A1: overexpression in *Escherichia coli*, purification, and characterization of alginate lyase IV (A1-IV). Protein Expr Purif 29:33–41. https://doi.org/10.1016/s1046-5928(03)00018-4
68. Jagtap SS, Hehemann JH, Polz MF, Lee JK, Zhao H (2014) Comparative biochemical characterization of three exolytic oligoalginate lyases from *Vibrio splendidus* reveals complementary substrate scope, temperature, and pH adaptations. Appl Environ Microbiol 80:4207–4214. https://doi.org/10.1128/AEM.01285-14
69. Yang J, Cui D, Ma S, Chen W, Chen D, Shen H (2021) Characterization of a novel PL 17 family alginate lyase with exolytic and endolytic cleavage activity from marine bacterium *Microbulbifer* sp. SH-1. Int J Biol Macromol 169:551–563. https://doi.org/10.1016/j.ijbiomac.2020.12.196
70. Kraiwattanapong J, Ooi T, Kinoshita S (1997) Cloning and sequence analysis of the gene (alyII) coding for an alginate lyase of *Pseudomonas* sp. OS-ALG-9. Biosci Biotechnol Biochem 61:1853–1857. https://doi.org/10.1271/bbb.61.1853
71. Kim HT, Chung JH, Wang D, Lee J, Woo HC, Choi IG et al (2012) Depolymerization of alginate into a monomeric sugar acid using Alg17C, an exo-oligoalginate lyase cloned from *Saccharophagus degradans* 2-40. Appl Microbiol Biotechnol 93:2233–2239. https://doi.org/10.1007/s00253-012-3882-x
72. Wang L, Li S, Yu W, Gong Q (2015) Cloning, overexpression and characterization of a new oligoalginate lyase from a marine bacterium, *Shewanella* sp. Biotechnol Lett 37:665–671. https://doi.org/10.1007/s10529-014-1706-z

73. Park HH, Kam N, Lee EY, Kim HS (2012) Cloning and characterization of a novel oligoalginate lyase from a newly isolated bacterium *Sphingomonas* sp. MJ-3. Mar Biotechnol (NY) 14:189–202. https://doi.org/10.1007/s10126-011-9402-7
74. Matsushima R, Danno H, Uchida M, Ishihara K, Suzuki T, Kaneniwa M et al (2010) Analysis of extracellular alginate lyase and its gene from a marine bacterial strain, *Pseudoalteromonas atlantica* AR06. Appl Microbiol Biotechnol 86:567–576. https://doi.org/10.1007/s00253-009-2278-z
75. Itoh T, Nakagawa E, Yoda M, Nakaichi A, Hibi T, Kimoto H (2019) Structural and biochemical characterisation of a novel alginate lyase from *Paenibacillus* sp. str. FPU-7 Sci Rep 9:14870. https://doi.org/10.1038/s41598-019-51006-1
76. Dong F, Xu F, Chen XL, Li PY, Li CY, Li FC et al (2019) Alginate lyase Aly36B is a new bacterial member of the polysaccharide lyase family 36 and catalyzes by a novel nechanism with lysine as both the catalytic base and catalytic acid. J Mol Biol 431:4897–4909. https://doi.org/10.1016/j.jmb.2019.10.023
77. Ji S, Dix SR, Aziz AA, Sedelnikova SE, Baker PJ, Rafferty JB et al (2019) The molecular basis of endolytic activity of a multidomain alginate lyase from *Defluviitalea phaphyphila*, a representative of a new lyase family, PL39. J Biol Chem 294:18077–18091. https://doi.org/10.1074/jbc.RA119.010716
78. Svanem BIG, Strand WI, Ertesvag H, Skjak-Braek G, Hartmann M, Barbeyron T et al (2001) The catalytic activities of the bifunctional *Azotobacter vinelandii* mannuronan C-5-epimerase and alginate lyase AlgE7 probably originate from the same active site in the enzyme. J Biol Chem 276:31542–31550. https://doi.org/10.1074/jbc.M102562200
79. Badur AH, Jagtap SS, Yalamanchili G, Lee JK, Zhao H, Rao CV (2015) Alginate lyases from alginate-degrading *Vibrio splendidus* 12B01 are endolytic. Appl Environ Microb 81:1856–1864. https://doi.org/10.1128/Aem.03460-14
80. Takagi T, Yokoi T, Shibata T, Morisaka H, Kuroda K, Ueda M (2016) Engineered yeast whole-cell biocatalyst for direct degradation of alginate from macroalgae and production of non-commercialized useful monosaccharide from alginate. Appl Microbiol Biotechnol 100:1723–1732. https://doi.org/10.1007/s00253-015-7035-x
81. Yoon HJ, Hashimoto W, Miyake O, Murata K, Mikami B (2001) Crystal structure of alginate lyase A1-III complexed with trisaccharide product at 2.0 angstrom resolution. J Mol Biol 307:9–16. https://doi.org/10.1006/jmbi.2000.4509
82. Pandey S, Mahanta P, Berger BW, Acharya R (2021) Structural insights into the mechanism of pH-selective substrate specificity of the polysaccharide lyase Smlt1473. J Biol Chem 297. https://doi.org/10.1016/j.jbc.2021.101014
83. Wang B, Dong S, Li FL, Ma XQ (2021) Structural basis for the exolytic activity of polysaccharide lyase family 6 alginate lyase BcAlyPL6 from human gut microbe *Bacteroides clarus*. Biochem Bioph Res Co 547:111–117. https://doi.org/10.1016/j.bbrc.2021.02.040
84. Violot S, Galisson F, Carrique L, Jugnarain V, Conchou L, Robert X et al (2021) Exploring molecular determinants of polysaccharide lyase family 6-1 enzyme activity. Glycobiology 31:1557–1570. https://doi.org/10.1093/glycob/cwab073
85. Dong S, Wei TD, Chen XL, Li CY, Wang P, Xie BB et al (2014) Molecular insight into the role of the N-terminal extension in the maturation, substrate recognition, and catalysis of a bacterial alginate lyase from polysaccharide lyase family 18. J Biol Chem 289:29558–29569. https://doi.org/10.1074/jbc.M114.584573
86. Park D, Jagtap S, Nair SK (2014) Structure of a PL17 family alginate lyase demonstrates functional similarities among exotype depolymerases. J Biol Chem 289:8645–8655. https://doi.org/10.1074/jbc.M113.531111
87. Yonemoto Y, Tanaka H, Hisano T, Sakaguchi K, Abe S, Yamashita T et al (1993) Bacterial alginate lyase gene: nucleotide sequence and molecular route for generation of alginate lyase species. J Ferment Bioeng 75:336–342. https://doi.org/10.1016/0922-338x(93)90130-Z
88. Li Q, Zheng L, Guo ZL, Tang TC, Zhu BW (2021) Alginate degrading enzymes: an updated comprehensive review of the structure, catalytic mechanism, modification method and

applications of alginate lyases. Crit Rev Biotechnol 41:953–968. https://doi.org/10.1080/07388551.2021.1898330
89. Ogura K, Yamasaki M, Mikami B, Hashimoto W, Murata K (2008) Substrate recognition by family 7 alginate lyase from *Sphingomonas* sp. A1. J Mol Biol 380:373–385. https://doi.org/10.1016/j.jmb.2008.05.008
90. Xu F, Chen XL, Sun XH, Dong F, Li CY, Li PY et al (2020) Structural and molecular basis for the substrate positioning mechanism of a new PL7 subfamily alginate lyase from the arctic. J Biol Chem 295:16380–16392. https://doi.org/10.1074/jbc.RA120.015106
91. Qin HM, Miyakawa T, Inoue A, Nishiyama R, Nakamura A, Asano A et al (2018) Structural basis for controlling the enzymatic properties of polymannuronate preferred alginate lyase FlAlyA from the PL-7 family. Chem Commun (Camb) 54:555–558. https://doi.org/10.1039/c7cc06523j
92. Hu F, Li Q, Zhu BW, Ni F, Sun Y, Yao Z (2019) Effects of module truncation on biochemical characteristics and products distribution of a new alginate lyase with two catalytic modules. Glycobiology 29:876–884. https://doi.org/10.1093/glycob/cwz064
93. Yang M, Yang SX, Liu ZM, Li NN, Li L, Mou HJ (2019) Rational design of alginate lyase from *Microbulbifer* sp. Q7 to improve thermal stability. Mar Drugs 17. https://doi.org/10.3390/md17060378
94. Wang B, Ji SQ, Ma XQ, Lu M, Wang LS, Li FL (2018) Substitution of one calcium-binding amino acid strengthens substrate binding in a thermophilic alginate lyase. FEBS Lett 592:369–379. https://doi.org/10.1002/1873-3468.12965
95. Li S, Wang Y, Li X, Lee BS, Jung S, Lee MS (2019) Enhancing the thermo-stability and anti-biofilm activity of alginate lyase by immobilization on low molecular weight chitosan nanoparticles. Int J Mol Sci 20. https://doi.org/10.3390/ijms20184565
96. Kunjukunju S, Roy A, Shekhar S, Kumta PN (2018) Cross-linked enzyme aggregates of alginate lyase: a systematic engineered approach to controlled degradation of alginate hydrogel. Int J Biol Macromol 115:176–184. https://doi.org/10.1016/j.ijbiomac.2018.03.110
97. Koti BA, Shinde M, Lalitha J (2013) Repeated batch production of agar-oligosaccharides from agarose by an amberlite IRA-900 immobilized agarase system. Biotechnol Bioproc E. 18:333–341. https://doi.org/10.1007/s12257-012-0237-5
98. Xiao AF, Xiao Q, Lin Y, Ni H, Zhu YB, Cai HN (2016) Immobilization of agarase from marine Vibrio onto carboxyl-functioned magnetic nanoparticles. J Nanosci Nanotechno 16:10048–10055. https://doi.org/10.1166/jnn.2016.12359
99. Zhang L, Guo X, Song Y, Tang N, Cheng P, Xiang J et al (2018) Bioadhesive immobilize agarase on magnetic ferriferous by polydopamine. Mater Sci Eng C Mater Biol Appl 93:218–225. https://doi.org/10.1016/j.msec.2018.07.068
100. Wang Q, Sun J, Liu Z, Huang W, Xue C, Mao X (2018) Coimmobilization of beta-agarase and alpha-neoagarobiose hydrolase for enhancing the production of 3,6-anhydro-l-galactose. J Agric Food Chem 66:7087–7095. https://doi.org/10.1021/acs.jafc.8b01974
101. Kornberg A (2009) Chapter 1 Why purify enzymes? In: Burgess RR, Deutscher MP (eds) Guide to protein purification. Academic Press, Exeter, pp 3–6
102. Scott JE, Williams KP (2004) Validating identity, mass purity and enzymatic purity of enzyme preparations. In: Markossian S, Grossman A, Brimacombe K, Arkin M, Auld D, Austin CP et al (eds) Assay Guidance Manual (Internet). Eli Lilly & Company and the National Center for Advancing Translational Sciences, Bethesda (MD)
103. Ramos OS, Malcata FX (2011) Food-grade enzymes. In: Moo-Young M (ed) Comprehensive biotechnology. Pergamon Press, Oxford, pp 555–569
104. Davidson IW, Sutherland IW, Lawson CJ (1976) Purification and properties of an alginate lyase from a marine bacterium. Biochem J 159:707–713. https://doi.org/10.1042/bj1590707
105. Swift SM, Hudgens JW, Heselpoth RD, Bales PM, Nelson DC (2014) Characterization of AlgMsp, an alginate lyase from *Microbulbifer* sp. 6532A. PLoS One 9:e112939. https://doi.org/10.1371/journal.pone.0112939

106. Zhang W, Xia X, Zhang Z (2019) Alginate lyase of a novel algae fermentation strain. Chem Biochem Eng Q 33:125–131. https://doi.org/10.15255/Cabeq.2018.1291
107. Ghadam P, Akhlaghi F, Ali AA (2017) One-step purification and characterization of alginate lyase from a clinical *Pseudomonas aeruginosa* with destructive activity on bacterial biofilm. Iran J Basic Med Sci 20:467–473. https://doi.org/10.22038/IJBMS.2017.8668
108. Wang Y, Guo EW, Yu WG, Han F (2013) Purification and characterization of a new alginate lyase from a marine bacterium *Vibrio* sp. Biotechnol Lett 35:703–708. https://doi.org/10.1007/s10529-012-1134-x
109. Peng C, Wang Q, Lu D, Han W, Li F (2018) A novel bifunctional endolytic alginate lyase with variable alginate-degrading modes and versatile monosaccharide-producing properties. Front Microbiol 9:167. https://doi.org/10.3389/fmicb.2018.00167
110. Shehadul Islam M, Aryasomayajula A, Selvaganapathy P (2017) A review on macroscale and microscale cell lysis methods. Micromachines 8. https://doi.org/10.3390/mi8030083
111. Stoll VS, Blanchard JS (2009) Chapter 6 Buffers. In: Burgess RR, Deutscher MP (eds) Guide to protein purification. Academic Press, London, pp 43–56
112. Gimmestad M, Ertesvag H, Heggeset TM, Aarstad O, Svanem BI, Valla S (2009) Characterization of three new *Azotobacter vinelandii* alginate lyases, one of which is involved in cyst germination. J Bacteriol 191:4845–4853. https://doi.org/10.1128/JB.00455-09
113. Novák P, Havlíček V (2016) Protein extraction and precipitation. In: Ciborowski P, Silberring J (eds) Proteomic profiling and analytical chemistry. Elsevier, Oxford, pp 51–62
114. Cao LX, Xie LJ, Xue XL, Tan HM, Liu YH, Zhou SN (2007) Purification and characterization of alginate lyase from *Streptomyces* species strain A5 isolated from banana rhizosphere. J Agr Food Chem 55:5113–5117. https://doi.org/10.1021/jf0704514
115. Singh RP, Gupta V, Kumari P, Kumar M, Reddy CRK, Prasad K et al (2011) Purification and partial characterization of an extracellular alginate lyase from *Aspergillus oryzae* isolated from brown seaweed. J Appl Phycol 23:755–762. https://doi.org/10.1007/s10811-010-9576-9
116. Zilda DS, Yulianti Y, Sholihah RF, Subaryono S, Fawzya YN, Irianto HE (2019) A novel *Bacillus* sp. isolated from rotten seaweed: identification and characterization alginate lyase its produced. Biodiversitas 20:1166–1172. https://doi.org/10.13057/biodiv/d200432
117. Burgess RR (2009) Chapter 20 Protein precipitation techniques. In: Burgess R, Deutscher M (eds) Guide to protein purification. Academic Press, London, pp 331–342
118. Tseng CH, Yamaguchi K, Kitamikado M (1992) Isolation and some properties of alginate lyase from a marine bacterium *Vibrio* sp. Al-128. Nippon Suisan Gakk 58:533–538
119. Lee CH (2017) A simple outline of methods for protein isolation and purification. Endocrinol Metab (Seoul) 32:18–22. https://doi.org/10.3803/EnM.2017.32.1.18
120. Westphal AH, Berkel vWJH. (2021) Techniques for enzyme purification. In: Gonzalo DG, Lavandera I (eds) Biocatalysis for practitioners: techniques, reactions and applications. Wiley-VCH, Baden-Württemberg, pp 3–33
121. Fu XT, Lin H, Kim SM (2007) Purification and characterization of a Na+/K+ dependent alginate lyase from turban shell gut *Vibrio* sp. YKW-34. Enzyme Microb Tech 41:828–834. https://doi.org/10.1016/j.enzmictec.2007.07.003
122. Falke JJ, Corbin JA (2013) Affinity tags for protein purification. In: Lennarz JW, LM D (eds) Encyclopedia of biological chemistry. Academic Press, London, pp 61–65
123. Lewkowicz N, Piatek P, Namiecinska M, Domowicz M, Bonikowski R, Szemraj J et al (2019) Naturally occurring nervonic acid ester improves myelin synthesis by human oligodendrocytes. Cells-Basel 8:786. https://doi.org/10.3390/cells8080786
124. He M, Guo M, Zhang X, Chen K, Yan J, Irbis C (2018) Purification and characterization of alginate lyase from *Sphingomonas* sp. ZH0. J Biosci Bioeng 126:310–316. https://doi.org/10.1016/j.jbiosc.2018.01.017
125. Beltagy EA, El-Borai A, Lewiz M, ElAssar SA (2016) Purification and characterization of alginate lyase from locally isolated marine *Pseudomonas stutzeri* Msea04. Acta Biol Hung 67:305–317. https://doi.org/10.1556/018.67.2016.3.8

126. Yonemoto Y, Murata K, Kimura A, Yamaguchi H, Okayama K (1991) Bacterial alginate lyase – characterization of alginate lyase-producing bacteria and purification of the enzyme. J Ferment Bioeng 72:152–157. https://doi.org/10.1016/0922-338x(91)90208-X
127. Coskun O (2016) Separation techniques: chromatography. North Clin Istanb 3:156–160. https://doi.org/10.14744/nci.2016.32757
128. Kong SG, Kagawa T, Wada M, Nagatani A (2013) A C-terminal membrane association domain of phototropin 2 is necessary for chloroplast movement. Plant Cell Physiol 54:57–68. https://doi.org/10.1093/pcp/pcs132
129. Otter D (2003) Protein/determination and characterization. In: Caballero B (ed) Encyclopedia of food sciences and nutrition. Academic Press, Maryland, pp 4824–4830
130. Wang M, Chen L, Lou Z, Yuan X, Pan G, Ren X et al (2022) Cloning and characterization of a novel alginate lyase from *Paenibacillus* sp. LJ-23. Mar Drugs 20:66. https://doi.org/10.3390/md20010066
131. Sun X, Shen W, Gao Y, Cai M, Zhou M, Zhang Y (2019) Heterologous expression and purification of a marine alginate lyase in *Escherichia coli*. Protein Expr Purif 153:97–104. https://doi.org/10.1016/j.pep.2018.09.002
132. Tavafi H, Ali AA, Ghadam P, Gharavi S (2018) Screening, cloning and expression of a novel alginate lyase gene from *P. aeruginosa* TAG 48 and its antibiofilm effects on *P. aeruginosa* biofilm. Microb Pathog 124:356–364. https://doi.org/10.1016/j.micpath.2018.08.018
133. Wang ZP, Cao M, Li B, Ji XF, Zhang XY, Zhang YQ et al (2020) Cloning, secretory expression and characterization of a unique pH-stable and cold-adapted alginate lyase. Mar Drugs 18:189. https://doi.org/10.3390/md18040189
134. Liu L, Wang Z, Zheng Z, Li Z, Ji X, Cong H et al (2021) Secretory expression of an alkaline alginate lyase with heat recovery property in *Yarrowia lipolytica*. Front Microbiol 12:710533. https://doi.org/10.3389/fmicb.2021.710533
135. Liu GL, Yue LX, Chi Z, Yu WG, Chi ZM, Madzak C (2009) The surface display of the alginate lyase on the cells of *Yarrowia lipolytica* for hydrolysis of alginate. Mar Biotechnol 11:619–626. https://doi.org/10.1007/s10126-009-9178-1
136. Madzak C (2015) *Yarrowia lipolytica*: recent achievements in heterologous protein expression and pathway engineering. Appl Microbiol Biot 99:4559–4577. https://doi.org/10.1007/s00253-015-6624-z
137. Alkawash MA, Soothill JS, Schiller NL (2006) Alginate lyase enhances antibiotic killing of mucoid *Pseudomonas aeruginosa* in biofilms. APMIS 114:131–138. https://doi.org/10.1111/j.1600-0463.2006.apm_356.x
138. Ochi Y, Takeuchi T, Murata K, Kawabata Y, Kusakabe I (1995) A simple method for preparation of poly-mannuronate using poly-guluronate lyase. Biosci Biotech Bioch 59:1560–1561. https://doi.org/10.1271/bbb.59.1560
139. Heyraud A, Colin-Morel P, Gey C, Chavagnat F, Guinand M, Wallach J (1998) An enzymatic method for preparation of homopolymannuronate blocks and strictly alternating sequences of mannuronic and guluronic units. Carbohydr Res 308:417–422. https://doi.org/10.1016/S0008-6215(98)00094-9

Agarase

Chengcheng Jiang and Xiangzhao Mao

1 Sources and Classification of Agarases

Sea is the main source of agarases, and agarases can be found in the seawater, marine sediments, marine algae, and marine mollusks. Agarases have been obtained from the marine microorganisms [1], such as *Alteromonas* sp. [2, 3], *Pseudomonas* sp. [4], *Vibrio* sp. [5], *Cytophaga* sp. [6], *Agarivorans* sp. [7], *Thalassomonas* sp. [5], *Bacillus* sp. [8], and *Acinetobacter* sp. [9], by the direct isolation or heterologous expression. Except for these marine microorganisms, agarase also has been obtained from the *Streptomyces coelicolor* A3(2) that isolated from soil [10], as well as the human intestinal bacterium *Bacteroides uniformis* (Bu) NP1 was found that can express agarases [11].

Agarases can be divided into α-agarase and β-agarase depending on the cleavage of glycosidic bonds. α-Agarase acts on the α-1,3-glycosidic bonds of agar or agarose to produce even-numbered agarooligosaccharides (EAOSs) with the D-galactose (D-Gal) as non-reducing end. β-Agarase acts on the β-1,4-glycosidic bonds of agar or agarose to produce even-numbered neoagarooligosaccharides (ENAOSs) with the 3,6-anhydro-α-L-galactose (L-AHG) as non-reducing end [12]. According to primary structure analysis, the characterized β-agarases were classified into four glycoside hydrolase (GH) families, including GH16, GH50, GH86, and GH118 (Fig. 1). All the characterized α-agarases belong to GH96 family till date. Additionally, the agarases could also be divided into endo-type hydrolases and exo-type hydrolases based on their action mode. Endo-type agarases cleave the inside-chain glycosidic bonds of agarose to create ENAOSs or EAOSs with various degree of polymerization (DP), while exo-type agarases can produce specific

C. Jiang · X. Mao (✉)
State Key Laboratory of Marine Food Processing and Safety Control, College of Food Science and Engineering, Ocean University of China, Qingdao, China
e-mail: xzhmao@ouc.edu.cn

X. Mao (ed.), *Degrading Enzymes for Marine Polysaccharides and Their Applications*, https://doi.org/10.1007/978-981-96-9517-1_7

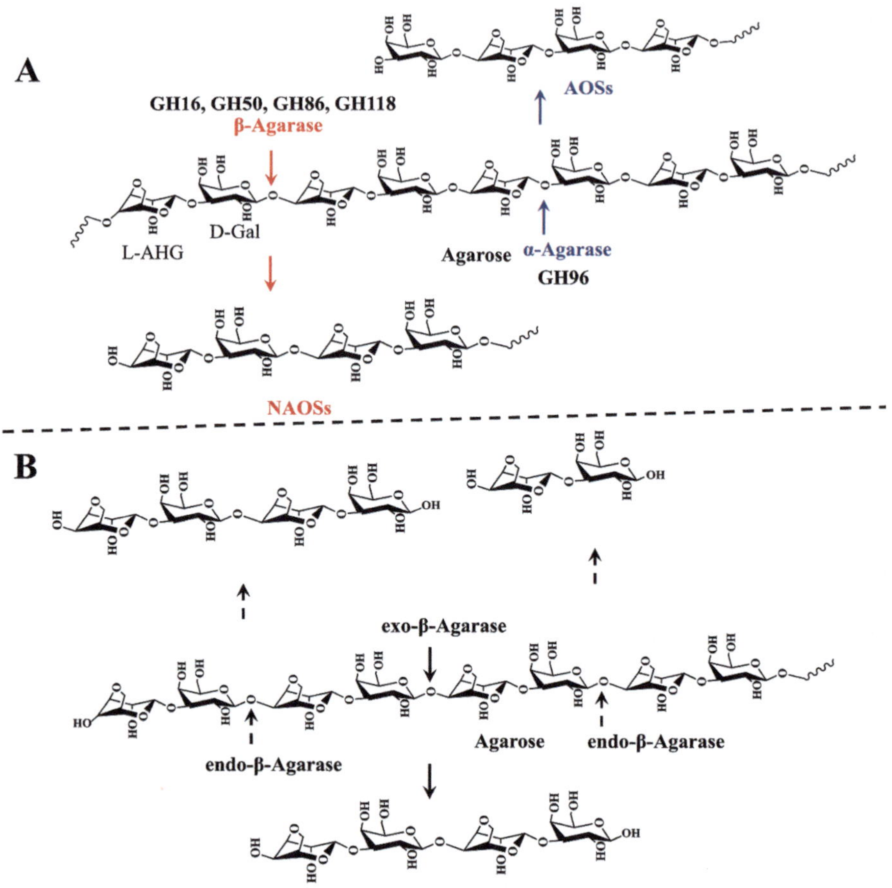

Fig. 1 (**a**) The cleavage site of α-agarase and β-agarase. (**b**) The action mode of endo- and exo-agarase

product by gradual degradation from the non-reducing end of agarose [1] (Fig. 1). Therefore, the exo-type agarases have great significance for preparation of agaro-oligosaccharides with specific DP.

2 Enzyme Activity Assay of Agarases

The Lugol's iodine solution has always been applied for visual detection of agarases activity on an agar solid plate [8]. It can stain agar into a dark brown color but cannot stain the agaro-oligosaccharide. Thus, the agarase-producing monoclonal cells are surrounded by transparent rings. Moreover, the Lugol's iodine solution can be used for identification of the molecular mass of the agarase [7]. The method of Lugol's iodine solution only preforms qualitative analysis of agarase activity.

DNS method can be used [13] to further quantify the agarase activity. The agaro-oligosaccharides product from enzymatic hydrolysis of agarose has reducibility and can react with DNS to appear red. Then, the amount of reducing sugar product can be determined by measuring the absorbance at 540 nm using D-Gal as the standard. One unit of enzymatic activity (U) was defined as the amount of enzyme required to liberate 1 μmol of reducing sugar per min from hydrolysis of agarose.

3 Biochemical Properties of Agarases

β-Agarases are distributed in various GH families, including GH16, GH50, GH86, and GH118. Therein, GH16 and GH50 family β-agarases have a thumping majority (Table 1). Agarase plays a significant role in functional agaro-oligosaccharide preparation due to its product specificity.

Moreover, the combined applications of different types of agarolytic enzymes can prepare different types of agaro-oligosaccharides [14]. In addition, the products of agarases show a relative relation with GH families. GH16 family β-agarases tend to produce neoagarobiose (NA2), neoagarotetraose (NA4), and neoagarohexaose (NA6) from the hydrolysis of agarose. GH50 family β-agarases act on agarose to obtain NA2 or NA4, while GH86 family β-agarases hydrolyze agarose to produce NA2. Especially, GH118 family β-agarases tend to create long-chain NAOSs, including neoagarooctaose (NA8), neoagarodecaose (NA10), and neoagarododecaose (NA12), from depolymerization of agarose. Moreover, five A4-producing α-agarases have been characterized, and all of them belong to GH96 family.

The higher optimal temperature and good thermal stability of agarase are very important for industrial application in agaro-oligosaccharide production because of the gel properties of agarose or agar [15]. As shown in Table 1, the optimal temperature of GH16 family β-agarases is in a range of 40–60 °C, and most of them exhibit a good thermal stability under above 40 °C. For example, the β-agarase AgaDL6 from *Flammeovirga* sp. HQM9 still retains more than 98% of residual activity after pre-incubation at its optimum temperature (50 °C) for 2 h [16]. GH16 β-agarase Aga3027 from *Flammeovirga* sp. OC4 could retain more than 80% of residual activity after incubating at its optimum temperature (40 °C) for 48 h [17]. Moreover, their optimal pHs were in a range of pH 6.0–7.5 apart from AgaDL6, which show its highest hydrolysis activity at pH 3.0 [16]. In addition, the catalytic activity has significance for the industry applications of agarases. Among the agarases, the β-agarase YM01–3 from *Catenovulum agarivorans* YM01 exhibited the highest hydrolysis activity with a V_{max} value of 1.14 × 10^4 μmol min^{-1} mg^{-1} [18]. As for GH50 β-agarases, their products are more specific (Table 1). One part of them produces NA2, and others produce NA4. NA4-producing β-agarase AgaWH50B, cloned from marine microorganism *Agarivorans gilvus* WH0801, shows a good hydrolysis efficiency with V_{max} values of 1.7 × 10^3 μmol min^{-1} mg^{-1} [19]. Only five GH86 β-agarases have been reported until now. Among them, NA2-producing β-agarase Aga3463 from *Pseudoalteromonas* sp. NJ21 displays the highest

Table 1 Biochemical properties of agarases

GH family	Protein	Bacterial species	Molecular mass (kDa)	Optimal temperature (°C)	Optimal pH	Main Products	Co-factor	V_{max} (μmol min^{-1} mg^{-1})	Thermostability	References
GH16	Gaa16A	*Gilvimarinus agarilyticus* JEA5	49	55	7.0	NA4	Ca^{2+}, Mn^{2+}, Fe^{2+}, Na^{+}	103	85% residual activity after incubating at 50 °C for 1 h	[23]
GH16	Aga091	*Pseudoalteromonas fuliginea*	32.5	60	6.0	NA4	Co^{2+}, Mg^{2+}, K^{+}, Na^{+}	334	57% residual activity after incubating at 50 °C for 1 h	[24]
GH16	Aga4436	*Flammeovirga* sp.OC4	54	55	6.5	NA4, NA6	Co^{2+}, Mn^{2+}	Not available	80% residual activity after incubating at 50 °C for 48 h	[25]
GH16	AgaA	*Microbulbifer* sp. JAMB-A94	48	55	7.0	NA4	No	517.3	Retaining more than 80% of maximum activity after pre-incubation at 60 °C for 15 min	[26]
GH16	AgaP	*Pseudoalteromonas* sp. AG4	72.5	55	5.5	NA4, NA6	Not available	207.5	Retaining more than 50% of maximum activity after pre-incubation at 55 °C for 60 min	[27]
GH16	Agy1	*Saccharophagus* sp. AG21	69	55	7.6	NA4, NA6	Fe^{2+}, Na^{+}, K^{+}	85	Good thermostability at 45 °C	[28]

GH16	RagaA7	*Microbulbifer* sp. JAMB-A7	49	50	7.0	NA4	No	398	Having a half-life of 502 min at 50 °C	[29]
GH16	AgaDL6	*Flammeovirga* sp. HQM9	52.8	50	3.0	NA4, NA6	Mg^{2+}, Ba^{2+}, Ca^{2+}, Co2+, Na^{+}, Zn^{2+}	Not available	Retaining more than 98% of maximum activity after pre-incubation at 50 °C for 2 h	[16]
GH16	Aga4	*Flammeovirga* sp. MY04	Not available	50	7.5	NA4, NA6	Na^{+}, DTT	Not available	Retaining more than 80% of maximum activity after pre-incubation at 50 °C for 1 h	[30, 31]
GH16	Aga672	*Aquimarina agarilytica* ZC1	98	25	7.0	NA4 and NA6 (main), NA8	No	154.3		[32]
GH16	Aga3027	*Flammeovirga* sp. OC4	93	40	9.0	NA4, NA6	Mn^{2+}, Sr^{2+}	Not available	Retaining more than 80% of maximum activity after pre-incubation at 40 °C for 48 h	[17]

(continued)

Table 1 (continued)

GH family	Protein	Bacterial species	Molecular mass (kDa)	Optimal temperature (°C)	Optimal pH	Main Products	Co-factor	V_{max} (μmol min^{-1} mg^{-1})	Thermostability	References
GH16	AgaB	*Flammeovirga* sp. OC4	106	45	8.0	NA4, NA6	No	700	Retaining more than 80% of maximum activity after pre-incubation at 45 °C for 1 h	[33]
GH16	AgaA	*Agarivorans* sp. LQ48	51	40	7.0	NA4, NA6	Mn^{2+}	909.1	Retaining more than 80% of maximum activity after pre-incubation at 40 °C for 1 h	[34]
GH16	Aga672	*Aquimarina agarilytica* ZC1	98	25	7.0	NA4 and NA6 (main), NA8	No	154.3	80% residual activity after incubating at 40 °C for 1 h	[32]
GH16	Aga862	*Pseudoalteromonas* sp. Q30F	50	45	6.5	NA4, NA6	$Co2^{+}$, Cu^{2+}, Zn^{2+}	256.41		[35]
GH16	AgaB34	*Agarivorans albus* YKW-34	49	30	7.0	NA4	No	50	80% residual activity after incubating at 40 °C for 1 h	[36]
GH16	β-agarase	*Microbulbifer* sp. Q7	65	40	6.0	NA4, NA6	No	Not available	90% residual activity after incubating at 37 °C for 2 h	[37]

GH16	CaAga1	*Cellulophaga algicola* DSM14237	39	40	7.0	NA4, NA6	Mn^{2+}, Fe^{2+}, Ca^{2+}, Co^{2+}, Mg^{2+}	36.21	80% residual activity after incubating at 70 °C for 2 h	[38]
GH16	GH16B	*Bacteroides uniformis* NP1	Not available	Not available	Not available	NA4, NA6	Not available	Not available	Not available	[11]
GH16	DagA	*Streptomyces coelicolor* A3(2)	32	40	7.0	NA4, NA6	Co^{2+}	39.06	Good thermostability at 40 °C	[10, 39]
GH16	Aga16B	*Saccharophagus degradans* 2–40	64	Not available	Not available	NA4	Not available	Not available	Not available	[12]
GH16	AgaA	*Zobellia galactanivorans*	60	40	Not available	NA4, NA6	Not available	Not available	Not available	[40, 41]
GH16	AgaA	*Marinimicrobium* sp. H1	57.5	40	6.0	NA4,	Na^{+}, K^{+}, Mg^{2+}, Ca^{2+}, Ba^{2+}		Good thermostability at 50 °C	[42]
GH16	YM01–3	*Catenovulum agarivorans* YM01	46.9	60	6.0	NA4, NA6	Na^{+}, K^{+}	1.14×10^{4}	Good thermostability at 50 °C	[18]
GH16	AgaML	*Metagenomic library*	71.6	50	7.0	NA4, NA6	Ca^{2+}, Mn^{2+}, Ba^{2+}, Co^{2+}, Na^{+}, K^{+}	967.5		[43]

(continued)

Table 1 (continued)

GH family	Protein	Bacterial species	Molecular mass (kDa)	Optimal temperature (°C)	Optimal pH	Main Products	Co-factor	V_{max} (μmol min^{-1} mg^{-1})	Thermostability	References
GH16	AgaL1	*Microbulbifer pacificus* LD25	64.3	50	7.0	NA2	SDS	16.8		[44]
GH16	PdAgaC	*Persicobacter* sp. CCB-QB2	30	60	7.5	NA2	No	2.9×10^3	Good thermostability at 55 °C	[45]
GH16	YM01–1	*Catenovulum agarivorans* YM01	37.3	50	7.0	NA2	Ca^{2+}, Mg^{2+}, K^+	4.35×10^3	Retaining more than 80% of maximum activity after pre-incubation at 50 °C for 1 h	[46]
GH16	N3–1	*Microbulbifer* sp. BN3	34.3	50	6.0	NA2, NA4	Na^+, K^+	Not available	Retaining more than 90% of maximum activity after pre-incubation at 50 °C for 1 h	[47]
GH16	ZgAgaC#	*Zobellia galactanivorans*	37.6	50 (agar)	6.5 (agarose) 9 (agar)	NA2, NA4	Not available	Not available	Not available	[48]
GH16	AgaB#	*Zobellia galactanivorans*	40	Not available	Not available	NA2, NA4 (main)	Not available	Not available	Not available	[40]
GH50	AgaB-4	*Paenibacillus agarexedens* BCRC17346	97	55	6.0	NA4	Co^{2+}, Mn^{2+}	874.61	Good thermostability at 45 °C	[49]

GH50	AgaW	*Cohnella* sp. LGH	97	50	7.0	NA4 (main), NA6	Ca^{2+}, Mg^{2+}, Na^+, K^+	387.11	Good thermostability at 55 °C	[50]
GH50	Aga_{ACN41}	*Vibrio* sp. strain CN41	110	40	7.5	NA4	Not available	5	Not available	[51]
GH50	AgWH50B	*Agarivorans gilvus* WH0801	105	40	7.0	NA4	No	1.7×10^3	90% residual activity after incubating at 50 °C for 1 h	[19]
GH50	AgWH50A	*Agarivorans gilvus* WH0801	105	30	6.0	NA4	No	0.781		[52]
GH50	VejGH50B	*Vibrio* sp. EJY3	111	65	7.0	NA4	Not available	43.5		[53]
GH50	VejGH50C	*Vibrio* sp. EJY3	107	45	9.0	NA4	Not available	229.8		[53]
GH50	VejGH50A	*Vibrio* sp. EJY3	108	45	7.0	NA2	Not available	144.7		[53]
GH50	EXB3	*Gilvimarinus chinensis*	84	30	6.0, 7.0	NA2	Ca^{2+}	126.8	Not available	[54]
GH50	Sco3487	*Streptomyces coelicolor* A3(2)	88.5	40	7.0	NA2	Co^{2+}	10.75	Good thermostability at 40 °C	[55]
GH50	β-agarase	*Agarivorans* sp. JA-1	109	40	8.0	NA2, NA4	$Na^{+,}$ Ca^{2+}	Not available		[56]
GH50	Aga50D	*Saccharophagus degradans* 2–40	84	30	7.0	NA2	Ca^{2+}	17.9	Not available	[57]

(continued)

Table 1 (continued)

GH family	Protein	Bacterial species	Molecular mass (kDa)	Optimal temperature (°C)	Optimal pH	Main Products	Co-factor	V_{max} (μmol min^{-1} mg^{-1})	Thermostability	References
GH50	AgaA11	*Agarivorans* sp. JAMB-A11	105	40	7.5–8.0	NA2	Cu^{2+}, Pb^{2+}, EDTA, SDS	371		[58]
GH50	β-agarase	*Alteromonas* sp. E-1	89	40	7.5	NA2	K^+, Na^+, EDTA			[3]
GH50	AgaA[#]	*Marinimicrobium* sp. H1	88.8	60	7.0	NA2	Ba^{2+}	Not available		[42]
GH50	AgaL4	*Microbulbifer pacificus* LD25	84.6	60	7.0	NA2	Ca^{2+}	9.6	93% residual activity after incubating at 70 °C for 180 min	[44]
GH50	VejGH50D	*Vibrio* sp. EJY3	85.5	30	7.0	NA2	Not available	3.07	Not available	[53]
GH50	AgWH50C	*Agarivorans gilvus* WH0801	83.7	30	6.0	NA2	Mg^{2+}, Mn^{2+}	1.17		[59]
GH50	AgaB	*Vibrio* sp. Strain JT0107	106	Not available	Not available	NA2	Not available	Not available	Not available	[60]
GH86	Aga86A_Wa	*Wenyingzhuangia aestuarii* OF219	79.73	30	6.5	NA2	Na^+, Mg^{2+}, Cd^{2+}	136.99		[61]
GH86	AgaJ5	*Gayadomonas joobiniege* G7	89	30	4.5	NA2, NA4, NA6 (main)	Na^+, K^+	188.6		[62]

GH86	Aga1	*Paenibacillus* sp. SSG-1	165	40	5.0	NA4, NA6	Not available	Not available		[63]
GH86	Aga3463	*Pseudoalteromonas* sp. NJ21	40	50	7.0	NA2	Ca^{2+}, Cd^{2+}, Mn^{2+}, Fe^{2+}	557		[20]
GH86	Aga86E	*Saccharophagus degradans* 2–40	146	Not available	Not available	NA2	Not available	Not available	Not available	[12]
GH118	AgaXa	*Catenovulum* sp. X3	52	52	7.4	NA6, NA8, NA10, NA12, the minimal substrate was NA8	Mg^{2+}, Ca^{2+}, Na^{+}, K^{+}	588.2	90% residual activity after incubating at 40 °C for 1 h	[21]
GH118	AgaC	*Vibrio* sp. PO-303	47	35	6.0	NA4, NA6, NA8, NA10, NA12, the minimal substrate was NA8	Not available	Not available		[64]

(continued)

Table 1 (continued)

GH family	Protein	Bacterial species	Molecular mass (kDa)	Optimal temperature (°C)	Optimal pH	Main Products	Co-factor	V_{max} (μmol min^{-1} mg^{-1})	Thermostability	References
GH118	AgaB	*Pseudoalteromonas* sp. CY24	51	40	6.0	NA4, NA6, NA8, NA10, the minimal substrate was NA8	Mg^{2+}, Ba^{2+}	Not available		[65]
GH118	AgaJA2	*Agarivorans* sp. JA-1	52	35	7.0	NA8	Not available	Not available		[66]
GH96	AgaA33	*Thalassomonas* sp. Strain JAMB-A33	158	45	8.5	A4	Ca^{2+}	Not available	Not available	[67]
GH96	AgaA	*Alteromonas agarilytica*	155	Not available	7.2	A4, A6	Ca^{2+}	Not available	Not available	[2]
GH96	AgaD	*Thalassomonas* sp. LD5	180	40	8.0	A4	Ca^{2+}	Not available	Not available	[68]
GH96	CaLJ96	*Catenovulum agarivorans*	200	37	7.0	A4	Ca^{2+}	12.65		[69]
GH96	AgaWS5	*Catenovulum sediminis* WS1-A	140	40	8.0	A4	Ca^{2+}	714.3	Not available	[22]

optimum reaction temperature (50 °C) and catalytic activity with V_{max} values of 557 μmol min^{-1} mg^{-1} [20]. Meanwhile, only four β-agarases belonging to GH118 family were identified. AgaXa from *Catenovulum* sp. X3 exhibited the best biochemical properties [21]. Besides, the α-agarase AgaWS5 from *Catenovulum sediminis* WS1-A showed a good catalysis ability toward agarose with a V_{max} value of 714.3 μmol min^{-1} mg^{-1} [22].

4 Degradation Patterns of Agarases

Agarases can be divided into endo-type hydrolases and exo-type hydrolases based on their action mode toward agarose. Either endo-type hydrolases or exo-type hydrolases, they all have the main products and corresponding degradation patterns. The analysis of agrases degradation patterns toward agaroase and NAOSs or AOSs with different DPs are critical for further study and application of agarases.

Until now, all the characterized α-agarases belong to endo-type hydrolases, and their main product from hydrolysis of agarose is agarotetraose (A4) [22, 67–70]. As for degradation patterns toward oligosaccharides, Liu et al. (2019) proved the activity of CaLJ96 to hydrolyze agaropentaose (A5) into agarotriose (A3) and A2, while it cannot act on the A3. It was speculated that A5 was the smallest substrate of all the A4-producing α-agarases [69]. Furthermore, Wang et al. (2020) illustrated the two different hydrolysis modes of a-agarase AgaD toward agarohexaose (A6). AgaD not only can act on the first α-1,3-glycosidic bond of A6 to release A2 from the non-reducing end but also can act on the second α-1,3-glycosidic bond of A6 to release A4 from the non-reducing end [70].

GH16 family β-agarases can be divided into >NA2-producing and NA2-producing β-agarase according to their products from degrading agarose with an endo-type hydrolysis mode. Take NA4, NA6-producing β-agarase Aga3027 for an example. Aga3027 acts on agarose to produce NA4 and NA6 simultaneously. The smallest substrate was agaroheptaose (A7), and minimum recognition unit was NA4 or NA6 from the non-reducing end of NAOSs [17].

Unlike GH16 family β-agarases, most of the GH50 family β-agarases belong to exo-type hydrolase. Therefore, their products were more specific than GH16 family β-agarases. For example, NA2-producing β-agarase AgWH50C could form one spot of NA2 at a short time (15 min) by degrading agarose, and the amount of NA2 gradually increased during the reaction. The results suggested that AgWH50C was an exo-type hydrolase with NA2 as the main product [59]. Another GH50 family β-agarase AgWH50B was proved; it is an NA4-producing exo-type hydrolase. The smallest cutting unit was NA4, while it could release NA4 from the non-reducing end of NA6 or NA8 [19].

The β-agarases from GH86 and GH118 families were relatively less reported. Among GH86 family β-agarases, an exo-type hydrolase Aga3463 from an Antarctic *Pseudoalteromonas* sp. strain has been reported [20]. Its degradation modes toward agarose and ENAOSs with different DPs were the same as that of NA2-producing

AgWH50C. Moreover, GH86 family β-agarase Aga86A_Wa from the marine bacterium *Wenyingzhuangia aestuarii* OF219 was proved to be the methyl group tolerating toward agarose to produce methylated ENAOSs [61]. As for GH118 family β-agarases, these enzymes exhibited endo-type activity toward agarose. Moreover, the smallest substrates were NA8. Thereby, the products of GH118 family β-agarases were larger than β-agarases from other GH families.

5 Structure and Action Mechanism of Agarases

Till date, some three-dimensional structures of β-agarases that belong to GH16, GH50, and GH86 families have been obtained, and the crystal structures of GH16 family β-agarases were the most reported. However, the structures of α-agarases have not been solved. The first two crystal structures resolved by Allouch et al. in 2003 are those of GH16 family β-agarases ZgAgaA and ZgAgaB from the marine bacterium *Zobellia galactanivorans* Dsij (PDB IDs: 1O4Y and 1O4Z) [71]. As shown in Fig. 2, 1O4Y and 1O4Z can be defined as a β-jelly roll fold, which is mainly composed of two antiparallel β-sheets, and each β-sheet was formed by seven β-strands. Meanwhile, there are several loops between these β-strands for connection. In addition, the research proved that ZgAgaA and ZgAgaB exist a long and open active site cleft that can accommodate at least eight sugar monomers. Moreover, the research further illustrated the catalytic machinery of these two β-agarases. The breakage of β-1,4-glycosidic bond by GH16 family β-agarases could be divided into two steps. The first step was the formation of glycosyl-enzyme intermediate, one of the glutamate residues (E189 in ZgAgaA) works as a general acid to make the glycosidic oxygen to promote aglycone departure. Meanwhile, the other nucleophile glutamate residue (E184 in ZgAgaA) is covalently bound to the anomeric carbon C1 of the sugar moiety. The second step was deglycosylation, and E189 in ZgAgaA works as a general base, hydrolyzing the glycosyl-enzyme intermediate to release the products [71] (Fig. 3). Next, the three-dimensional structure of *Mt*AgaA from *Microbulbifer thermotolerans* JAMB-A94 (PDB ID: 3WZ1) in GH16 family was resolved and reported. Allouch et al. (2004) illustrated the crystal structure of the inactive mutant ZgAgaA E147S in complex with NA8 (PDB ID: 1URX) (Fig. 2), which is also from GH16 family.

The first three-dimensional structure of GH50 family β-agarases was determined by Pluvinage et al. (2013) (PDB ID: 4BQ2) [72]. It is composed of a C-terminal $(\alpha/\beta)_8$ barrel fold single domain with a β-sandwich CBM-like domain at the N terminus (Fig. 2). An active channel of about 25–30 Å in length was existed for hydrolyzing substrates. The research revealed that it is consistent with a retaining catalytic mechanism whereby E695 acts as the nucleophile that attacks the anomeric carbon C1 of the sugar moiety, while E534 acts as a general acid [72]. Besides, the crystal structures of GH50 family β-agarases AgWH50C and PfGH50B have also been solved, and their PDB IDs were 5Z6P and 6XJ9 [73, 74]. There was only one

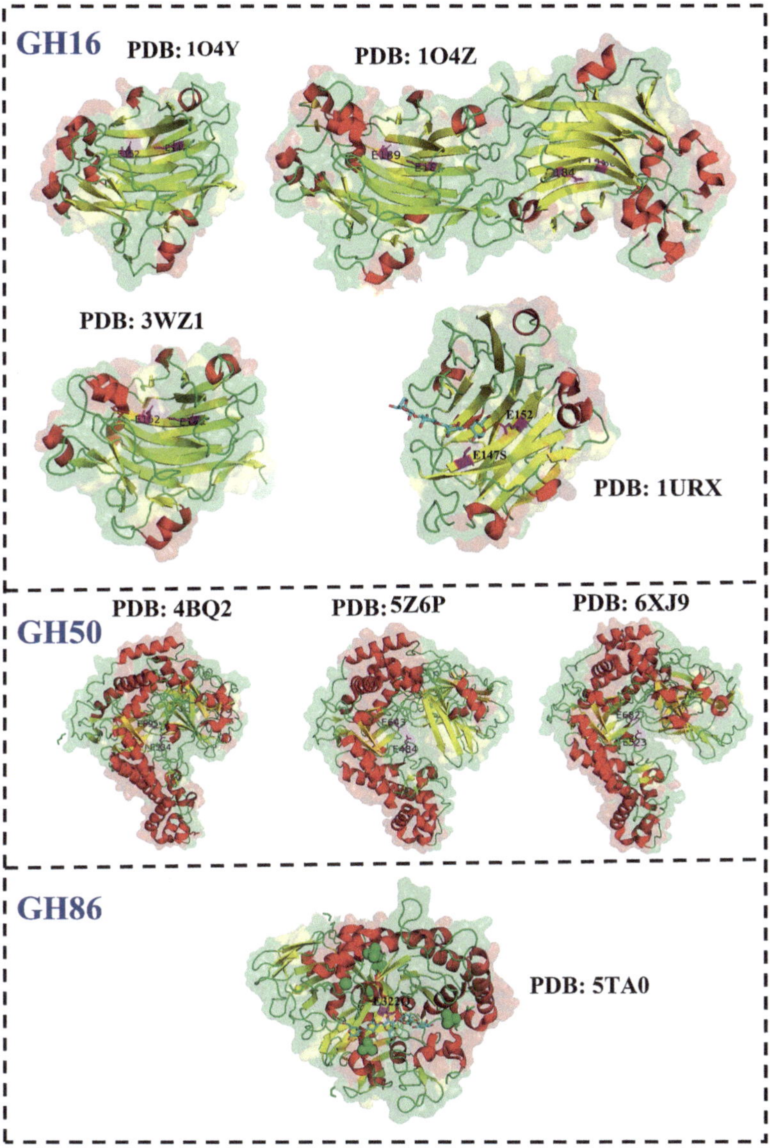

Fig. 2 The crystal structures of β-agarases from different GH families. Images were created using PDB data and PyMOL software. PDB IDs: 1O4Y, 1O4Z, 3WZ1, 1URX, 4BQ2, 5Z6P, 6XJ9, 5TA0

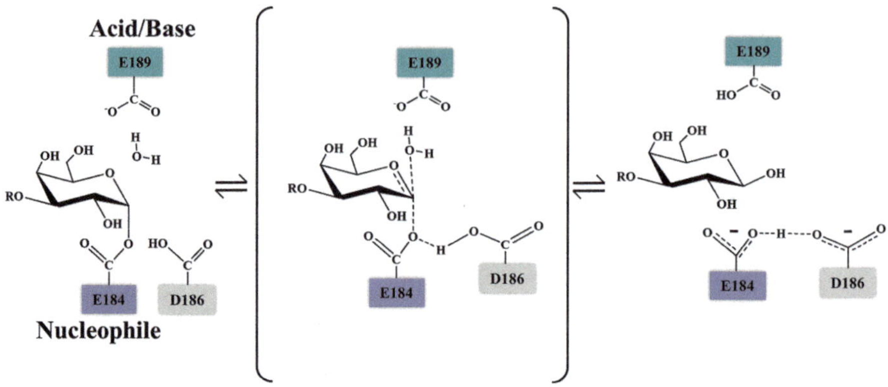

Fig. 3 The catalytic machinery of ZgAgaA

three-dimensional structure in GH86 family β-agarase from the human intestinal bacterium *B. uniformis* NP1 was determined [11]. This β-agarase is composed of a $(\alpha/\beta)_8$ core barrel with two Ig-like domains (5TA0) (Fig. 2).

6 Modification Strategies for Agarases

To enhance the application performance of agarases, many research studies have been devoted to improve the enzymatic properties of agarases, such as catalytic activity and stability. In general, the strategies of the enzyme immobilization (EI) and protein engineering (PE) were of great significance for improving enzyme properties during the past decades [75].

6.1 Enzyme Immobilization Strategies Used in Agarases

EI technology has always been utilized to improve the application stability and increase the reusability of agarases due to the unstable and incapable of recycling of free enzyme (Table 2). Basawaraj et al. (2013) reported an effective method to immobilize β-agarases by using amberlite IRA-900 as support material. Compared with the free enzyme, the optimum temperature of the immobilized enzyme has increased by 5 °C (from 40 to 45 °C). Besides, immobilized β-agarase showed better stability at high temperature than free β-agarase. The immobilized β-agarase retained 74% and 50.84% of residual activity at 55 and 60 °C, while the free β-agarase retained only 40% and 16.79% of residual activity. Moreover, the immobilized enzyme displayed good repeated utilization, which still retained around 60% of residual activity after eight cycles of use [76]. For better to isolate the

Table 2 The immobilized methods were applied in agarase

Immobilized agarase	Immobilized material	Optimal pH of free enzyme/immobilized enzyme	Optimal temperature (°C) of native enzyme/immobilized enzyme	Comparison of pH stability	Comparison of temperature stability	References
β-agarase from *Pseudomonas aeruginosa* AG LSL-11	Amberlite IRA-900	8.0/8.0	40/45	The immobilized enzyme retained 97% and 85% of its initial activity at pH 9.0 and 10.0, while free enzyme retained 80% and 60%	The immobilized enzyme retained 74% and 50.84% of its initial activity at 55 and 60 °C, while free enzyme retained 40% and 16.79%.	[76]
β-agarase from *Vibrio natriegens* CICC 23820	Carboxyl-functioned magnetic Fe_3O_4 nanoparticles	8.0/7.0	50/40	The immobilized enzyme retained 93.2% of its initial activity at pH 9.0, while free enzyme only retained 28.5%	The immobilized enzyme retained 48% of its initial activity at temperatures higher than 50 °C, while free enzyme lost its activity.	[77]
β-agarase from *Marinobacter adhaerens* G4	Magnetic ferriferous oxide	No data	No data	The pH stability of immobilized enzyme was enhanced mildly	The temperature stability of immobilized enzyme was enhanced mildly.	[83]
β-agarase AgWH50B from *Agarivorans gilvus* WH0801	Functionalized silica-coated modified magnetite nanoparticles	7.0/7.0	40/35	No data	–	[82]
NABH AgaWH117 from *Agarivorans gilvus* WH0801	Functionalized silica-coated modified magnetite nanoparticles	6.0/7.0	27.5/22.5	No data	–	[82]
β-agarase from *Vibrio natriegens* CICC 23820	Tannic acid modified-Fe_3O_4 magnetic nanoparticles	6.5/6.5	50/50	The pH stability of immobilized enzyme was better than free enzyme	The immobilized enzyme retained 77% of its initial activity at 60 °C, while free enzyme retained 70%.	[78]

immobilized enzyme from the reaction system, the carboxyl-functioned magnetic Fe_3O_4 nanoparticles were developed to immobilize β-agarase via covalent binding. The research data suggested that the immobilized β-agarase achieved better thermal and pH stability. The immobilized enzyme still retained 93.2% of residual activity after an incubation at pH 9.0 for 1 h, whereas that of the free enzyme only retained 28.5% of residual activity. As for thermal stability, the free enzyme was completely inactivated after an incubation at 50 °C for 1 h, while the immobilized enzyme still retained approximately 48% of residual activity. Moreover, the immobilized enzyme retained approximately 40% of residual activity after seven cycles of use [77]. Furthermore, the researchers explored another involved Fe_3O_4 support material (tannic acid–modified Fe_3O_4 magnetic nanoparticles) for immobilizing β-agarase. The residual activity of immobilized enzyme still maintained 77% after incubating at 60 °C for 1 h, while free enzyme retained 70% of its initial activity [78].

L-AHG, one monomer of agarose, exhibits some biological activities, such as anti-inflammatory, skin-whitening and anticariogenic activities [79–81]. Its enzymatic preparation process is usually divided into two steps by using two different types of enzymes, including β-agarase and α-neoagarobiose hydrolase (NABH). Wang et al. (2018) developed a coimmobilization system to coimmobilze an NA4-producing β-agarase AgWH50B and an NABH K134D (the mutant of AgaWH117) for one-step production of L-AHG and A3 from agarose by using trichlorotriazine-functionalized silica-coated magnetite nanoparticles as supports. The coimmobilized AgWH50B and K134D showed a yield of 40.6% for producing L-AHG from agarose, whereas that of the yield of 34.1% by directly using free enzyme [82].

6.2 Protein Engineering Strategies Used in Agarases

The various PE strategies (including random mutagenesis, directed evolution, and rational design) could be applied in improvement of the biological properties of target enzyme. As shown in Table 3, Shi et al. (2008) adopted directional evolution technology via error-prone polymerase chain reaction (PCR) and deoxyribonucleic acid (DNA) shuffling to increase the thermostability of β-agarase AgaB. This made an increment of 4.6 °C of the melting temperature (*T*m) compared with the wild-type AgaB [84]. The random mutagenesis method has been performed on NA2- and NA4-producing β-agarase AgaB from *Z. galactanivorans* to enhance its catalytic efficiency and thermostability. The results suggested that the activities of three mutants E99K, T307I, and E99K-T307I were reached approximately 140%, 190%, and 200% of the initial AgaB, respectively. All of them were more stable than wild AgaB in which E99K-T3071 exhibited the best thermostability, and its *T*m was increased by 5.2 °C [85]. The same strategy was applied in another GH16 family β-agarase AgaA from the same bacterium *Z. galactanivorans*, and three mutants (S63K, C253I, and S63K-C253I) were obtained. Therein, only S63K achieved an improvement of catalytic activity, which reached approximately 126% of the wild-type enzyme. Besides, the thermostability of the mutant S63K reached 125% of the

Table 3 The different technologies for enhancing enzyme activity or temperature stability of agarase

Wild-type enzyme	Mutant enzyme	Bacterial source	Modification strategy	Improvement optimal Temperature (°C)	Change of degradation products	Improvement of catalytic activity	Improvement of thermostability	References
AgaB	S2	*Pseudoalteromonas* sp. CY24	Error-prone PCR and DNA-shuffling	Not available	No obvious change	Not available	The half-life of S2 was 18.4-fold longer than AgaB at 40 °C	[84]
AgaB	E99K	*Zobellia galactanivorans*	Random mutagenesis	+5	Not available	140% at 40 °C	T_m + 2.6 °C	[85]
AgaB	T307I	*Zobellia galactanivorans*	Random mutagenesis	+5	Not available	190% at 40 °C	T_m + 3.8 °C	[85]
AgaB	E99K–T307I	*Zobellia galactanivorans*	Random mutagenesis	+5	Not available	200% 40 °C 260% with 1 mM $CaCl_2$ at 40 °C	T_m + 5.2 °C	[85]
AgaA	S63K	*Zobellia galactanivorans*	Site-Directed Mutagenesis	+5	166%, 257%, and 220% NA6, and 230%, 427%, and 350% more NA4 than the wild-type in sol, gel, and nonmelted powder agar, respectively, at 45 °C over 24 h	126% at 40 °C	128% higher than the wild-type enzyme when treated at 45 °C for 1 h	[41]

(continued)

Table 3 (continued)

Wild-type enzyme	Mutant enzyme	Bacterial source	Modification strategy	Improvement optimal Temperature (°C)	Change of degradation products	Improvement of catalytic activity	Improvement of thermostability	References
AgaG4	AgaG4-T140	*Flammeovirga* sp. MY04	Deletion of a non-catalytic region	Not change	Not change	35 times increase	Not available	[31]
rAgaM1	trAgaM1	Mangrove sediment microbiome	Deletion of a non-catalytic region from the C-terminal	Not available	Not available	Not available	Not available	[86]
AgWH50C	K621F	*Agarivorans gilvus* WH0801	Multiple cross-linked rational design Protocol	+8	Not available	K621F has five times higher activity than AgWH50C at 41 °C	K621F retained 38% residual activity at 40 °C for 30 min, while the AgWH50C totally inactivation	[74]

wild-type enzyme at 45 °C. Moreover, the ability of NAOSs production was tested, and the result indicated that S63K produced 166%, 257% and 220% more NA6, and 230%, 427% and 350% more NA4 than the wild-type AgaA from sol, gel, and powder agar, respectively [41].

Apart from the random mutagenesis, the rational design methods also have been applied in modifying agarases. Han et al. (2015) reported that the truncated β-agarase rAgaG4-T140 made an approximately 35 times increase over than that of wild-type enzyme, which was constructed by deleting the noncatalytic region (NCR) fragment (S364-H503) from the wild-type enzyme [31]. Furthermore, a mutant K621F with increased enzyme activity and stability was obtained by using rational design tactic based on the crystal structure of NA2-producing β-agarase AgWH50C. K621F achieved an approximately 45% improvement as well as the optimal temperature increased by 8 °C (from 30 to 38 °C) [74].

7 Preparation Methods for Agarases

7.1 *Isolation and Purification of Agarases from the Wild Bacteria*

In order to isolate and purify agarases from the wild agarases-producing bacteria, several purification processes are indispensable (Table 4). The first step is ultrafiltration and ammonium sulfate precipitation for removing the medium solution and condensing crude enzyme [2, 87–89], and the ammonium sulfate can be replaced by ethanol-precipitating protein [6]. then followed by anion-exchange chromatography and gel filtration chromatography for obtaining pure agarase [7, 90].

Apart from the anion-exchange chromatography and gel filtration chromatography, the affinity chromatography also has been applied in purifying agarase. Potin et al. (1993) adopted Sepharose CL-6B to isolate agarase, which used the agaro-oligosaccharides solution as the elution buffer to specifically elute the agarase bound in the filler [2]. Besides, the native-polyacrylamide gel electrophoresis (native-PAGE) was applied to purify agarase and to identify the molecular weight (Mw) [89].

7.2 *Heterologous Expression of Agarase*

Although the agarase can be obtained by isolating from the wild agrases-producing bacteria, but the deficiency of expression level may bring a lot of work. Moreover, it is only suitable for the preparation of extracellular enzymes. Thereby, the heterologous expression has been applied in preparing agarases. The expression system of recombinant *Escherichia coli* has been wildly used in expressing agarase,

Table 4 The agarases purified from wild bacteria

Enzyme	Bacterial source	Molecular mass (kDa)	Medium for enzyme production	Change of degradation products	References
α-agarase	*Alteromonas agarlyticus* GJ1B	180 (SDS-PAGE); 360 (gel filtration)	Y-2 modified medium was composed of A/B/C (v/v/v = 1000/60/0.33), A: 25 g/L NaC1, 5.0 g/L $Mg_2SO_4 \cdot 7H_20$, 0.2 g/L $CaC1_2 \cdot 2H_20$, 0.1 g/L KCl, 2.0 g/L NH_4NO_3, 0.2 g/L casamino acids, 0.2 g/L yeast extract and 2.5 g/L agar, B: 0.2 M sodium phosphate, pH 7.2, C: 10% $FeSO_4 \cdot 7H_20$	Ultrafiltration → Ammonium sulfate precipitation → Affinity chromatography by with Sepharose CL-6B column → Fast performance liquid chromatography (FPLC) with Mono Q column	[2]
β-agarase	*Vibrio* sp. JT0107	107	The medium was composed of 0.3% (w/v) agar, 0.02% (w/v) K_2HPO_4 and 0.1% (w/v) yeast extract, which were dissolved in seawater, pH 8.0	Ammonium sulfate precipitation → Dialysis → QAE-Toyopearl chromatography for removing insoluble particles and proteins →Fast performance liquid chromatography (FPLC) with Mono Q column	[87]
β-agarase	*Alteromonas* sp. C-1	52	The medium was composed of 0.25% casein hydrolysate, 0.05% yeast extract, 2.5% NaCl, 0.06% NaH_2PO_4, 0.5% $MgSO_4$, 0.002% $FeSO_4 \cdot 7H_20$, 0.01% $CaCl_2$, and 0.15% agar, pH 7.25	Ammonium sulfate precipitation → Dialysis → DEAE-cellulose column → Sephadex G-75 column	[88]
Two of β-agarases	*Agarivorans* sp. HZ105	54 and 58	The agarases solution was prepared by using agar plate	Native-polyacrylamide gel electrophoresis (Native-PAGE)	[89]

β-agarase	*Alteromonas* sp. SY 37–12	39.5	The medium was composed of 2.0% NaCl, 0.1% K_2HPO_4, 0.05% $MgSO_4$, 0.1% NH_4Cl, 0.01% $CaCl_2$, and 0.2% agar	Ammonium sulfate precipitation → Dialysis → DEAE-Sepharpse column → Sephacryl S-100HR	[91]
β-agarase	*Pseudoalteromonas antarctica* strain N-1	33	The medium was composed of 0.25% casein hydrolysate, 0.05% yeast extract, 2.5% NaCl, 0.06% NaH_2PO_4, 0.5% $MgSO_4$, 0.002% $FeSO_4 \cdot 7H_20$, 0.01% $CaCl_2$, and 0.2% agar	Ammonium sulfate precipitation → Dialysis → Sephadex G-75 column → DEAE-cellulose column	[92]
β-agarase	*Vibrio* sp. AP-2	20	The medium was composed of 0.5% (w/v) peptone, 0.1% (w/v) yeast extract and 3% (w/v) NaCl in the algal extract, pH 7.4	Ammonium sulfate precipitation → Carboxymethyl-Sephadex C-50 chromatography →Affinity chromatography by with Sepharose CL-6B column →Fractogel H W-55 chromatography →DEAE-Fractogel65UM chromatogvaphy	[90]
β-agarase AgaA34	*Agarivorans albus* YKW-34	50	The medium was composed of 2.5% NaCl, 0.01% K_2HPO_4, 0.5% $MgSO_4 \cdot 7H_20$, 0.1% KCl, 0.02% $CaCl_2$, 0.002% $FeSO_4 \cdot 7H_20$, and 0.23% agar with 0.27% yeast extract, pH 7.8	Ultrafiltration→ Dialysis → DEAE-Sepharose FF column → Sephacryl S-100 HR	[7]

(continued)

Table 4 (continued)

Enzyme	Bacterial source	Molecular mass (kDa)	Medium for enzyme production	Change of degradation products	References
β-agarase	*Cytophaga flevensis*	26	The medium was composed of 1.0 g/L NH_4Cl, 1 g/L K_2HPO_4, 0.5 g/L $MgSO_4 \cdot 7H_20$, 0.02 g/L $CaCl_2 \cdot 2H_20$, and 0.3% agar	Alcohol precipitation → Sephadex G-100 → DEAD-Sephadex A-50	[6]
β-agarase	*Alteromonas* sp. E-1	82 (SDS-PAGE); 180 (Superdex 200 gel filtration)	The medium was composed of 1.45 g/L NH_4Cl, 0.75 g/L KCl, 0.075 g/L K_2HPO_4, 12.35 g/L $MgSO_4 \cdot 7H_20$, 1.45 g/L $CaCl_2 \cdot 2H_20$, 1.25 g/L casamino acids, 1.25 g/L yeast extract in 10 mm Tris-HCl buffer (pH 8.0) with 0.1% (w/v) agar	Ultrafiltration→ Dialysis → DEAE-Toyopearl 650 M column →FPLC with Mono Q column	[3]
β-agarase LSL-1	*Acinetobacter* sp. AG LSL-1	100	Czapek-Dox medium with 0.3% (w/v) agar as the sole carbon source	Ammonium sulfate precipitation → DEAE Sephacel column → Sephacryl S-200 column	[9]
β-agarase	*Pseudomonas atlantica*	Not determined	The medium was composed of 25 g/L NaCl, 0.1 g/L K_2HPO_4, 0.1 g/L $MgSO_4$ $7H_20$, 0.2 g/L KCl, 5.0 g/L $CaCl_2$, 0.3% (w/v) $FeSO_4$ $7H_20$, 2.5 g/L casamino acids and 2.5 g/L Bacto agar	Ammonium sulfate precipitation → Sepharose 6B	[4]

including β-agarase and α-agarase [19, 69]. The plasmid for carrying target gene contains pEASY-E2, pET22b, pET28a(+), pET151/D-TOPO, and so on [19, 44, 68, 69], which can be induced by specific compounds. Such as the plasmid with T7 promoter can be induced by isopropyl β-D-Thiogalactoside (IPTG) or its structural analogues α-lactose. The advantages of *Escherichia coli* expression system include easy operation, high expression level, and stable expression. However, it is not very convenient to secrete agarase that limits its industry application. Thus, the recombinant *Bacillus subtilis* system was developed for secretion of an NA2, NA4, and NA6-producing β-agarase AgaA7 with a signal peptides (SPs) library; therein, the SP lipA could achieve the most secretory effect with a 1.44-fold increase than the initial engineered strain [93] (Table 5). In addition, there are several other β-agarases have been expressed in different types of *B. subtilis* hosts, including *B. subtilis* WB800n, *B. subtilis* ISW1214, and *B. subtilis* RIK1285 [35, 93, 94]. Recently, Jeon et al. (2021) explored the *Corynebacterium glutamicum* for secreting β-agarase [95].

After successful expression, the purification was necessary for obtaining pure agarases. The affinity chromatography can be used for purification of agarase. For example, the recombinant NA4-producing β-agarase AgWH50B with a 6 × His tag protein can be purified by using Ni^{2+}-NTA resin [19]. The target β-agarase and Ni^{2+}-NTA medium can be specifically combined, then eluting with different concentrations of imidazole solution. Moreover, the size-exclusion chromatography can be furtherly used to obtain purer enzyme for illustrating the crystal structure of agarase [96].

Table 5 The different expression system used in agarases production

Enzyme	Bacterial source	Expression plasmid	Expression host	References
Aga862	*Pseudoalteromonas* sp. Q30F	pHT43	*Bacillus subtilis* WB800n	[35]
rAgaO	*Microbulbifer*	pOAG1	*Bacillus subtilis* ISW1214	[94]
RagaA7	*Microbulbifer* JAMB-A7	pUC18	*Bacillus subtilis* ISW1214	[26]
AgaA	*Microbulbifer* JAMB-A94	pBAG1	*Bacillus subtilis* ISW1214	[29]
RagaA11	*Agarivorans* sp. JAMB-A11	Not available	*Bacillus subtilis* ISW1214	[58]
AgaA7	*Pseudoalteromonas hodonsis* sp. nov	pGEX-5X-1	*Bacillus subtilis* RIK1285	[93]
DagA	*Streptomyces coeruleus*	Phcp-CgR	*Corynebacterium glutamicum*	[95]
EXB3	*Gilvimarinus chinensis*	Phcp-CgR	*Corynebacterium glutamicum*	[95]

8 Applications for Agarases

8.1 Agarases for Agaro-Oligosaccharides Preparation

As aforementioned, the agarase could cleave the glycosidic bond of agarose to produce different types of agaro-oligosaccharides. This was of great significant for applying in food, pharmaceutical, and cosmetic industries due to its good physiological activities. For example, Wang et al. (2017) proved the anti-inflammatory activity of NA4 toward to lipopolysaccharide (LPS)-induced RAW264.7 cells [97]. ENAOS mixtures with different average DPs (DP3, DP5, and DP8) exhibited different degrees of antioxidant activities in vitro tests [98]. In addition, the research illustrated the potential prebiotic activity of ENAOSs [98]. Wang et al. (2020) reported the effect of AOSs with DPs from 3 to 9 on alleviating type 2 diabetes mellitus (T2DM) mice induced by high-fat diet (HFD) [99]. Apart from aforementioned bioactivities, antifatigue [100, 101], anti-nerve injury [102], anti-hepatic injury [103], antitumor [104], and antiviral [105] bioactivities of agaro-oligosaccharides have been wildly reported and studied. L-AHG, the monomer of agarose, has proved the anti-cavities and skin-whitening effects [81, 106]. Moreover, some research studies demonstrated that the physiological activities of agaro-oligosaccharides correlated with their structures and DPs [14, 107].

Considering the abundant physiological activities of agaro-oligosaccharides, the development of effective agaro-oligosaccharides preparation methods is of great importance. The present methods mainly contain chemical (acidolysis), enzymatic, and fermentation [14]. The enzymatic method by using agarases can exactly cleave agar or agarose polymers to produce ENAOSs and EAOSs, leading to the products are more specific than acidolysis. In addition, the reaction conditions of enzymatic method are milder. Thus, the enzymatic method based on agarases was of important role to produce agaro-oligosaccharides with specific DP for further revealing the structure–function relationship. Ma et al. (2019) used the truncated β-agarase (Aga16A-ΔCBM), cloned from the marine bacterium *Microbulbifer* sp. BH-1 to hydrolyze agarose for NA4 production, which was performed in 100 mL of 2 g agarose with 200 U Aga16A-ΔCBM at 45 °C for 24 h. The research data suggested an achieved high yield of NA4 (93.2%, w/w) (Table 6) [108]. Except for NA4-producing β-agarase, the β-agarases with other products also have been utilized in NAOSs preparation. For example, NA4 and NA6-producing β-agarase AgaA was applied to decompose 2 g agarose dissolved in 1 L buffer for simultaneously obtaining NA4 and NA6, and their yields were 43.35% (w/w) and 41.65% (w/w), respectively (Table 6) [109]. Another NA4 and NA6-producing β-agarase AgaG1 from *Alteromonas* sp. GNUM1 achieved a high conversion of agarose, which led to 3.8 g/L NA2 and 6.4 g/L NA4 production from agarose with a concentration of 10 g/L (w/v) [110].

To achieve more efficient production of NA2, Yan et al. (2020) exploited a two-step enzymatic hydrolysis process. The process was composed of an NA2, NA4-producing GH16 family β-agarase AgaA and an NA2-producing GH50 family

Table 6 The enzymatic method used in agaro-oligosaccharides preparation

Enzymatic products	Bioactivities of products	Enzymatic substrate	β-Agarases for enzymatic hydrolysis	Purity	Yield w/w (%)	References
NA4 NA6	Antioxidant, potential prebiotic activity, anti-inflammatory, anti-diabetic, and anti-obesity, antifatigue, anti-hepatic injury, antitumor, anti-melanogenic	0.2% (w/v) agarose	β-Agarase AgaA cloned from marine *Pseudoalteromonas* sp. CY24	–	43.35% 41.65%	[109]
NA2 NA4 NA6	Antioxidant, potential prebiotic activity, anti-inflammatory, anti-diabetic, and anti-obesity, antifatigue, anti-hepatic injury, antitumor, anti-melanogenic	9.63% (w/v) agar	Agarases produced by *Pseudomonas vesicularis* MA103 and *Aeromonas salmonicida* MAEF108, which have activity of 250 U/mL and 278 U/mL respectively.	100% 100% 100%	5.8% 4.6% 7.2%	[112]
NA2 NA4	Antioxidant, potential prebiotic activity, anti-inflammatory, anti-diabetic, and anti-obesity, antifatigue, anti-hepatic injury, anti-melanogenic	1% (w/v) agarose	β-Agarase cloned from marine from *Alteromonas* sp. *GNUM1*	–	38% 64%	[110]

(continued)

Table 6 (continued)

Enzymatic products	Bioactivities of products	Enzymatic substrate	β-Agarases for enzymatic hydrolysis	Purity	Yield w/w (%)	References
NA2 NA4 NA6 NA8	Antioxidant, potential prebiotic activity, anti-inflammatory, anti-diabetic, and anti-obesity, antifatigue, anti-hepatic injury, antitumor, anti-melanogenic	0.5% (w/v) agarose	Agarases from *Pseudomonas atlantica*	63.6% 66.6% 71.9% 64.8%	5.8% 13% 17% 15%	[113]
NA4	Antioxidant, potential prebiotic activity, anti-inflammatory, anti-diabetic, and anti-obesity, antifatigue, anti-hepatic injury, anti-melanogenic	0.5% (w/v) agar	BN3 β-agarase cloned from *Micrococcaceae*. Ssp.	100%	44.8%	[114]
NA4	Antioxidant, potential prebiotic activity, anti-inflammatory, anti-diabetic, and anti-obesity, antifatigue, anti-hepatic injury, anti-melanogenic	2% (w/v) agarose	β-Agarase Aga16A-ΔCBM cloned from *Microbulbifer* sp. BH-1	–	93.2%	[108]

(continued)

Table 6 (continued)

Enzymatic products	Bioactivities of products	Enzymatic substrate	β-Agarases for enzymatic hydrolysis	Purity	Yield w/w (%)	References
NA2 NA4 NA6 NA8 NA10 NA12	Antioxidant, potential prebiotic activity, anti-inflammatory, anti-diabetic, and anti-obesity, antifatigue, anti-hepatic injury, anti-melanogenic	0.3% (w/v) agar	β-Agarase purified from *Vibrio natriegens*	99.3% 98.9% 98.0% 97.6% 97.3% 97.4%	23.2% 35.8% 10.2 7.5% 4.2% 3.2%	[111]
NA2	Anti-diabetic and anti-obesity, antifatigue, anti-hepatic injury, anti-melanogenic	1% (w/v) agar	β-Agarase A and B cloned from *Marinimicrobium* sp. H1	94%	>89.2%	[42]

β-agarase AgaB, to produce NA2 with an extremely high yield (94%, w/w) [42]. Moreover, an endo-type NA4, NA6-producing β-agarase DagA and an exo-type NA2-producing β-agarase EXB3 were coexpressed in extracellular of *C. glutamicum*. The crude fermentation enzyme solution was used to hydrolyze agar for NA2 production with a two-stage hydrolysis process, while almost all the agar was transformed into NA2 in 8 h [95]. NAOSs with different DP can produced from decomposing agarose by using different β-agarases. Surprisingly, Lin et al. (2019) developed the method for obtaining NAOSs with DPs from 2 to 12 by using the same β-agarase via controlling the reaction time [111].

Although several α-agarases have been successfully cloned and expressed [2, 22, 67–69], but their ability for AOSs preparation have not been performed.

8.2 *Agarases for Red Algae Biorefinery*

Another important usage of agarases is to saccharify red algae for bioethanol or other chemicals production (Table 7). In 2012, the agarose transformation process combining chemical liquefaction with enzyme saccharification was established for ethanol production [115]. Firstly, the acetic acid was used for liquefaction of agarose and achieved a degree of liquefaction of 95.6% (w/w). Then, the simultaneous saccharification and fermentation (SSF) process was progressed at 30 °C for 72 h.

Table 7 The biochemicals production based on enzymatic saccharification by using agarase

Product	Method	Agarses used	Yield of product	References
Ethanol	Combining chemical liquefaction with enzymatic saccharification	NA4 and NA6-producing β-agarase DagA, Aga16B, NA2-producing β-agarase Aga50D	35.9 (%theoretical max. ethanol from galactose)	[115]
Ethanol	Combining chemical liquefaction with enzymatic saccharification	NA2-producing β-agarase Aga50D	37.1 (%theoretical max. ethanol from galactose)	[116]
Biohydrogen	Combining fermentation and enzymatic saccharification	NA8, NA10, NA12 and NA14-producing β-agarase AgaXa	114.2 mL/g	[117]
Ethanol	Combining enzymatic saccharification and fermentation	NA4 and NA6-producing β-agarase AgaH71, NA2-producing β-agarase AgaG1	1.97 g/L ethanol from 10 g/L agarose	[118]

Herein, the saccharification was performed with three agarases (NA4 and NA6-producing β-agarase DagA, Aga16B, NA2-producing β-agarase Aga50D). And an NABH (EMBL id: ABD81917) was used for decomposing liquefied agarose to mono sugars, including D-Gal and L-AHG. Then, the *Saccharomyces cerevisiae* D5A (ATCC 200062) was used for metabolizing D-Gal to produce ethanol. By calculation, 4.4 g/L ethanol was obtained from totally 30 g/L agarose [115] (Fig. 4). In 2013, the researchers further optimized the conditions of chemical liquefaction for reducing the acid concentration and liquefaction time and for improving the reducing sugar production. Using 3.0% (w/v) acetic acid to act on 7% (w/v) agarose at 130 °C for 0.5 h reached a yield of 30.6% (w/v) of reducing sugar production, while the yield of reducing sugar was only 16.7% (w/v) in the previous study. Furthermore, SSF was carried out at 30 °C for 72 h. Different from the previous study, the saccharification was performed with the only NA2-produicng β-agarase Aga50D and an NABH [116] (Fig. 4).

Apart from producing ethanol, Wu et al. (2017) developed the enzymatic saccharification with agarolytic enzymes (NA8, NA10, NA12, and NA14-producing β-agarase AgaXa and the NABH NH852) and fermentation by *Enterobacter* sp. CN1 to obtain biohydrogen from agar. After calculation, 5047 ± 228 mL/L biohydrogen was obtained from 50 g/L agar [117].

The aforementioned saccharification processes were carried out in vitro. Lee et al. (2019) constructed the engineered *S. cerevisiae* harboring *AGAH71* and *AGAG1* genes encode NA4 and NA6-producing β-agarase AgaH71 and NA2-producing β-agarase AgaG1, respectively, and *NABH558* gene encoding NABH to degrade agarose for ethanol production. The engineered stain was cultured in two stages, which can be divided at 40 °C for 48 h (for saccharification) and then 30 °C

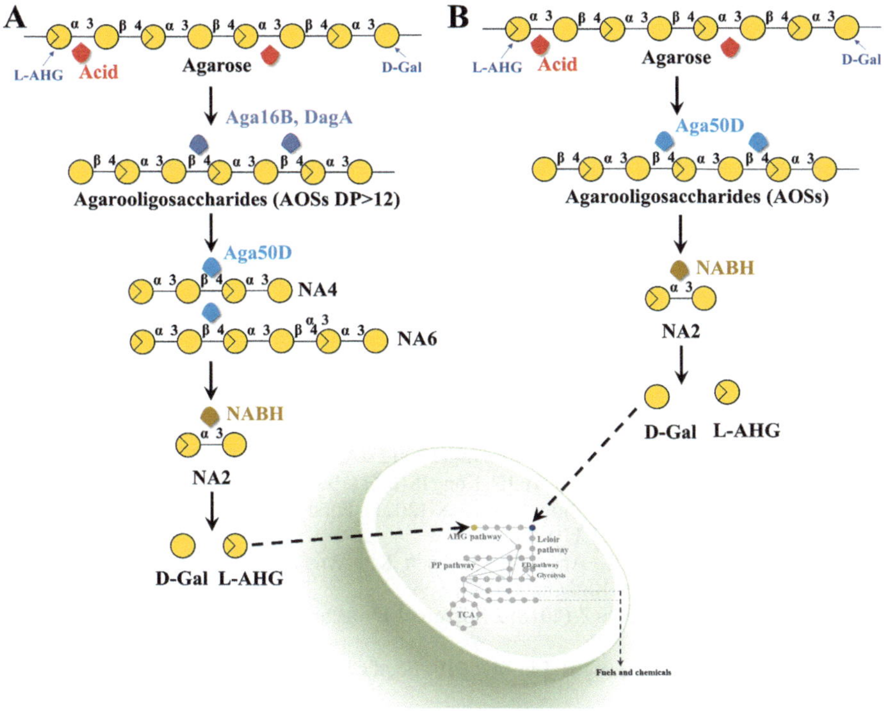

Fig. 4 Ethanol production from saccharified agarose combing with acidolysis and enzymolysis based on agarases

for 72 h to obtain ethanol. After calculation, 1.97 g/L ethanol was produced from metabolism of agarose with a concentration of 10 g/L agarose [118].

8.3 *Agarases for Recovery of DNA from Agarose Gel*

Agarose gel has been wildly used in identification of DNA length. Some agarases could degrade agarose at a high temperature, so the agarases with good temperature stability should be a powerful tool for the recovery of DNA. For example, Takara Company applied the β-agarase to dissolve DNA agarose gel at 60 °C.

References

1. Fu XT, Kim SM (2010) Mar Drugs 8:200. https://doi.org/10.3390/md8010200
2. Potin P, Richard C, Rochas C, Kloareg B (1993) Eur J Biochem 214:599

3. Kirimura K, Masuda N, Iwasaki Y, Nakagawa H, Kobayashi R, Usami S (1999) J Biosci Bioeng 87:436
4. Groleau D, Yaphe W (1977) Can J Microbiol 23:672
5. Ohta Y, Hatada Y, Miyazaki M, Nogi Y, Ito S, Horikoshi K (2005) Curr Microbiol 50:212. https://doi.org/10.1007/s00284-004-4435-z
6. Van der Meulen HJ, Harder W (1975) Antonie Van Leeuwenhoek 41:431
7. Fu XT, Lin H, Kim SM (2008) Appl Microbiol Biotechnol 78:265. https://doi.org/10.1007/s00253-007-1303-3
8. Suzuki H, Sawai Y, Suzuki T, Kawai K (2003) J Biosci Bioeng 93:456. https://doi.org/10.1016/S1389-1723(02)80092-5
9. Lakshmikanth M, Manohar S, Lalitha J (2009) Process Biochem 44:999. https://doi.org/10.1016/j.procbio.2009.04.025
10. Bibb MJ, Jones GH, Joseph R, Buttner MJ, Ward JM (1987) J Gen Appl Microbiol 133:2089. https://doi.org/10.1099/00221287-133-8-2089
11. Pluvinage B, Grondin JM, Amundsen C et al (2018) Nat Commun 9:1043. https://doi.org/10.1038/s41467-018-03366-x
12. Ekborg NA, Taylor LE, Longmire AG, Bernard H, Weiner RM, Hutcheson SW (2006) Appl Environ Microbiol 72:3396. https://doi.org/10.1128/AEM.72.5.3396-3405.2006
13. Ha JC, Kim GT, Kim SK, Oh TK, Yu JH, Kong ISBAB (1997) Biotechnol Appl Biochem 26:1–6
14. Jiang C, Cheng D, Liu Z, Sun J, Mao X (2021) Food Res Int 145:110408. https://doi.org/10.1016/j.foodres.2021.110408
15. Normand V, Lootens DL, Amici E, Plucknett KP, Aymard P (2000) Biomacromolecules 1:730. https://doi.org/10.1021/bm005583j
16. Liu Y, Tian X, Peng C, Du Z (2018) J Microbiol Biotechnol 29:235. https://doi.org/10.4014/jmb.1810.09065
17. Chen X, Lin H, Jin M, Zeng R, Lin M (2019) Food Chem 295:311. https://doi.org/10.1016/j.foodchem.2019.05.132
18. Cui F, Dong S, Shi X, Zhao X, Zhang XH (2014) Mar Drugs 12:2731. https://doi.org/10.3390/md12052731
19. Liang Y, Ma X, Zhang L, Li F, Liu Z, Mao X (2017) J Agric Food Chem 65:7982. https://doi.org/10.1021/acs.jafc.7b01533
20. Li J, Xie M, Gao Y (2019) Int J Biol Macromol 129:162. https://doi.org/10.1016/j.ijbiomac.2019.01.204
21. Xie W, Lin B, Zhou Z et al (2013) Appl Microbiol Biotechnol 97:4907. https://doi.org/10.1007/s00253-012-4385-5
22. Lee CH, Lee CR, Hong SK (2019) Appl Microbiol Biotechnol 103:8403. https://doi.org/10.1007/s00253-019-10056-1
23. Lee Y, Jo E, Lee YJ et al (2018) J Microbiol Biotechnol 28:776. https://doi.org/10.4014/jmb.1709.09050
24. Wang Y, Liu T, Guo S et al (2018) J Gen Appl Microbiol 64:276. https://doi.org/10.2323/jgam.2018.02.002
25. Chen X-L, Hou Y-P, Jin M, Zeng R-Y, Lin H-T (2016) J Agric Food Chem 64:7251. https://doi.org/10.1021/acs.jafc.6b02998
26. Ohta Y, Nogi Y, Miyazaki M et al (2004) Biosci Biotechnol Biochemist 68:1073. https://doi.org/10.1271/bbb.68.1073
27. Oh C, Nikapitiya C, Lee Y et al (2010) J Ind Microbiol Biotechnol 37:483. https://doi.org/10.1007/s10295-010-0694-9
28. Lee Y, Oh C, De Zoysa M et al (2013) J Microbiol Biotechnol 23:913. https://doi.org/10.4014/jmb.1209.09009
29. Ohta Y, Hatada Y, Nogi Y et al (2004) Appl Microbiol Biotechnol 64:505. https://doi.org/10.1007/s00253-004-1573-y
30. Han W, Gu J, Yan Q et al (2012) J Ocean U China 11:375. https://doi.org/10.1007/s11802-012-1929-3

31. Han W, Gu J, Liu H, Li F, Wu Z, Li Y (2015) J Ocean U China 14:841. https://doi.org/10.1007/s11802-015-2800-0
32. Lin B, Liu Y, Lu G, Zhao M, Hu Z (2017) FEMS Microbiol Lett 364:12. https://doi.org/10.1093/femsle/fnx012
33. Dong Q, Ruan L, Shi H (2016) Carbohydr Res 432:1. https://doi.org/10.1016/j.carres.2016.05.002
34. Long M, Yu Z, Xu X (2010) Mar Biotechnol 12:62. https://doi.org/10.1007/s10126-009-9200-7
35. Cui X, Jiang Y, Chang L et al (2018) Int J Biol Macromol 120:657. https://doi.org/10.1016/j.ijbiomac.2018.07.118
36. Fu XT, Pan CH, Lin H, Kim SM (2009) J Microbiol Biotechnol 19:257. https://doi.org/10.4014/jmb.0801.026
37. Su Q, Jin T, Yu Y, Yang M, Mou H, Li L (2017) AMB Expr 7:220. https://doi.org/10.1186/s13568-017-0525-8
38. Han Z, Zhang Y, Yang J (2019) Int J Mol Sci 20:2143. https://doi.org/10.3390/ijms20092143
39. Temuujin U, Chi WJ, Lee SY, Chang YK, Hong SK (2011) Appl Microbiol Biotechnol 92:749. https://doi.org/10.1007/s00253-011-3347-7
40. Jam M, Flament D, Allouch J et al (2005) Biochem J 385:703. https://doi.org/10.1042/BJ20041044
41. Lee S, Lee DG, Jang MK, Jeon MJ, Jang HJ, Lee SH (2011) J Microbiol Biotechnol 21:1116. https://doi.org/10.4014/jmb.1107.07001
42. Yan J, Chen P, Zeng Y et al (2020) Int J Biol Macromol 160:288. https://doi.org/10.1016/j.ijbiomac.2020.05.206
43. Mai Z, Su H, Zhang S (2016) Int J Mol Sci 17:1360. https://doi.org/10.3390/ijms17081360
44. Chen YP, Wu HT, Wang GH et al (2018) J Biosci Bioeng 127:403. https://doi.org/10.1016/j.jbiosc.2018.10.001
45. Hafizah NF, Teh AH, Furusawa G (2018) Appl Biochem Biotechnol 187:770. https://doi.org/10.1007/s12010-018-2849-5
46. An K, Shi X, Cui F et al (2018) Protein Expr Purif 143:1. https://doi.org/10.1016/j.pep.2017.10.002
47. Li RK, Chen Z, Ying XJ, Ng TB, Ye XY (2018) Int J Biol Macromol 119:1164. https://doi.org/10.1016/j.ijbiomac.2018.08.053
48. Naretto A, Fanuel M, Ropartz D et al (2019) J Biol Chem 294:6923. https://doi.org/10.1074/jbc.RA118.006609
49. Chen ZW, Lin HJ, Huang WC, Hsuan SL, Lin JH, Wang JP (2018) AMB Express 8:49. https://doi.org/10.1186/s13568-018-0581-8
50. Li G, Sun M, Wu J et al (2015) Appl Microbiol Biotechnol 99:10019. https://doi.org/10.1007/s00253-015-6869-6
51. Li L, Xu XW, Jiang XW et al (2011) Appl Environ Microbiol 77:7077. https://doi.org/10.1128/AEM.05364-11
52. Liu N, Mao X, Du Z, Mu B, Wei D (2014) Carbohydr Res 388:147. https://doi.org/10.1016/j.carres.2014.02.019
53. Yu S, Yun EJ, Kim DH, Park SY, Kim KH (2020) Appl Environ Microbiol 86:e02724. https://doi.org/10.1128/AEM.02724-19
54. Jeong YJ, Choi JW, Cho MS, Jeong KJ (2019) Biotechnol Bioprocess Eng 24:250. https://doi.org/10.1007/s12257-018-0362-x
55. Temuujin U, Chi WJ, Chang YK, Hong SK (2012) J Bacteriol 194:142. https://doi.org/10.1128/JB.05978-11
56. Lee DG, Park GT, Kim NY et al (2006) Biotechnol Lett 28:1925. https://doi.org/10.1007/s10529-006-9171-y
57. Kim HT, Lee S, Lee D et al (2010) Appl Microbiol Biotechnol 86:227. https://doi.org/10.1007/s00253-009-2256-5
58. Ohta Y, Hatada Y, Ito S, Horikoshi K (2005) Biotechnol Appl Biochem 41:183. https://doi.org/10.1042/BA20040083

59. Liu N, Mao X, Yang M, Mu B, Wei D (2014) World J Microbiol Biotechnol 30:1691. https://doi.org/10.1007/s11274-013-1591-y
60. Yasushi S, Takashi M, Masana N (1994) Biochim Biophys Acta 1218:105
61. Cao S, Shen J, Zhang Y, Chang Y, Xue C (2020) J Agric Food Chem 68:7678. https://doi.org/10.1021/acs.jafc.0c02672
62. Lee YR, Jung S, Chi WJ et al (2018) J Microbiol Biotechnol 28:284. https://doi.org/10.4014/jmb.1710.10011
63. Song T, Xu H, Wei C et al (2016) Sci Rep 6:34103. https://doi.org/10.1038/srep34103
64. Dong J, Hashikawa S, Konishi T, Tamaru Y, Araki T (2006) Appl Environ Microb 72:6399. https://doi.org/10.1128/AEM.00935-06
65. Ma C, Lu X, Shi C et al (2007) J Biol Chem 282:3747. https://doi.org/10.1074/jbc.M607888200
66. Lee DG, Jeon MJ, Lee SH (2012) J Microbiol Biotechnol 22:1692. https://doi.org/10.4014/jmb.1209.09033
67. Hatada Y, Ohta Y, Horikoshi K (2006) J Agric Food Chem 54:9895. https://doi.org/10.1021/jf0613684
68. Zhang W, Xu J, Liu D, Liu H, Lu X, Yu W (2018) Appl Microbiol Biotechnol 102:2203. https://doi.org/10.1007/s00253-018-8762-6
69. Liu J, Liu Z, Jiang C, Mao X (2019) J Agric Food Chem 67:10373. https://doi.org/10.1021/acs.jafc.9b03073
70. Wang H, Zhang W, Cui Z, Lu Z, Lu X (2020) Biotechnol Lett 42:1919. https://doi.org/10.1007/s10529-020-02901-5
71. Allouch J, Jam M, Helbert W et al (2003) J Biol Chem 278:47171. https://doi.org/10.1074/jbc.M308313200
72. Pluvinage B, Hehemann J-H, Boraston AB (2013) J Biol Chem 288:28078. https://doi.org/10.1074/jbc.M113.491068
73. Giles K, Pluvinage B, Boraston AB (2017) Proteins 85:182. https://doi.org/10.1002/prot.25189
74. Zhang P, Zhang J, Zhang L et al (2018) Appl Microbiol Biotechnol 103:1289. https://doi.org/10.1007/s00253-018-9540-1
75. Bernal C, Rodriguez K, Martinez R (2018) Biotechnol Adv 36:1470. https://doi.org/10.1016/j.biotechadv.2018.06.002
76. Koti BA, Shinde M, Lalitha J (2013) Biotechnol Bioprocess Eng 18:333. https://doi.org/10.1007/s12257-012-0237-5
77. Xiao A, Xiao Q, Lin Y, Ni H, Zhu Y, Cai H (2016) J Nanosci Nanotechnol 16:10048. https://doi.org/10.1166/jnn.2016.12359
78. Xiao Q, Liu C, Ni H, Zhu Y, Jiang Z, Xiao A (2019) Food Chem 272:586. https://doi.org/10.1016/j.foodchem.2018.08.017
79. Kim JH, Yun EJ, Yu S, Kim KH, Kang NJ (2017) Mar Drugs 15:321. https://doi.org/10.3390/md15100321
80. Yun EJ, Lee S, Kim HJ et al (2013) Appl Microbiol Biotechnol 97:2961. https://doi.org/10.1007/s00253-012-4184-z
81. Yun EJ, Lee AR, Kim JH, Cho KM, Kim KH (2017) Food Chem 221:976. https://doi.org/10.1016/j.foodchem.2016.11.066
82. Wang Q, Sun J, Liu Z, Huang W, Xue C, Mao X (2018) J Agric Food Chem 66:7087. https://doi.org/10.1021/acs.jafc.8b01974
83. Zhang L, Guo X, Song Y et al (2018) Mater Sci Eng C 93:218. https://doi.org/10.1016/j.msec.2018.07.068
84. Shi C, Lu X, Ma C, Ma Y, Fu X, Yu W (2008) Appl Biochem Biotechnol 151:51. https://doi.org/10.1007/s12010-008-8169-4
85. Jang MK, Lee SW, Lee DG et al (2010) Biotechnol Lett 32:943. https://doi.org/10.1007/s10529-010-0237-5

86. Qu W, Wang D, Wu J et al (2020) Front Microbiol 11:574771. https://doi.org/10.3389/fmicb.2020.574771
87. Sugano Y, Terada I, Arita M, Noma M (1993) Appl Environ Microbiol 59:1549
88. Leon O, Quintana L, Peruzzo G (1992) JC Slebe Appl Environ Microbiol 58:4060
89. Hu Z, Lin BK, Xu Y, Zhong MQ, Liu GM (2009) J Appl Microbiol 106:181. https://doi.org/10.1111/j.1365-2672.2008.03990.x
90. Aoki T, Araki T, Kitamikado MP (1990) Eur J Biochem 187:461
91. Wang J, Mou H, Jiang X, Guan H (2006) Appl Microbiol Biotechnol 71:833. https://doi.org/10.1007/s00253-005-0207-3
92. Vera J, Alvarez R, Murano E, Slebe JC, Leon O (1998) Appl Environ Microbiol 64:4378
93. Ramos KR, Valdehuesa KN, Cabulong RB et al (2016) Enzym Microb Technol 90:19. https://doi.org/10.1016/j.enzmictec.2016.04.009
94. Ohta Y, Hatada Y, Nogi Y, Li Z, Ito S, Horikoshi K (2004) Appl Microbiol Biotechnol 66:266. https://doi.org/10.1007/s00253-004-1757-5
95. Jeon EJ, Choi JW, Cho MS, Jeong KJ (2021) Microb Biotechnol 14:2164. https://doi.org/10.1111/1751-7915.13899
96. Zhang P, Rui J, Du Z, Xue C, Li X, Mao X (2016) J Biotechnol 219:22. https://doi.org/10.1016/j.jbiotec.2015.12.007
97. Wang W, Liu P, Hao C, Wu L, Wan W, Mao X (2017) Sci Rep 7:44252. https://doi.org/10.1038/srep44252
98. Zhang YH, Song XN, Lin Y et al (2019) Int J Biol Macromol 137:177. https://doi.org/10.1016/j.ijbiomac.2019.06.207
99. Wang X, Yang Z, Xu X, Jiang H, Cai C, Yu G (2020) Carbohydr Polym:240. https://doi.org/10.1016/j.carbpol.2020.116261
100. Zhang N, Mao X, Li RW et al (2017) Mol Nutr Food Res 61:1600585. https://doi.org/10.1002/mnfr.201600585
101. Chen X, Yu J, Xue C, Wang Y, Tang Q, Mao X (2019) Biosci Biotechnol Biochemist 83:1227. https://doi.org/10.1080/09168451.2019.1607246
102. Ye Q, Wang W, Hao C, Mao X (2019) J Funct Foods 57:222. https://doi.org/10.1016/j.jff.2019.04.017
103. Jin M, Liu H, Hou Y et al (2017) Food Res Int 100:186. https://doi.org/10.1016/j.foodres.2017.08.032
104. Lee MH, Jang JH, Yoon GY et al (2017) BMB Rep 50:263. https://doi.org/10.5483/bmbrep.2017.50.5.014
105. Kim M, Lee JE, Cho H et al (2020) Biomaterials 263:120391. https://doi.org/10.1016/j.biomaterials.2020.120391
106. Lee JE, Kim YA, Yu S, Park SY, Kim KH, Kang NJ (2019) J Dermatol Sci 96:90. https://doi.org/10.1016/j.jdermsci.2019.10.005
107. Chen X, Fu X, Huang L, Xu J, Gao X (2021) Carbohydr Polym 265:118076. https://doi.org/10.1016/j.carbpol.2021.118076
108. Ma J, Yan Q, Yi P, Yang S, Liu H, Jiang Z (2019) Process Biochem 87:119. https://doi.org/10.1016/j.procbio.2019.08.021
109. Hu B, Gong Q, Wang Y, Ma Y, Li J, Yu W (2006) Anaerobe 12:260. https://doi.org/10.1016/j.anaerobe.2006.07.005
110. Seo YB, Lu Y, Chi WJ et al (2014) Process Biochem 49:430. https://doi.org/10.1016/j.procbio.2013.12.014
111. Lin F, Ye J, Huang Y, Yang Y, Xiao M (2019) PRO 7:267. https://doi.org/10.3390/pr7050267
112. Pan C, Kazlowski B, Ko Y, Kong Z, Wu S, Wang F (2010) US patent: 0323407 A1
113. Kazłowski B, Pan CL, Ko YT (2015) Carbohydr Polym 122:351. https://doi.org/10.1016/j.carbpol.2014.09.003
114. Xu XQ, Su BM, Xie JS et al (2018) Food Chem 240:330. https://doi.org/10.1016/j.foodchem.2017.07.036

115. Kim HT, Lee S, Kim KH, Choi IG (2012) Bioresour Technol 107:301. https://doi.org/10.1016/j.biortech.2011.11.120
116. Kim HT, Yun EJ, Wang D, Chung JH, Choi IG, Kim KH (2013) Bioresour Technol 136:582. https://doi.org/10.1016/j.biortech.2013.03.038
117. Wu YR, Zhang M, Zhong M, Hu Z (2017) Bioresour Technol 241:369. https://doi.org/10.1016/j.biortech.2017.05.117
118. Lee JS, Hong SK, Lee CR, Nam SW, Jeon SJ, Kim YH (2019) J Microbiol Biotechnol 29:625. https://doi.org/10.4014/jmb.1902.02012

Chondroitinase

Changliang Zhu and Haijin Mou

1 Chapter Overview

Chondroitin sulfate (CS) and dermatan sulfate (DS), which belong to a class of natural complex polysaccharides called glycosaminoglycans (GAGs), are constituents of cellular membranes and the extracellular matrix of mammalian tissues [1]. The carbohydrate components of proteoglycans (PGs) are thought to be involved in the regulation of various cellular processes, such as adhesion, differentiation, migration, and proliferation [2]. Accordingly, the preparation of CS oligosaccharides has become a popular research topic, and chondroitinase is a powerful tool for producing CS oligosaccharides. Different CS lyase activities have been reported for chains with O-sulfo groups at various positions [3]. In addition, chondroitinases are used in the treatment of lumbar disc herniation, functional recovery of injured spinal cord, tumor therapy, and cystic fibrosis via the CSPGs [4–6]. This chapter catalogs cleavage enzymes, sources, physical and chemical properties, structure, preparation methods, and other topics to guide increased awareness and application of chondroitinase.

2 Enzyme Classification and Source

Chondroitinase (ChSase), also known as chondroitin sulfate (CS) lyase, can be used to catalyze the degradation of CS. The various CS structures comprise linear polymers of 40–100 repeating disaccharide units containing a hexosamine, N-acetyl

C. Zhu · H. Mou (✉)
State Key Laboratory of Marine Food Processing and Safety Control, College of Food Science and Engineering, Ocean University of China, Qingdao, China
e-mail: mousun@ouc.edu.cn

X. Mao (ed.), *Degrading Enzymes for Marine Polysaccharides and Their Applications*, https://doi.org/10.1007/978-981-96-9517-1_8

galactosamine (GalNAc), and a glucuronic acid (GlcA) joined by β(1, 3) and β (1, 4) linkages, respectively (Fig. 1).

Based on their action on different CS/dermatan sulfate (DS) chains, ChSases are classified into four distinct classes of enzymes.

1. **ChSase ABC**

ChSase ABC (EC 4.2.2.4) acts on CSA, CSB, CSC, and hyaluronic acid (HA) substrates and catalyzes the generation of unsaturated disaccharides and other oligosaccharides by β-eliminating reaction. A variety of microorganisms can synthesize ChSase ABC. These include *Proteus vulgaris* and *Bacteroides thetaiotaomicron*. Conventional preparations from these bacteria contain two distinct enzymes, an endo- and an exolyase [7]. In recent years, ChSase ABC derived from *P. vulgaris* has been the most frequently studied enzyme in medicine.

2. **ChSase AC**

ChSase AC (EC 4.2.2.5) also acts on the aforementioned substrates and catalyzes unsaturated disaccharides and other oligosaccharides by the β-eliminating reaction. However, it cannot degrade keratin sulfate, heparin, or CS B. Strains that produce ChSase AC include *Flavobacterium heparinum* [8], *Arthrobacter aurescens* [9], and *Aeromonas liquefaciens* [10]. ChSase AC has attracted considerable attention in medicine.

3. **ChSase B**

ChSase B (EC 4.2.2.19) can specifically degrade HA and DS in glycosaminoglycans and generate unsaturated disaccharides and other oligosaccharides by

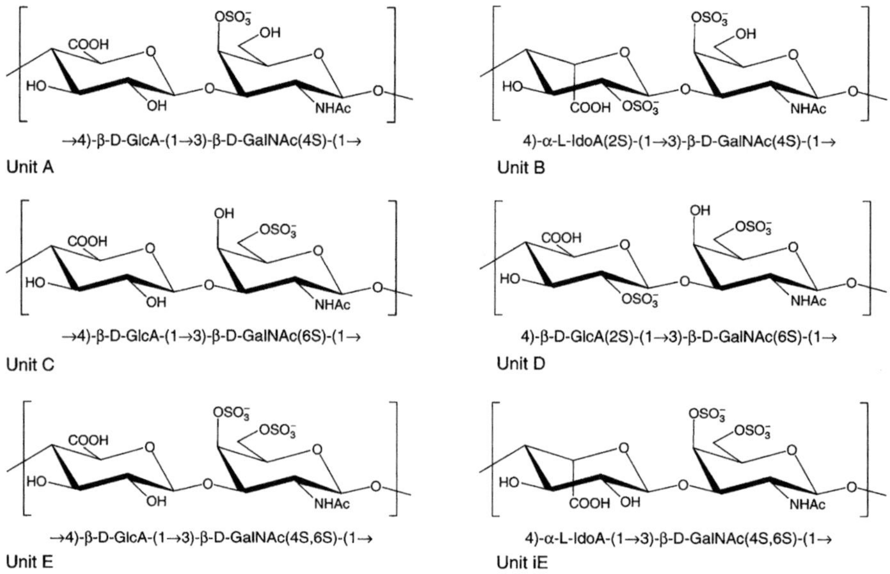

Fig. 1 Six typical disaccharide units in chondroitin sulfate/dermatan sulfate (CS/DS) chains

β-eliminating reaction. *Flavobacterium heparinum* produces this enzyme [8]. The medicinal value of ChSase B does not appear compelling.

4. **ChSase C**

   ChSase C (no EC number) specifically degrades CS C and HA, generating unsaturated disaccharides and other oligosaccharides by a β-eliminating reaction. This enzyme is a new type of chondroitinase sulfate found in the purified ChSase B produced by *F. heparinum* [11].

   Classification and source information of ChSase enzymes are provided in Table 1.

3 Detection of Enzyme Activity

To study the enzymology of ChSases, a robust and reliable measure of the progress of enzyme-catalyzed reactions is required. ChSase activity can be measured using ultraviolet (UV) light at 232 nm. The unsaturated double bond formed at the C4-C5 position of the nonreducing end of the product by cleaving the link between D-glucuronic acid (GlcA) and N-acetylgalactosamine (GalNAc) β-1,4-glycosidic bond is a characteristic absorption peak at 232 nm. The enzyme activity is calculated based on the molar extinction coefficient of the unsaturated products [12].

The enzymatic reaction is performed using 2 mg/mL CS as substrate in 20 mM Tris-HCl, pH 7.5 at 37 °C. In total, 20 μL enzyme sample is added to a reaction mixture. The reaction is terminated by a 5-min boiling water bath treatment after 20 min of process, following by detecting the absorbance value of the reaction solution at a wavelength of 232 nm, and using 20 μL of inactivated enzyme solution as a blank control.

One enzyme activity unit (*U*) is defined as the amount of enzyme required to degrade CS and produce 1 μmol of unsaturated disaccharide per minute at 37 °C. The millimolar extinction coefficient is 5.1 during the calculation process [13].

Table 1 Classification and sources of ChSases

Name	Substrate	Enzymolysis products	Source
Chondroitinase ABC	CS-A, CS-B, CS-C, HA	Unsaturated disaccharides, oligosaccharides	*Proteus vulgaris* and *Bacteroides thetaiotaomicron*
Chondroitinase AC	CS-A, CS-B, CS, HA	Unsaturated disaccharides, oligosaccharides	*Flavobacterium heparinum*, *Arthrobacter aurescens*and *Aeromonas liquefacien*
Chondroitinase B	HA, DS	Unsaturated disaccharides, oligosaccharides	*Flavobacterium heparinum*
Chondroitinase C	CS-C, HA	Unsaturated disaccharides, oligosaccharides	*Flavobacterium heparinum*

$$U/L = \frac{A}{\varepsilon \cdot d} \times \frac{1}{t} \times \frac{V_t}{V_s} \times 1000$$

where ε represents the millimolar absorption coefficient, d represents the thickness of the quartz cuvette, t represents the action time, A represents absorbance at 232 nm, V_t represents the total volume of the reaction, and V_s represents the sample volume.

4 Physical and Chemical Properties

Temperature, pH, molecular weight, and substrate specificity are important parameters affecting the hydrolysis reaction selectivity of microbial ChSases. Table 2 summarizes a few studies on chondroitinase sulfate and its enzymatic properties.

5 Degradation Modes

5.1 *Endolytic Mode*

CS-AC lyases cleave the glycosidic bond on the nonreducing end of a uronic acid, catalytic substrate of [3)-β GalNAc(4S)-(1, 4)-β GlcA(2x,3x)-(1], or [3)-β GalNAc(6S)-(1, 4)-β GlcA(2x,3x)-(1] and [3)-β GluNAc-(1, 4)-β-D-GlcA-(1,] in CS-A, CS-C, and HA, respectively [22]. These enzymes have random endolytic modes of action. ChSase B is a periplasmic O-glycosylated protein that cleaves the

Table 2 Biochemical properties of ChSases from various species

Enzyme types	species	Mol mass (kDa)	pH_{opt}	T_{opt}	pI	References
Chase AC	*A.aurescens*	79	6.0	50	5.5	[14, 15]
	Aurebacterium	81	7.5	37	/	[16]
	A.liquefaciens	/	6.6	<46	/	[10]
	Arthrobacter. sp	84	6.5	37	/	[17]
	B.stercoris	170	5.7–6.0	45–50	8.3	[18]
	F. heparinum	74	6.6	40	8.85	[12]
Chase ABC	*B. stercoris*	83	7.0	40	7.9	[18]
	Bacillus sp.	53	5.5–6.0	75–80	/	[19]
	P. vulgaris	100	8.0	37	8.25/8.50	[7]
	P. vulgaris	105	8.0	40	8.45	[7]
	Acinetobacter sp.	76	5–6	37	/	[20]
Chase B	*F. heparinum*	55	/	/	9.05	[12]
Chase C	*F. heparinum*	/	/	/	/	[21]

[3)-β GalNAc(4 × 6x)-(1,4)-aIdoA(2x)-(1] glycosidic bond in a random endolytic manner to produce unsaturated disaccharides [12]. ChSases ABC catalyzes the endolytic cleavage of CSs and their oligosaccharides [7].

5.2 *Exolytic Mode*

ChSases ABC and AC exhibit endonuclease activity. However, ChSase ABC I catalyzes the endolytic and exolytic cleavage of disaccharide residues from the nonreducing ends of both polymeric CSs and their oligosaccharides [7]. In addition, ChSase AC from *A. aurescens* displays higher activity toward HA than CS-A and CS-C, acting as an exolyase to release disaccharides [14, 15] (Table 3).

6 Structure and Catalytic Mechanism

6.1 *ChSase ABC*

The crystal structure of the ChSase ABC I from *P. vulgaris* (Fig. 2) features three domains arranged linearly to form a catalytic domain in a wide-open cleft [22]. Substrate binding may provoke the closure of the binding site within a flexible side chain of an arginine (Arg) residue and interact with the carboxylic group of GlcA or IdoA. This scenario explains why this enzyme can cleave the two substrates of CS-A [3)-β GalNAc(4S)-(1, 4)-β GlcA(2x,3x)-(1] (A-unit) and CS-C ([3)-β GalNAc(6S)-(1, 4)-β GlcA(2x,3x)-(1]) (C-unit). According to molecular docking data, the Arg500 amino acid residue interacts with 4-O-sulfated and 6-O-sulfated

Table 3 Cleavage modes of ChSases

Enzyme types	species	Cleavage mode	References
Chase AC	*A.aurescens*	Exolytic mode	[14, 15]
	Aurebacterium	Endolytic mode	[16]
	A.liquefaciens	/	[10]
	Arthrobacter. sp.	Exolytic mode	[17]
	B.stercoris	Endolytic mode	[18]
	F. heparinum	/	[12]
Chase ABC	*B. stercoris*	Exolytic mode	[18]
	Bacillus sp.	/	[19]
	P. vulgaris	Endolytic mode	[7]
	P. vulgaris	Exolytic mode	[7]
	Acinetobacter sp.	Exolytic mode	[20]
Chase B	*F. heparinum*	Endolytic mode	[12]
Chase C	*F. heparinum*	/	[21]

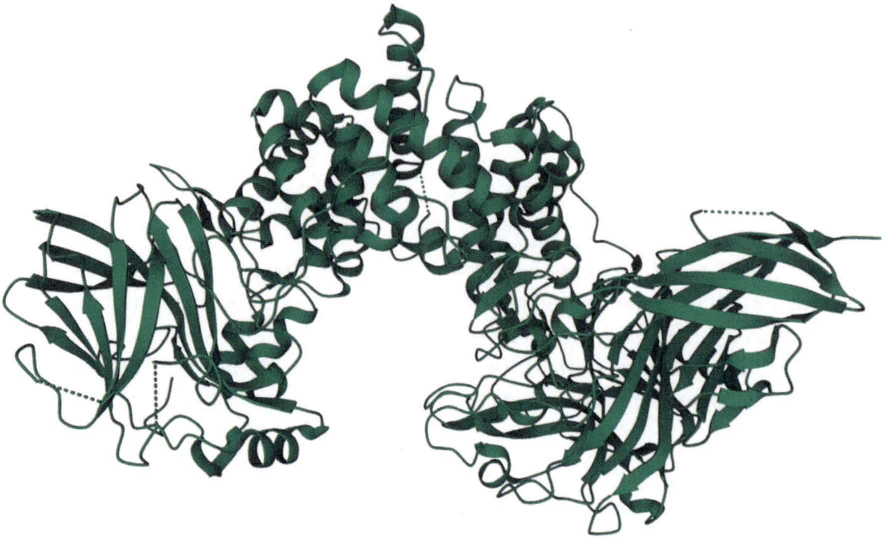

Fig. 2 Structure of ChSase ABC from (7E1Q)

GalNAc residues. The distance between Arg500 and the 4-O-sulfate group is 0.8 Å, which is shorter than the distance between Arg500 and the 6-O-sulfated group. The 6-O-sulfated group is electrostatically repelled by the nearby Asp490. Thus, cABC-I has the highest affinity for the CSA that is richest in 4-O-sulfated GalNAc residues among all CS isomers. The C-terminal domain is composed of a stack of four anti-parallel β-sheets. Moreover, it is structurally similar to the ChSase AC sequence homology from *F. heparinum*. Recombinant cABC expressed in *Escherichia coli* is similar to that of *P. vulgaris*. Mutagenesis has revealed that a tetrad residue (His501, Tyr508, Arg560, and Glu653) is essential for enzyme activity. According to the previously observed GAG lyases, this group of residues is capable of stabilizing the proton shuffling required for the degradation of GAGs [23].

6.2 ChSase AC

Three-dimensional crystal structure of ChSase AC from *A. aurescens* (1RW9, Fig. 3) in its native form and complex with its substrates (chondroitin 4-sulfate tetrasaccharide, CS (tetra), and hyaluronan tetrasaccharide) at resolutions of 1.25–1.9 Å [15]. Overall α + β architecture of ChSase AC supports Supporting Tyr242 as the cleavage mechanism for extracting protons from the C-5 position of GlcA, and the charge on the carboxylate group is neutralized by Asn183 and His233. Thirteen alpha helices exist in the N-terminal alpha helix domain, among which 10 participate in the formation of incomplete bilayer (alpha/alpha) 5-ring structures. The long

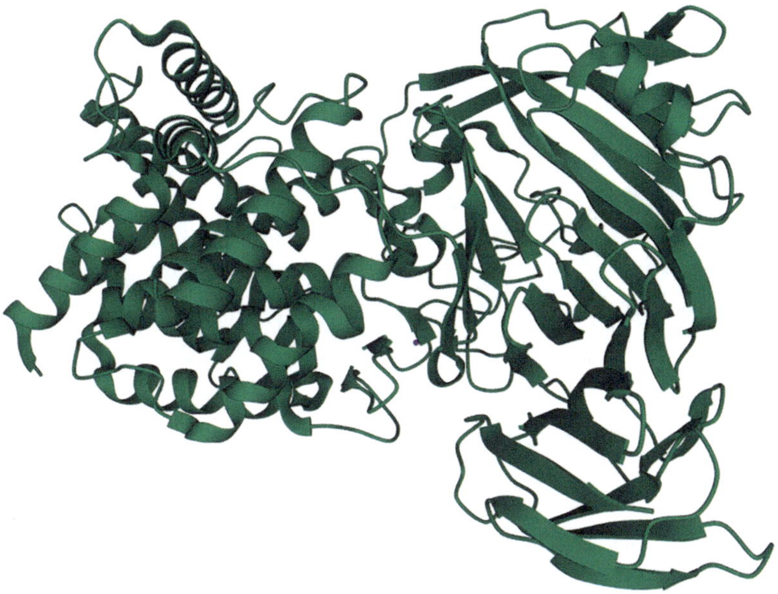

Fig. 3 Structure of ChSase AC from *A. aurescens* (1RW9)

deep groove on one side of the circular object contains active and substrate-binding sites. In addition, Arg, the Ala292 mutation in ChSase AC from *F. heparinum*, leads to a subsequent stepwise exolytic mechanism due to a loss of high-affinity contact [23], similar to ChSase AC from *A. aurescens*. Both closely resemble HA lyase from *Streptococcus pneumoniae* [24] .

6.3 ChSase B

The crystal structure of ChSase B adopts a right-handed parallel β-helical fold. This fold was originally found in pectate lyases and subsequently in several polysaccharide lyases and hydrolases. The binding of the disaccharide product occurs within a positively charged cleft formed by loops extending from the surface of the β-helix. Amino acid residues responsible for the recognition of disaccharides and potential catalytic residues have been identified. Two Arg residues, Arg318 and Arg364, interact with the carboxyl groups of unsaturated IdoA [25]. Moreover, DS is mostly sulfated at C4 of GalNAc. Arg 364 interacts directly with this C-4 sulfate group to reinforce specificity. Cleavage of DS likely occurs at the reducing end of the disaccharide, with Glu333 possibly acting as a general base (Fig. 4).

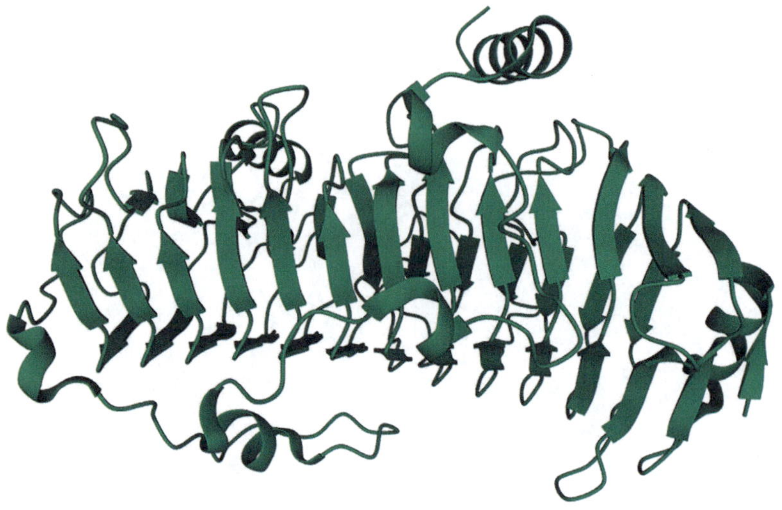

Fig. 4 Structure of ChSase B (1DBO)

7 Structural and Molecular Modification of Enzymes

A fusion protein expression strategy has been used to improve the enzymatic properties of ChSases. Zhu et al. used a gene encoding glyceraldehyde-3-phosphate dehydrogenase (GAPDH) in front of the ChSase AC gene to form the GAPDH-ChSase fusion protein. After adding the GAPDH tag, the molecular weight of ChSase increased from 84 to 118 kDa, specific activity increased from 19.8 to 25.74 U/mg, and K_m changed from 0.483 to 0.380 mg/mL, which enhanced the affinity for the substrate. The optimum reaction pH changed from 6.5 to 7.0, and the optimum reaction temperature changed from 37 °C to 40 °C. Temperature stability was greatly enhanced; GAPDH-ChSase retained 69.8% relative enzyme activity after incubation at 40 °C for 120 min, while ChSase retained only 32.1% relative enzyme activity under the same conditions [17].

8 Preparation Methods

8.1 Genetic Engineering

The ChSase content secreted by natural bacteria is generally low, and the study of enzyme structure generally requires high-purity protein crystallization for structural analysis. Therefore, many enzymatic proteins have been synthesized via heterologous expression using genetic engineering. ChSase AC and ABC have been expressed in *E. coli* and yeast [17, 22, 23]. ChSase AC II enzyme activity increased with the chaperone GroEs-GroEL, and the specific activity increased from 19.8 to 25.74 U/mg with the GAPDH tag [17].

8.2 *Fermentation Engineering*

Low expression increases the cost of chondroitinase sulfate and does not meet industrial production requirements. The biodiversity of microorganisms, their adaptability to efficient genetic manipulation, and their ease of growth make them versatile sources of ChSases. ChSase production depends on the selected microorganism, composition of the growth medium, physicochemical growth parameters, and method of enzyme-producing fermentation. Nutritional and culture conditions, including temperature, pH, substrate water content, aeration, and incubation period, significantly influence ChSase production. ChSase AC II enzyme activity has been increased from 3.12 to 9.15 U/mL by optimizing nutritional and culture conditions [17].

9 Applications of ChSases

9.1 *Medicinal*

CSPGs form an integral part of the extracellular matrix. Their various biological applications include mechanical strength, cell migration, cell adhesion, and proliferation, via the degradation of CS in the lesion location [26]. An increase in CSPG levels is associated with numerous debilitating conditions. The following subsections discuss the role of CSPGs in various pathological conditions and the use of ChSases to treat these conditions (Table 4 and Fig. 5) [27].

9.2 *Structural Analysis of CS*

The structural diversity of CS/DS is the main reason for its functional diversity. The highly complex sugar chain structure significantly hinders the study of the structure and function of CS/DS. Traditional analytical methods, such as high-resolution mass spectrometry and magnetic resonance spectroscopy, have important applications in the study of the structure of CS/DS oligosaccharides with relatively simple structures. However, the analysis of complex CS/DS polysaccharides and relatively high-molecular-weight oligosaccharides is challenging. Analysis of the structure and function of complex CS/DS sugar chains mainly involves a combination of enzymatic degradation and spectral analysis. Among these, CS/DS-specific enzymes play an important role. Homogeneous structurally characterized CS/DS-oligosaccharides standards are useful to facilitate the development of new GAG sequencing technologies. These standards help to better understand the structure and size specificities of the binding sites within the CS for biologically important proteins, chondroitin lyases, and hydrolases [25].

Table 4 The mechanism of therapeutic effects on various pathologies by ChSases

Indication(s)	Mechanism
Spinal cord injury	Axon regeneration by removing inhibitory glial scars
Chemonucleolysis	Reduce intervertebral disc pressure in the lumbar spine
Keloid scars	The reduction of scar stiffness relies on the decomposition of chondroitin sulfate
Tumor	Inhibition of angiogenesis and metastasis of cancer cells through surface removal of CSPG
Vitreolysis	Enzymatic vitreous detachment to reduce surgical complications
Visual acuity/ amblyopia	Improvement of plasticity in the visual cortex by removing the neural network associated with CSPGs
Analytical tool	Detection and quantification of end products produced by chondroitin sulfate action through enzymes, serving as markers for analyzing cancer progression and treatment

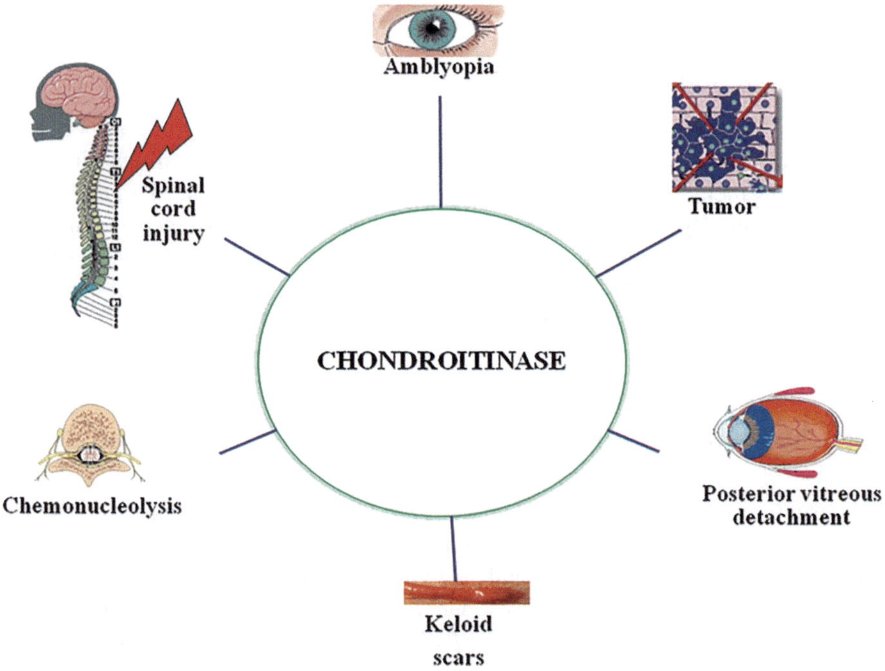

Fig. 5 Therapeutic applications of chondroitinase. The information is from [27]

9.3 Preparation of Oligosaccharides

Low-molecular-weight CS (LMWCS) has the characteristics of easy absorption, low viscosity, and good solubility, and has broad application prospects in the prevention of Alzheimer's disease and atherosclerosis, inhibiting oxidation, and protecting articular cartilage.

The enzymatic hydrolysis method for preparing LMWCS has the advantages of a simple process, easy control, fast reaction speed, mild conditions, no damage to the sulfuric acid group on the CS molecule, and no environmental pollution. The method is more suitable for the green industrial production of LMWCS. Many reports exist on the enzymatic preparation of CS oligosaccharides. An enzymatic hydrolysis method for preparing LMWCS would be significant.

References

1. Volpi N (2006) Chondroitin sulfate: structure, role and pharmacological activity. Academic, Amsterdam
2. Petit E, Delattre C, Papy-Garcia D, Michaud P (2006) Chondroitin sulfate Lyases: applications in analysis and Glycobiology. Adv Pharmacol 53:167–186. https://doi.org/10.1016/S1054-3589(05)53008-4
3. Schiraldi C, Cimini D, De Rosa M (2010) Production of chondroitin sulfate and chondroitin. Appl Microbiol Biotechnol 87:1209–1220. https://doi.org/10.1007/s00253-010-2677-1
4. Kriks S, Shim JW, Piao J, Ganat YM, Wakeman DR, Xie Z, Carrillo-Reid L, Auyeung G, Antonacci C et al (2011) Dopamine neurons derived from human ES cells efficiently engraft in animal models of Parkinson's disease. Nature 480:547–551. https://doi.org/10.1038/nature10648
5. Bradbury EJ, Moon LD, Popat RJ, King VR, Bennett GS, Patel PN, Fawcett JW, McMahon SB (2001) Chondroitinase ABC promotes functional recovery after spinal cord injury. Nature 416:636–640. https://doi.org/10.1038/416636a
6. García-Alías G, Barkhuysen S, Buckle M, Fawcett JW (2009) Chondroitinase ABC treatment opens a window of opportunity for task-specific rehabilitation. Nat Neurosci 12:1145–1151. https://doi.org/10.1038/nn.237
7. Hamai A, Hashimoto N, Mochizuki H, Kato F, Makiguchi Y, Horie K, Suzuki S (1997) Two distinct chondroitin sulfate ABC lyases: an endoeliminase yielding tetrasaccharides and an exoeliminase preferentially acting on oligosaccharides. J Biol Chem 272(14):9123–9130. https://doi.org/10.1074/jbc.272.14.9123
8. Michelacci YM, Horton DS, Población CA (1987) Isolation and characterization of an induced chondroitinase ABC from *Flavobacterium heparium*. Biochim Biophys Acta 923:291–301. https://doi.org/10.1016/0304-4165(87)90016-X
9. Hiyama K, Okada S (1975) Crystallization and some properties of chondroitinase from *Arthrobacter aurescens*. J Biol Chem 250:1824–1828. https://doi.org/10.1016/S0021-9258(19)41768-7
10. Kitamikado M, Lee YZ (1975) Chondroitinase-producing bacteria in natural habitats. Appl Environ Microbiol 29:414–421. https://doi.org/10.1128/am.29.3.414-421.1975
11. Huang WJ, Matte A, Li Y, Kim YS, Linhardt RJ, Su HS, Cygler M (1999) Crystal structure of chondroitinase B from *Flavobacterium heparinum* and its complex with a disaccharide product at 1.7 a resolution. J Mol Biol 294(1):1257–1269. https://doi.org/10.1006/jmbi.1999.3292
12. Gu K, Linhardt RJ, Laliberte M (1995) Purification, characterization and specificity of chondroitin lyases and glycuronidase from *Flavobacterium heparinum*. J Biochem 312:569–577. https://doi.org/10.1042/bj3120569
13. Yamagata T, Saito H, Habuchi O, Suzuki S (1968) Purification and properties of bacterial chondroitinases and chondrosulfatases. J Biol Chem 243:1523–1535. https://doi.org/10.1016/S0021-9258(18)93574-X

14. Hiyama K, Okada S (1977) Action of chondroitinases: III. Ionic strength effects and kinetics in the action of chondroitinase AC. Action of chondroitinases III. J Biochem 82:429–436. https://doi.org/10.1093/oxfordjournals.jbchem.a131710
15. Lunin VV, Li Y, Linhardt RJ, Miyazono H, Kyogashima M, Kaneko T, Bell AW, Cygler M (2004) High-resolution crystal structure of *Arthrobacter aurescens* chondroitin AC lyase: an enzyme-substrate complex defines the catalytic mechanism. J Mol Biol 337:367–386. https://doi.org/10.1016/j.jmb.2003.12.071
16. Takegawa K, Iwahara K, Iwahara S (1991) Purification and properties of chondroitinase produced by a bacterium isolated from soil. J Ferment Bioeng 72:128–131. https://doi.org/10.1016/0922-338X(91)90323-9
17. Fang YT, Yang SX, Fu XD, Xie WC, Li L, Liu ZM, Mou HJ, Zhu CL (2019) Expression, purification and characterization of chondroitinase AC II from marine bacterium *Arthrobacter* sp. CS01. Mar Drugs 17:185. https://doi.org/10.3390/md17030185
18. Hong SW, Kim BT, Shin HY, Kim WS, Lee KS, Kim YS, Kim DH (2002) Purification and characterization of novel chondroitin ABC and AC lyases from *Bacteroides stercoris* HJ-15, a human intestinal anaerobic bacterium. Eur J Biochem 269:2934–2940. https://doi.org/10.1046/j.1432-1033.2002.02967.x
19. Bellamy WR (1990) A novel *Bacillus* sp. capable of degrading sulfated glycosaminoglycans. In: Horikoshi K, Grant WD (eds) Superbugs, microorganisms in extreme environments. Japan Scientific Societies Press, Tokyo, pp 143–157
20. Zhu CL, Zhang JL, Li LY, Zhang J, Jiang YH, Shen ZP, Guan HS, Jiang XL (2017) Purification and characterization of hyaluronate lyase from *Arthrobacter globiformis* A152. Appl Biochem Biotechnol 182(1):216–228. https://doi.org/10.1007/s12010-016-2321-3
21. Ototani N, Yosizawa Z (1979) Purification of chondroitinase B and chondroitinase C using glycosaminoglycan-bound AH-Sepharose 4B. Carbohydr Res 70:295–306. https://doi.org/10.1016/S0008-6215(00)87109-8
22. Takashima M, Watanabe I, Miyanaga A, Eguchi T (2021) Substrate specificity of Chondroitinase ABC I based on analyses of biochemical reactions and crystal structures in complex with disaccharides. Glycobiology. https://doi.org/10.1093/glycob/cwab086
23. Prabhakar V, Capila I, Bosques CJ, Pojasek K, Sasisekharan R (2005) Chondroitinase ABC I from *Proteus vulgaris*: cloning, recombinant expression and active site identification. Biochem J 386(1):103–112. https://doi.org/10.1042/BJ20041222
24. Jedrzejas MJ, Mello LV, De Groot BL, Li S (2002) Mechanism of hyaluronan degradation by *Streptococcus pneumoniae* hyaluronate lyase: structures of complexes with the substrate. J Biol Chem 277:28287–28297. https://doi.org/10.1074/jbc.M112009200
25. Capila I, Wu Y, Rethwisch DW, Matte A, Cygler M, Linhardt RJ (2002) Role of arginine 292 in the catalytic activity of chondroitin AC lyase from *Flavobacterium heparinum*. Biochim Biophys Acta 1597:260–270. https://doi.org/10.1016/S0167-4838(02)00304-7
26. Kjellen L, Lindahl U (1991) Proteoglycans: structures and interactions. Annu Rev Biochem 60:443–475. https://doi.org/10.1146/annurev.bi.60.070191.002303
27. Kasinathan N, Volety SM, Josyula VR (2016) Chondroitinase: a promising therapeutic enzyme. Crit Rev Microbiol 42(3):474–484. https://doi.org/10.3109/1040841X.2014.959893

Recent Progress of Carrageenan-Degrading Enzymes

Hui Ni

1 Classification and Origin of Carrageenase

Carrageenan is a high-molecular-weight polysaccharide, one of the main constituents of the cell wall of red algae (Rhodophyta) [18, 71]. It consists of disaccharides of α-1,3 linked 3,6-anhydro-galactose (A unit) and galactose (G unit) or two G units, which are connected through β-1,4 linkage to form the linear structure, with different degrees of sulfation on G and A units. Accordingly, carrageenan can be classified into different isoforms including κ-, ι-, and λ-carrageenans based on the location and number of sulfate substitutions and the presence of 3,6-anhydro-D-galactose (Fig. 1). The different forms of carrageenans display distinct gelling characteristics, and the κ-form is the most used in industrial food production.

Carrageenases are the enzymes that are capable of cleaving the internal β-1,4 link in the carrageenan polymer, releasing the oligo-carrageenans with different units of digalactose repeats. In light of their substrate specificity, carrageenases can be divided into κ-, ι-, and λ-carrageenases, corresponding to the main types of carrageenans, respectively. κ-Carrageenase specifically hydrolyzes the 1,4-β-D-linkages between 3,6-anhydro-α-D-galactose and β-4-sulfate-D-galactose in κ-carrageenans, producing neocarrabiose-sulfate and neocarratetraose-sulfate. Similarly, ι-carrageenase acts on the bonds between 3,6-anhydro-D-galactose-2-sulfate and β-D-galactose-4-sulfate in ι-carrageenans to produce ι-neocarratetraose sulfate and ι-neocarrahexaose sulfate. λ-Carrageenase functions on the β-1,4-glycosidic bonds in the backbone of λ-carrageenans producing tetrasaccharides. Most of the carrageenases that have been characterized so far are κ-carrageenases.

H. Ni (✉)
College of Ocean Food and Biology Engineering, Jimei University,
Xiamen, Jimei District, China
e-mail: nihui@jmu.edu.cn

X. Mao (ed.), *Degrading Enzymes for Marine Polysaccharides and Their Applications*, https://doi.org/10.1007/978-981-96-9517-1_9

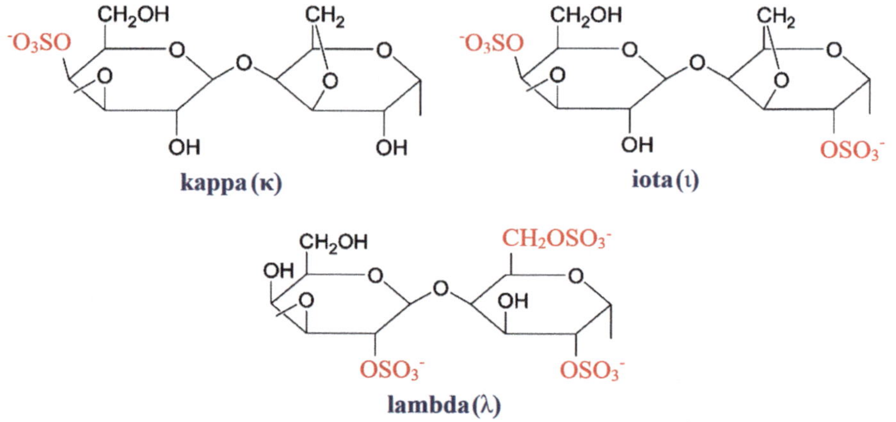

Fig. 1 Structure of kappa, iota, and lambda carrageenans [4]

Carrageenases exist in a wide range of sources, including microorganisms and marine animals. The bacteria that can secrete carrageenases mainly include *Pseudoalteromonas, Pseudomonas*, *Shewanella*, *Vibrio*, *Cellulophaga*, *Cytophaga*, *Tamlana*, *Microbulbifer*, *Alteromonas*, *Pedobacter*, and *Zobellia* [69]. Bacteria use carrageenase as an enzymatic tool to degrade carrageenan for invasion, and the degradation products sometimes can act as an inducer for more carrageenase production [8]. In addition, carrageenase has also been reported in marine animals that feed on algae, such as conch. Moreover, the carrageenase enzymatic activity has also been detected in the gut secretion of grass carp [51].

2 Method for Detection of Carrageenase Activity

To detect the carrageenase activity rapidly and qualitatively in microorganisms, plate assays are developed for screening carrageenan-degrading strains. The enzymatic activity can be indicated by the formation of depressions on the plate surface or liquefaction or clearing zone of the opaque medium as the expansion of the growing colonies to metabolize the substrate. As cetypyridinium chloride (CPC) can react with sulfated galactans to form a precipitation of polysaccharides, 0.5% CPC was utilized to detect the carrageenase activity for screening carrageenase-producing microorganisms on a culture plate [22]. Moreover, a method using zymography was developed to show carrageenase activity, in which the patterns of carrageenase-mediated degradation were visualized by alcian blue staining on polyacrylamide gel electrophoresis (PAGE) gel [52]. For quantitative analysis of the carrageenase activity, the dinitrosalicylic (DNS) method was commonly used by spectrophotometric monitoring of the reducing sugar production as a result of the hydrolysis of carrageenan polysaccharide by carrageenanase [69]. In addition, the activity of carrageenase can also be determined by measuring the viscosity as a more direct and sensitive approach under defined conditions allowing the carrageenan to form a viscous solution [69].

3 Physical and Chemical Properties of Carrageenase

Carrageenase endolytically depolymerizes carrageenan by cleaving the β-1,4 glycosidic bonds to yield oligogalactans of either neocarrabiose or neoagarobiose series [54]. The identification and characterization of novel carrageenases from carrageenolytic microorganisms have drawn considerable research attention in recent years. A list of carrageenases and their important features is summarized in Tables 1, 2 and 3.

The molecular weight of carrageenase is normally estimated by sodium dodecyl sulfate–polyacrylamide gel electrophoresis (SDS-PAGE) through their migration with reference to the protein ladder on the gel [24, 60, 68]. In most cases, the purified carrageenase presented a single band. Interestingly, three isoforms of κ-carrageenase were discovered in *Cytophaga* sp., generating more than one protein band detected on SDS-PAGE [48]. The molecular mass of reported carrageenases can vary from 27.6 kDa to 128 kDa in different microbial species [54].

The optimum pH of most reported carrageenases is between 7.2 and 8.0, close to the pH of seawater [54]. Exceptions are observed on some carrageenases that favor acidic, neutral, or more alkaline conditions, such as the carrageenases from *Zobellia* sp. ZM-2 [31], *Cellulophaga lytica* strain N5-2 [60], and *Pseudoalteromonas tetraodonis* JAM-K142 [23], with optimum pH at 6.0, 7.0, and 8.8, respectively. Notably, a κ-carrageenase isolated from *P. elongata* MTCC5261 displayed optimum activity under acidic (pH 5.6) and alkaline (pH 7.7) conditions [40].

The optimum temperature for carrageenases can range from 30 °C to 55 °C with some exceptions. For example, a ι-carrageenase from *Wenyingzhuangia fucanilytica* showed its highest activity at lower temperature of 25 °C [49], and the λ-carrageenase from *Bacillus* sp. Lc50-1 favored 75 °C [25].

The Michaelis–Menten constant (K_m) and maximal reaction rate (V_{max}) are the most important enzymatic kinetics parameters. They are calculated from Lineweaver–Burk plot and can be used for guiding research and engineering design. The value of K_m is defined as a certain concentration of substrate in which the enzymatic reaction rate is half of the maximal reaction rate. The smaller the K_m value for an enzyme, the higher its affinity for the given substrate. The V_{max} value is affected by two factors including enzyme concentration and flexibility. The more flexible the enzyme, the faster it reaches its V_{max}. Calculated values of K_m and V_{max} for some carrageenases under the optimal temperature and pH are depicted in Table 1 and Table 3.

Thermal stability is also an important parameter of carrageenases, which greatly impacts their industrial applications. The advantage of excellent thermostability of carrageenase is supported by allowing high concentration of carrageenan to be

Table 1 Sources and biochemical properties of κ-carrageenases

Organism	Classification	Optimal pH	Optimal temperature (°C)	GH family	Kinetic parameters	Thermostability	pH stability	Enzymatic hydrolysis products	References
Zobellia sp. ZM-2	Recombinase	6	39	16	$K_m = 0.84$ mg/mL; $V_{max} = 2.92$ U/mL	50% residual activity after 40 °C for 1 h	Remained 85% activity after 20 °C for 2 h at pH 6.0–8.0	DPs 4 and 6	[31]
Cellulophaga lytica strain N5–2	Natural enzyme	7	35	16	$K_m = 1.647$ mg/mL; $V_{max} = 8.7$ μmol min^{-1} mg^{-1}	25% residual activity after 50 °C for 1 h	–	DPs 2, 6 and 8	[60]
Pseudoalteromonas tetraodonis JAM-K142	Recombinase	8.8	30	16	$K_m = 0.869$ mg/mL; $V_{max} = 66.4$ μmol min^{-1} mg^{-1}	–	–	–	[23]
Pseudomonas elongate MTCC5261	Recombinase	5.6/7.7	40	16	$K_m = 6.66$ mg/mL; $V_{max} = 4$ μmol min^{-1} mL^{-1}	–	–	–	[20]
Pseudoalteromonas sp. QY203	Natural enzyme	7.2	45	16	–	40% residual activity after 50 °C for 1 h	6.0 to 9.0	DPs 2 and 4	[24]
Pseudoalteromonas porphyrae LL-1	Natural enzyme	8	55	16	$K_m = 4.4$ mg/mL; $V_{max} = 66.4$ mg min^{-1} mL^{-1}	50% residual activity after 40 °C for 1 h	6.0 to 9.0	–	[30]
Cytophaga-like bacterium	Natural enzyme	7.2	40	–	$K_m = 0.244$ μM; $V_{max} = 0.28$ U	–	–	DP 2	[45]
Vibro sp. CA-1004	Natural enzyme	8.0	40	–	$K_m = 3.3$ mg/mL	10% residual activity after 50 °C for 10 min	5.0 to 11.0	DP 2	[1]

Cytuphaga MCA-2	Natural enzyme	7.2	–	–	–	–	–	DPs 4 and 6	[43]
Pseudoalteromonas WZUC10	Natural enzyme	7.5	30	–	$K_m = 0.56$ mg/mL; $k_{cat} = 125\ s^{-1}$	Completely inactivated after 7.5 min at 70 °C	–	DP 4	[67]
Pseudoalteromonas sp. AJ5–13	Natural enzyme	8.0	55	–	$K_m = 9.8$ mg/mL	5% residual activity after 50 °C for 30 min	6.5 to 9.0	DPs 2, 4, 6, 8 and 10	[33]
Tamlana sp. HC4	Natural enzyme	8.0	30	–	$K_m = 7.63$ mg/mL	10% residual activity after 50 °C for 30 min	7.2 to 8.6	DPs 2 and 4	[53]
Vibrio sp. NJ-2	Natural enzyme	8.0	40	16	$K_m = 2.54$ mg/mL; $V_{max} = 138.89$ mmol mg^{-1} min^{-1}	90% residual activity after 40 °C for 30 min	6.0 to 10.0	DPs 2–8	[68]
Bacillus sp. HT19	Natural enzyme	7.0	60	–	$K_m = 0.061$ mg/mL; $V_{max} = 115.13$ U/mg	90% residual activity after 60 °C for 24 h	–	DP 2	[28]
Cellulophaga lytica strain N5–2	Recombinase	7.0	35	–	$K_m = 0.94$ mg/mL; $V_{max} = 13.42$ U mg^{-1} min^{-1}	75% residual activity after 35 °C for 2 h	5.0 to 8.0	DPs 6 and 8	[6]
Pedobacter hainanensis NJ-02	Recombinase	8.0	40	16	$K_m = 2.4$ mg/mL; $V_{max} = 126$ mmol mg^{-1} min^{-1}	–	5.5 to 10.0	DPs 2, 4, 6 and 8	[70]
Pseudoalteromonas carrageenovora QY203	Recombinase	7.0	55	16	–	90% residual activity after 40 °C for 1 h	More than 80% activity after 4 °C for 24 h at pH 3.6–10.0	DPs 2 and 4	[57]

(continued)

Table 1 (continued)

Organism	Classification	Optimal pH	Optimal temperature (°C)	GH family	Kinetic parameters	Thermostability	pH stability	Enzymatic hydrolysis products	References
Pseudoalteromonas porphyrae LL1	Recombinase	8.0	40	16	–	20% residual activity after 45 °C for 1 h	70% residual activity at pH 11.0 and 4 °C for 12 h	DPs 2 and 4	[65]
Shewanella sp. Kz7	Recombinase	8.0	45	16	K_m = 0.15 mg/L; V_{max} = 807.6 U/mg	85% residual activity after 30 °C for 1 h	pH 6.0 to 9.0	DPs 2 and 4	[55]
Wenyingzhuangia aestuarii OF219	Recombinase	8.0	25	16	K_m = 0.17 μM; V_{max} = 24.81 U/mg	42% residual activity after 25 °C for 24 h	pH 6.0 to 10.5	DPs 2 and 4	[50]
Zobellia sp. ZM-2	Recombinase	7.0	50	16	K_m = 2.07 mg/mL; V_{max} = 0.77 mg ml^{-1} min^{-1}	89% residual activity after 55 °C for 10 min	–	DPs 4 and 6	[62]
Zobellia sp. ZL-4	Recombinase	7.0	55	16	K_m = 0.704 mg/mL	90% residual activity after 40 °C for 1 h	pH 5.0 to 8.0	DPs 2, 4, and 6	[64]
Pseudoalteromonas sp. ZDY3	Recombinase	8.0	45	16	K_m = 3.67 mg/mL; K_m/k_{cat} = 53.0 mL mg^{-1} s^{-1}	80% residual activity after 30 °C for 15 d	80% activity in pH 4.5–10.5 for 48 h	DPs 2 and 4	[66]
Polaribacter sp. NJDZ03	Recombinase	7.0	55	16	–	60–94% activity for 4–12 h at 55 °C	–	DP 2	[12]

Table 2 Sources and biochemical properties of λ-carrageenases

Organism	Classification	Optimal pH	Optimal temperature (°C)	GH family	Kinetic parameters	Thermostability	pH stability	Enzymatic hydrolysis products	References
Pseudoalteromonas carrageenovora ATCC43555	Recombinase	7.5	30	–	–	–	–	DP 6	[14]
Bacillus sp. Lc50–1	Natural enzyme	8	75	–	–	50% residual activity after 85 °C for 10 min	6.0 to 9.0	DP 2	[25]
Pseudoalteromonas CL19	Natural enzyme	7	35	–	–	Stable at 25 °C for 120 min	–	DP 4	[44]

Table 3 Sources and biochemical properties of ι-carrageenases

Organism	Classification	Optimal pH	Optimal temperature (°C)	GH family	Kinetic parameters	Thermostability	pH stability	Enzymatic hydrolysis products	References
Zobellia galactanivorans	Recombinase	–	–	82	–	–	–	–	[46]
Microbulbifer thermotolerans JAMB-A94^{T}	Recombinase	7.5	50	82	K_m = 1.2 mg/mL	–	–	DP 4	[17]
Wenyingzhuangia fucanilytica CZ1127^{T}	Recombinase	8	25	82	K_m = 1.12 μM; k_{cat} = 560.75 s^{-1}	84% residual activity after 25 °C for 24 h	5.0 to 9.0	DP 4	[49]
Cellulophaga sp. QY3	Natural enzyme	7.0	50	82	K_m = 3.686 mM; k_{cat} = 14.226 s^{-1}	80% residual activity after 50 °C for 1 h	More than 70% residual activity at pH 5.0–10.6 and 4 °C for 24 h	DPs 2 and 4	[34]
Pseudoalteromonas carrageenovora Psc^{T}	Recombinase	8.3	40	–	K_m = 13.4 μM; k_{cat} = 1.64 min^{-1}	–	–	–	[10]
Flavobacterium sp. YS-80-122	Recombinase	7.6	30	82	–	–	6.0 to 8.6	DP 4	[27]
Pseudoalteromonas carrageenovora ASY5	Natural enzyme	8.0	40	82	–	40% residual activity after 40 °C for 45 min	More than 60% residual activity at pH 8.0–9.5 and 4 °C for 75 min	DPs 2 and 4	[56]

processed as high temperature can effectively reduce the viscosity of carrageenan solution, facilitating the degradation of carrageenan by carrageenase [25]. In addition, the tolerance to high temperature is greatly favored for the depolymerization process because the high temperature promotes the polysaccharide's conformational change to be more accessible to carrageenase [34]. Notably, the purified ι-carrageenase from *Flavobacterium* sp. YS-80-122 demonstrated thermotolerance and cold adaptation, facilitating its storage and transportation and industrial application in production of ι-carrageenan-derived oligosaccharides from seaweed [27].

The activity of carrageenases can be affected by some cations, as a result of conformational changes in the enzyme as well as changes in the substrate. Some metal ions such as Na^+, K^+, Ca^{2+}, and Mg^{2+}, especially Na^+ at high concentration, can increase the activity of carrageenases [17, 24, 25, 30, 57, 61, 68]. In contrast, some heavy metal ions such as Co^{2+}, Zn^{2+}, Mn^{2+}, Cu^{2+}, Pb^{2+}, Hg^{2+}, and Ag^+ can inhibit the activity of carrageenases [4, 17, 24, 25, 30, 34, 53, 61, 68]. Reducing reagent such as DL-dithiothreitol (DTT) was reported to increase the activity of carrageenase isolated from *Zobellia* sp. ZM-2 [31], owing to its protective effect on a cysteine residue to maintain the enzyme conformation integrity and stability [59]. The chelating agent ethylenediaminetetracetic acid (EDTA) showed the inhibitory effects on some κ-carrageenases such as the enzymes from *Tamlana* sp. HC4 [53], *Pseudoalteromonas* sp. QY203 [24], and *Vibrio* sp. NJ-2 [68], suggesting that they are metalloenzymes. The detergent SDS had been reported to inhibit the activity of carrageenases [24], and the brominating agent N-bromosuccinimide can almost block the activity of the κ-carrageenase isolated from *Vibrio* sp. CA-1004. This finding indicates that an interacting tryptophan residue is involved in the active site or substrate recognition site [1].

4 Cleavage Manner of Carrageenase

Based on the versions of building blocks and the sulfation patterns, carrageenans can be categorized into three basic types of κ-, ι-, and λ-carrageenans. The diversity of carrageenan structure requires specific enzymes to recognize the context of the linear structure and to maintain the native structure of the backbone. Accordingly, the enzymes specifically functioning on the κ-, ι-, and λ-carrageenans are called κ-carrageenase (EC 3.2.1.83), ι-carrageenase (EC 3.2.1.157), and λ-carrageenase (EC 3.2.1.162), respectively. The carrageenases of different types all are endohydrolases, cleaving the internal β-1,4 bonds rather than breaking the bonds from the ends, releasing even-numbered oligosaccharides (Fig. 2). However, the products from the digestions by κ- and ι- carrageenases are normally short oligosaccharides of DP2 and DP4, suggesting that they hydrolyze the polysaccharide chain in a processive manner. In contrast, the action of λ-carrageenase results in oligosaccharides of DP6 or higher, indicating λ-carrageenase behaves more randomly compared with κ- and ι-carrageenases. The diversity in DPs of digestion products is a result of the heterogenous property of carrageenan primary structure and also the stereostructure

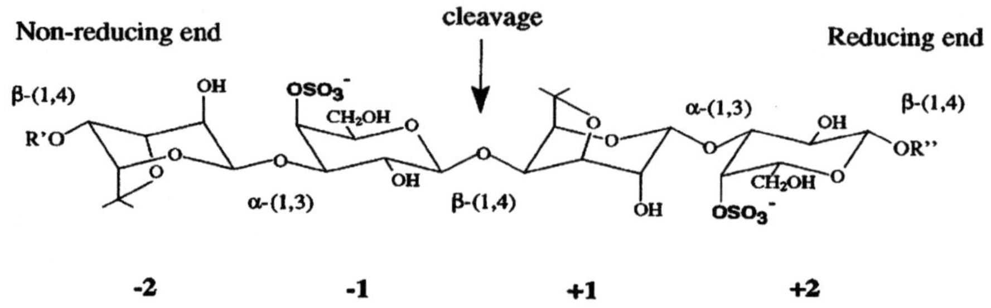

Fig. 2 Schematic of two disaccharide-repeating units of κ-carrageenan with reducing as well as non-reducing ends. Carrageenan cleavage site is indicated by arrow [37]

determined by the glycosidic linkages. The oligosaccharides derived from the enzymatic digestion of carrageenan were more uniform in molecular weights than the ones from acid hydrolysis [61].

5 Structural Characteristics and Catalytic Mechanism of Carrageenase

5.1 Structural Characteristics

The three-dimensional structures of carrageenases are not fully understood yet, with only few publicly available reports investigating their folding (Table 4). The structures of κ- carrageenases will be interpreted by CgkA from *Pseudoalteromonas carrageenovora* [37] and ZgCgkA from *Zobellia galactanivorans* Dsij^T [36]. Both CgkA and ZgCgkA fold into a β-jelly roll, which consists of six and seven strands, respectively, arranged into two closely pack and curved anti-parallel β-sheets (Fig. 3a, b). The structure is featured with a tunnel active site, implying a processive action mode of the enzyme. In light of their secondary structure analysis, the fold is composited of three minor antiparallel β-sheets and one α-helix. Some minor structural differences between CgkA and ZgCgkA can be revealed by their structural superposition, including the amino acid variations residing in the extremity of the catalytic channel, the presence of an additional finger in the extremity of the tunnel of ZgCgkA, and the presence of a helix on the very top of ZgCgkA [36]. The catalytic mechanism of CgkA and ZgCgkA was featured by the double-displacement reaction mechanism with the involvement of two catalytic carboxyl groups, E_{163}/E_{159} and E_{168}/E_{164}, as the nucleophile and acid/base residues, respectively [36] (Fig. 4a, b). In addition, the D_{165} and D_{161} residues were also involved in the enzymatic action to accelerate the deglycosylation step in CgkA and ZgCgkA, respectively [36] (Fig. 4a, b).

The structure of ι-carrageenase will be explained by CgiA from *Alteromonas fortis* [38]. The crystal structure of CgiA presents a right-handed β-helix fold and

Table 4 The structural features of different carrageenases

Enzyme	Source	Catalytic sites	Resolution	PDB	References
CgkA	*Pseudoalteromonas carrageenovora*	E_{163}, D_{165}, E_{168}	1.54 Å	1DYP	[37]
ZgCgkA	*Zobellia galactanivorans* Dsij^T	E_{159}, D_{161}, E_{164}	1.66 Å	5OCR	[36]
CgiA	*Alteromonas fortis*	E_{245}, D_{247}, E_{310}, H_{281}	1.60 Å	1H80	[38]

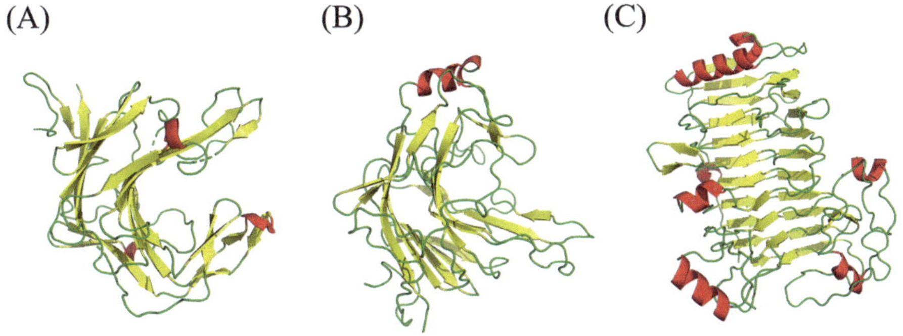

Fig. 3 Overall structure of CgkA (1DYP) from *Pseudoalteromonas carrageenovora* (**a**), ZgCgkA (5OCR) from *Zobellia galactanivorans* Dsij^T (**b**) and CgiA (1H80) from *Alteromonas fortis* (**c**)

possesses two additional domains (A and B) in the C-terminal region compared to the structure of κ-carrageenase (Fig. 3c). The domain A has an α/β fold, similar to the DNA/RNA binding protein, and the domain B is stabilized by a peptide loop with two disulfide bridges. The structural modeling of CgiA complexed with a substrate indicates that the substrate binds to the enzyme through bonding with several residues (Fig. 4c), and the interaction leads to dramatic conformational changes in the domain A. The changes in the domain A cause its shift toward the β-helix cleft and shape a tunnel to hold a tetraose substrate. The switch from an open conformation to a closed tunnel might interpret the processivity of ι-carrageenase [38].

5.2 Catalytic Mechanism

Carrageenan is hydrolyzed by carrageenase through the coordination of acid/base and nucleophile catalysts consisting of glutamate and/or aspartate residues. Different types of carrageenases employ different strategies for the hydrolysis of carrageenan polymer. The κ-carrageenases degrade the polysaccharide by a retaining mechanism via double-replacement reactions, in which the catalytic carboxylate residues distributed at the opposite site of sugar plane attack the glycosyl oxygen and anomeric caron atoms, respectively, completing the glycosylation and deglycosylation steps (Fig. 5) [3, 5]. Specifically, the proton donated to the glycosyl oxygen to

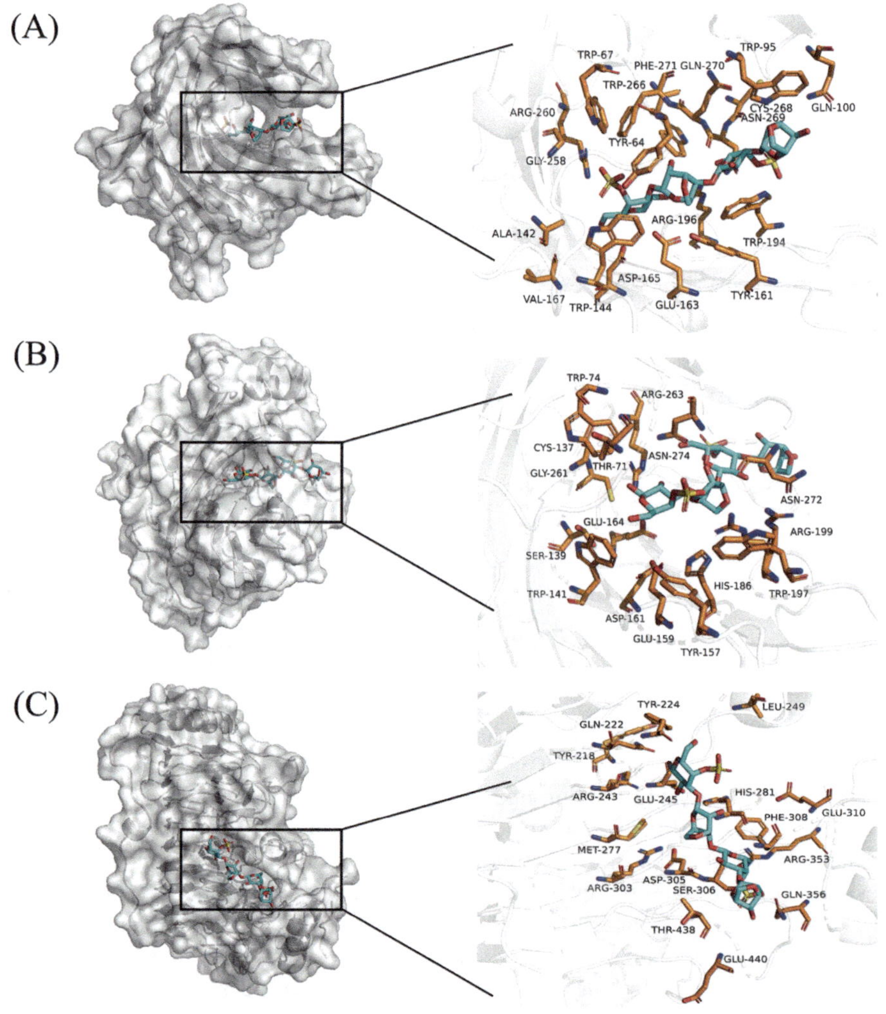

Fig. 4 Active sites and molecular docking of CgkA (1DYP) from *Pseudoalteromonas carrageenovora* (**a**), ZgCgkA (5OCR) from *Zobellia galactanivorans* Dsij^{T} (**b**), and CgiA (1H80) from *Alteromonas fortis* (**c**)

facilitate the departure of leaving group and the nucleophilic attack on the anomeric carbon by the carboxylate residue facilitates the formation of glycosyl–enzyme intermediate. The deprotonated acid/base then functions as a general base to activate the nucleophile, which in turn breaks the bonding of glycosyl–enzyme complex.

The ι-carrageenases hydrolyze their substrate by inverting mechanism, involving an overall inversion of the anomeric configuration at the cleavage site (Fig. 6). The two crucial carboxylic residues act as general acid and base catalysts, and these groups are circa 10.5 Å apart from each other. The protonation of the glycosidic

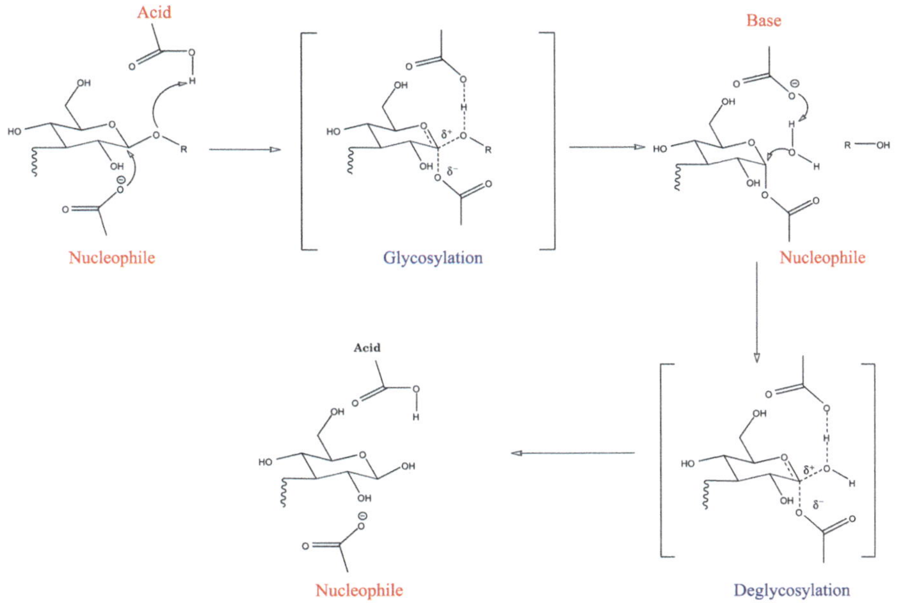

Fig. 5 Retaining mechanism for the degradation of carrageenan by κ-carrageenase via double-displacement reaction [5]

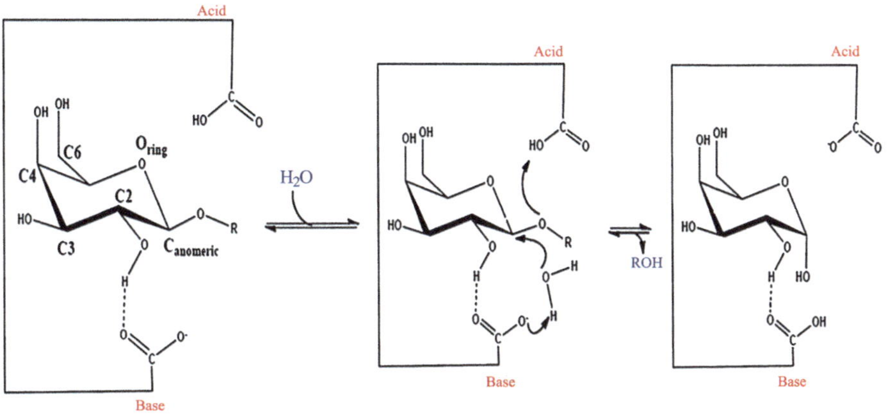

Fig. 6 Inverting mechanism for the degradation of carrageenan by ι-carrageenase via single displacement reaction [2]

bond by the attack of the acid/base residue liberates the aglycon. Meanwhile, the anomeric carbon is attacked by a water molecule activated by a base residue, yielding a stereochemically distinct anomer opposite to that of the substrate [38, 39].

Due to the limited studies on the structure of carrageenases and the availability of primary sequences of carrageenases, the overall picture of the structural contribution of carrageenase to its catalytic mechanism is still elusive. Advancements in exploring the microbial genome and developments in structural biology can deepen our understanding of the structure and reaction mechanism of these enzymes in the long run.

6 Molecular Modification and Immobilization of Carrageenase

6.1 Molecular Modification of Carrageenase

Enzymes are gaining popularity in diverse fields such as industry, agriculture, medicine, and environmental protection. However, enzymes have not reached the degree of large-scale application, mainly due to their properties such as instability, strict requirements for pH and temperature, and the high cost of their production and purification. As such, enzyme modification has been developed with the attempts to enhance their adaptation. At present, efforts have been invested on expression of the truncated or mutated versions of carrageenases and immobilization of the free enzyme on a supporting matrix.

The κ-carrageenase ZL-4 from *Zobellia* sp. ZL-4 was modified by removing the carbohydrate-binding module (CBM) and the type-IX secretion module, and the resultant κ-ZL-4-GH16 exhibited enhanced thermal and pH stability and more favored acidic conditions at pH 3.0–6.0 [64]. In addition, the truncated ZL-4 demonstrates higher substrate affinity, and the major degradation products had shifted to more disaccharides. Molecular investigation uncovered that the ZL-4 enzyme formed an active site cleft to process a κ-carrageenan hexose, whereas the catalytic channel of κ-ZL-4-GH16 was partially closed and that of κ-ZL-4 was slightly open, which might explain the differential enzyme–substrate affinity. The attempt on the κ-carrageenase cgkZ from *Zobellia* sp. ZM-2 was used to be referred as another trial on carrageenase modification by truncations [36]. The Por secretion tail (Pst) and the carbohydrate-binding module (CBM) were deleted to generate the truncated carrageenases of cgkZΔPst and cgkZΔCBM, respectively. The transcript abundance of *Pst* truncation was increased when the recombinant was expressed in *Pichia pastoris* and the cgkZΔPst exhibited the enhanced enzyme activity. Meanwhile, the *CBM* truncation promoted the transcription but reduced the hydrolytic activity of cgkZ, suggesting that CBM has an important role in maintaining the enzymatic activity of κ-carrageenase.

The strain designated *Pseudoalteromonas* sp. AJ5 was isolated from the intestine of *Apostichopus japonicus*, which demonstrated the lytic activity against κ-carrageenan. Through random mutagenesis using UV irradiation and ethyl methanesulphonate (EMS) treatment, the potential mutants of *Pseudoalteromonas* sp. AJ5 were obtained. AJ5-13 was identified from the mutant population showing increased carrageenan degradation activity by 2.9-fold compared to that of the parent culture, providing another approach to carrageenase enzyme engineering to improve enzyme function [33].

6.2 *Immobilization of Carrageenase*

In addition to the attempts of molecular modification of carrageenases through genetic engineering, the immobilization of carrageenase on a matrix provides another strategy to improve its reusability and tolerance to a wider range of reaction conditions. Generally, the immobilization can be achieved by adsorption, embedding, covalent bonding, and cross-linking methods. Currently, only a few reports describe the immobilization of carrageenase and characterizing its catalytic parameters. The κ-carrageenases from *Pseudoalteromonas* sp. ASY5 and *Pseudoalteromonas carrageenovora* were successfully covalently immobilized to carboxyl-functionalized magnetic nanoparticles (CMNP) [56, 58]. The catalytic domain of *Pseudoalteromonas purpurea* κ-carrageenase (named KCgCD) was covalently immobilized on the synthesized amine-modified zeolitic imidazolate framework-8 nanoparticles (An-ZIF-8) with the formation of cross-linked enzyme aggregates (CLEAs). The immobilized κ-carrageenase was named KCgCD-An-ZIF-8-CLEA, and its SEM image is shown in Fig. 7 [72]. These attempts have demonstrated that the immobilized carrageenases not only improve the stability and catalytic efficiency during the reaction but also have feasibility in the production of immobilized enzymes that can be reused several times to produce the carrageenan-derived oligosaccharides. Further optimization of the immobilization conditions and exploration of other carrageenase candidates would pave the way for the industrial utilization of carrageenases.

Fig. 7 SEM images of the carrier An-ZIF-8 (left) and the immobilized κ-carrageenase KCgCD-An-ZIF-8-CLEA (right) [72]

7 Methods for the Production and Purification of Carrageenase

Wild-type strains and engineered strains can be used for carrageenase production. The carrageenase-secreting bacterial strains have been screened through different approaches, and these carrageenases are normally inducible by elicitors such as the degradation products of carrageenan by the carrageenase [4, 26]. A few carrageenases have also been found to be localized intracellularly [24].

The mass production of carrageenase can be achieved through complex fermentation systems. Many attempts aimed to maximize the enzyme activity with the consideration of culture temperature, inducer concentration and induction time, medium constituents, and the pH [47]. A fed-batch fermentation system was set up, and the substitution of the casamino acids by an equal amount of glutamic acid significantly increased the carrageenase enzyme activity [8]. The optimization tests were performed on *Pseudomonas aeruginosa* ZSL-2 by varying single factors and with orthogonal array design, and the κ-carrageenase production through the optimized medium and growth conditions was enhanced [73]. Optimization efforts were also invested on *Pseudomonas elongate*, and the yield of carrageenase was significantly increased with the optimized ingredients of the culture medium [21]. Similarly, the growth medium was optimized for *Thalassospira* sp. fjfst-332 to achieve maximized production capacity of κ-carrageenase, and endeavors were made to provide guide for industrial utilization of this strain for enzyme production [15]. Moreover, the strain of *Thalassospira* sp. fjfst-332 was immobilized to magnetic Fe_3O_4-chitosan carrier, and the immobilized strain gains tolerance to higher temperature and wider pH range. The outcome is a more efficient processing of carrageenan into oligosaccharides with extended lifetime [16].

Except for using wild-type strains for carrageenase production, some carrageenases are heterologously expressed in engineered strains. By introducing a wild-type signal peptide of *Zobellia* sp. ZM-2 κ-carrageenase into the carrageenase-expressing vector *pProEX-HTa-cgkZ*, the secretion of the recombinant κ-carrageenase into extracellular space was greatly increased [32]. The catalytic domain of a κ-carrageenase from *Pseudoalteromonas porphyrae* LL1 was effectively expressed in *Brevibacillus choshinensis*, and the extracellular κ-carrageenase is highly tolerant to alkaline condition [65]. Additionally, the yeast strain *Pichia pastoris* was successfully employed to express full-length κ-carrageenase and its truncated versions, providing another approach to expressing carrageenase heterologously [62].

The purification strategy of expressed carrageenase may vary depending on the expression vehicles and the genetic elements associated with the enzymes. The purification procedures of carrageenase from extracellular secretion may include several successive operations [69]. Initially, ammonium sulfate precipitation is performed to obtain the proteins from fermentation system of wild-type strains. Ion-exchange chromatography is then used to further purify the enzyme of interest, using either cation exchanger to bind carrageenase with alkaline pI or anion exchanger to pull carrageenase with low pI. Additionally, gel filtration and hydrophobic interaction chromatography have also been employed during the purification procedures [24, 44]. For recombinant carrageenase enzymes, they are usually purified using affinity chromatography taking advantage of the epitope tags, such as Ni^{2+}-cheated resin interacting with hexahistidine-tags [6, 35, 49, 55, 70].

8 Application of Carrageenase

8.1 Preparation of Carrageenan Oligosaccharides

Carrageenase can efficiently and specifically degrade carrageenan to produce carrageenan oligosaccharides. Moreover, the enzymatic treatment is gentle and can minimize the modification to the saccharide molecules. Thus, the enzymatic digestion of carrageenan by carrageenase would be an ideal alternate to the traditional acid hydrolysis method.

Carrageenan oligosaccharides have a variety of biological activities, such as antitumor [42], immune regulation [19, 29], antioxidant activity [63], and promotion of plant growth [41]. The hydrolysate of κ-carrageenan by κ-carrageenase purified from *Bacillus* sp. HT19 confers cucumber-enhanced resistance to cucumber mosaic virus (CMV) and increased activities of antioxidant enzymes in infected plants [28]. By combining the cellulase and heterogeneously expressed κ-carrageenase originated from cgkZ in marine bacterium, a high-efficiency method is developed to prepare carrageenan oligosaccharides directly from *Eucheuma cottonii* [7]. The mixture of κ-neocarrabiose-sulfate, κ-neocarrahexaose-sulfate, and κ-neocarraoctaose-sulfate prepared from enzymatic digestion of κ-carrageenan demonstrates strong anti-tumor activity by repressing the transcription of the cell proliferation-promoting factors such as bFGF and bFGFR [61]. Therefore, carrageenan oligosaccharides not only play an important role in organism but also have important research value and broad development prospect in biomedicine and other aspects.

8.2 Study of Carrageenan Structure

The structure of carrageenan can be inferred from analysis of the structure of its hydrolysate by carrageenanase. The action manner of carrageenase over carrageenans determines the degradation pattern and the composition of the enzymatic products. The analysis of the components in the lysate would provide clue to drawing the structure of carrageenan polysaccharide. Spectroscopic and spectrometric methods are normally employed to dissect the polysaccharide structure, and the insight into the enzymatic fractionation offers great help to elucidate various structural elements in the intricate carrageenan polysaccharide molecules [11]. By analysis of the carrageenanase-digested carrageenan in combination with nuclear magnetic resonance technology, it revealed that λ-carrageenan unit may contain four sulfate groups instead of only three groups [13].

8.3 Application in Algal Genetic Engineering

Carrageenan is closely associated with the cell wall of red algae. The high viscosity of carrageenan confers difficulty in extracting genomic DNA from red algae. The specific activity of carrageenase imparts the advantage of using carrageenanase to digest the seaweed prior to DNA extraction. In addition, enzymatic digestion of algae cell wall can also facilitate protoplast preparation, so as to provide smooth progress of seaweed genetic engineering.

8.4 Study on Carrageenan Metabolism in Bacteria

The heterotrophic marine bacteria attached to the surface of red algae can secrete carrageenan-degrading enzymes and use carrageenan oligosaccharides to obtain energy. The knowledge about carrageenan degrading enzyme would support the study of carrageenan metabolism in bacteria. The identification of carrageenan-induced elements involved in carrageenan degradation including carrageenases contributes to a broad picture of genetic network deciphering carrageenan utilization in heterotrophic bacteria [9].

References

1. Araki T, Higashimoto Y, Morishita T (1999) Purification and characterization of κ-carrageenase from a marine bacterium, *Vibrio* sp. CA-1004. Fish Sci 65(6):937–942
2. Chang CF (ed) (2012) Carbohydrates-comprehensive studies on glycobiology and glycotechnology. InTech, pp 171–134
3. Chauhan PS, Gupta N (2016) Insight into microbial mannosidases: a review. Crit Rev Biotechnol 8:1–12
4. Chauhan PS, Saxena A (2016) Bacterial carrageenases: an overview of production and biotechnological applications. 3 Biotech 6(2):146
5. Chauhan PS, Puri N, Sharma P, Gupta N (2012) Mannanases: microbial sources, production, properties and potential biotechnological applications. Appl Microbiol Biotechnol 93(5):1817–1830
6. Cui HL, Peng YX, Zhao BW, Liu YG, Chen FJ, Wu HG, Yao Z (2017) Cloning, identification and characterization of a novel κ-carrageenase from marine bacterium *Cellulophaga lytica* strain N5-2. Int J Biol Macromol 105:509–515

7. Duan FY, Yu Y, Liu ZM, Tian L, Mou HJ (2016) An effective method for the preparation of carrageenan oligosaccharides directly from *Eucheuma cottonii* using cellulase and recombinant κ-carrageenase. Algal Res 15:93–99
8. Dyrset N, Lystad KQ, Levine DW (1997) Development of a fermentation process for production of a κ-carrageenase from *Pseudomonas carrageenovora*. Enzym Microb Technol 20(6):418–423
9. Ficko-Blean E, Préchoux A, Thomas F, Rochat T, Larocque R, Zhu Y, Stam M, Genicot S, Jam M, Calteau A (2017) Carrageenan catabolism is encoded by a complex regulon in marine heterotrophic bacteria. Nat Commun 8(1):1685
10. Genicot SM, Groisillier A, Rogniaux H, Meslet-Cladiere L, Barbeyron T, Helbert W (2014) Discovery of a novel iota carrageenan sulfatase isolated from the marine bacterium *Pseudoalteromonas carrageenovora*. Front Chem 2:67
11. Ghanbarzadeh M, Golmoradizadeh A, Homaei A (2018) Carrageenans and carrageenases: versatile polysaccharides and promising marine enzymes. Phytochem Rev 17(3):535–571
12. Gui YY, Gu XQ, Fu LP, Zhang Q, Zhang PY, Li J (2021) Expression and characterization of a thermostable carrageenase from an antarctic *Polaribacter* sp. NJDZO3 strain. Front Microbiol 12:631039
13. Guibet M, Kervarec N, Génicot S, Chevolot Y, Helbert W (2006) Complete assignment of 1H and 13C NMR spectra of Gigartina skottsbergii lambda-carrageenan using carrabiose oligosaccharides prepared by enzymatic hydrolysis. Carbohydr Res 341:1859–1869
14. Guibet M, Colin S, Barbeyron T, Genicot S, Kloareg B, Michel G, Helbert W (2007) Degradation of lambda-carrageenan by *Pseudoalteromonas carrageenovora* lambda-carrageenase: a new family of glycoside hydrolases unrelated to kappa- and iota-carrageenases. Biochem J 404:105–114
15. Guo JJ, Zhang LT, Lu X, Zeng SX, Zhang Y, Xu H, Zheng BD (2016) Medium optimization and fermentation kinetics for κ-carrageenase production by *Thalassospira* sp Fjfst-332. Molecules 21(11):1479
16. Guo JJ, Zheng ZC, Chen C, Lu X, Zhang Y, Zheng BD (2017) Enhanced production of κ-carrageenase and κ-carrageenan oligosaccharides through immobilization of *Thalassospira* sp Fjfst-332 with magnetic Fe_3O_4-chitosan microspheres. J Agric Food Chem 65(36):7934–7943
17. Hatada Y, Mizuno M, Li Z, Ohta Y (2011) Hyper-production and characterization of the ι-carrageenase useful for ι-carrageenan oligosaccharide production from a deep-sea bacterium, *Microbulbifer thermotolerans* JAMB-A94^T, and insight into the unusual catalytic mechanism. Mar Biotechnol 13(3):411–422
18. Hjerde T, Christensen BE, Smidsrod O (2017) The influence of the conformational state of κ- and τ-carrageenan on the rate of acid hydrolysis. J Acta Mater 128:249–257
19. Jagtap AS, Manohar CS (2021) Overview on microbial enzymatic production of algal oligosaccharides for nutraceutical applications. Mar Biotechnol 23(2):159–176
20. Khambhaty Y, Mody K, Jha B (2007a) Purification and characterization of κ-carrageenase from a novel γ-proteobacterium, *Pseudomonas elongata* (MTCC 5261) syn. *Microbulbifer elon* comb. nov. Biotechnol Bioprocess Eng 12(6):668–675
21. Khambhaty Y, Mody K, Jha B, Gohel V (2007b) Statistical optimization of medium components for κ-carrageenase production by *Pseudomonas elongata*. Enzym Microb Technol 40(4):813–822
22. Kobayashi T, Koide O, Mori K, Shimamura S, Matsuura T, Miura T, Takaki Y, Morono Y, Nunoura T, Imachi H, Inagaki F, Takai K, Horikoshi K (2008) Phylogenetic and enzymatic diversity of deep subseafloor aerobic microorganisms in organics- and methane-rich sediments off Shimokita Peninsula. Extremophiles 12:519–527
23. Kobayashi T, Uchimura K, Koide O, Deguchi S, Horikoshi K (2012) Genetic and biochemical characterization of the *Pseudoalteromonas tetraodonis* alkaline κ-carrageenase. Biosci Biotechnol Biochem 76(3):506–511

24. Li S, Jia P, Wang L, Yu W, Han F (2013) Purification and characterization of a new thermostable κ-carrageenase from the marine bacterium *Pseudoalteromonas* sp. QY203. J Ocean Univ China 12(1):155–159
25. Li J, Hu QS, Seswita-Zilda D (2014) Purification and characterization of a thermostable λ-carrageenase from a hot spring bacterium, *Bacillus* sp. Biotechnol Lett 36(8):1669–1674
26. Li YP, Huang ZJ, Qiao LK, Gao Y, Guan HS, Hwang HM, Aker WG, Wang P (2015) Purification and characterization of a novel enzyme produced by *Catenovulum* sp. LP and its application in the pre-treatment to *Ulva prolifera* for bio-ethanol production. Process Biochem 50(5):799–806
27. Li S, Hao J, Sun M (2017) Cloning and characterization of a new cold-adapted and thermotolerant ι-carrageenase from marine bacterium *Flavobacterium* sp. YS-80-122. Int J Biol Macromol 102:1059–1065
28. Li J, Pan AH, Xie MS, Zhang PP, Gu XQ (2019) Characterization of a thermostable κ-carrageenase from a hot spring bacterium and plant protection activity of the oligosaccharide enzymolysis product. J Sci Food Agric 99(4):1812–1819
29. Ling X, Yao Z, Wu H, Wang FF, Zhang SX (2012) The immune regulation of κ-carrageenan oligosaccharide and its desulfated derivatives on LPS-activated microglial cells. Neurochem Int 61:689–696
30. Liu GL, Li Y, Chi Z, Chi ZM (2011) Purification and characterization of κ-carrageenase from the marine bacterium *Pseudoalteromonas porphyrae* for hydrolysis of κ-carrageenan. Process Biochem 46(1):265–271
31. Liu ZM, Li GY, Mo ZL, Mou HJ (2013) Molecular cloning, characterization, and heterologous expression of a new κ-carrageenase gene from marine bacterium *Zobellia* sp. ZM-2. Appl Microbiol Biotechnol 97(23):10057–10067
32. Liu ZM, Tian L, Chen YL, Mou HJ (2014) Efficient extracellular production of kappa-carrageenase in *Escherichia coli*: effects of wild-type signal sequence and process conditions on extracellular secretion. J Biotechnol 185:8–14
33. Ma YX, Dong SL, Jiang XL, Mou HJ (2010) Purification and characterization of carrageenase from marine bacterium mutant strain *Pseudoalteromonas* sp. Aj5-13 and its degraded products. J Food Biochem 34:661–678
34. Ma S, Duan G, Chai W, Geng C, Tan Y, Wang L, Le Sourd F, Michel G, Yu W, Han F (2013a) Purification, cloning, characterization and essential amino acid residues analysis of a new ι-carrageenase from *Cellulophaga* sp. QY3. PLos One 8(5):e64666
35. Ma S, Tan YL, Yu WG, Han F (2013b) Cloning, expression and characterization of a new ι-carrageenase from marine bacterium, *Cellulophaga* sp. Biotechnol Lett 35(10):1617–1622
36. Matard-Mann M, Bernard T, Leroux C, Barbeyron T, Larocque R, Prechoux A, Jeudy A, Jam M, Nyvall Collen P, Michel G, Czjzek M (2017) Structural insights into marine carbohydrate degradation by family GH16 κ-carrageenases. J Biol Chem 292(48):19919–19934
37. Michel G, Chantalat L, Duee E, Barbeyron T, Kloareg B, Chantalat L, Henrissat B, Dideberg O (2001a) The κ-carrageenase of *P. carrageenovora* features a tunnel-shaped active site a novel insight in the evolution of clan-B glycoside hydrolases. Structure 9:513–525
38. Michel G, Chantalat L, Fanchon E, Henrissat B, Kloareg B, Dideberg O (2001b) The ι-carrageenase of *Alteromonas fortis*. A β-helix fold-containing enzyme for the degradation of a highly polyanionic polysaccharide. J Biol Chem 276(43):40202–40209
39. Michel G, Helbert W, Kahn R, Dideberg O, Kloareg B (2003) The structural bases of the processive degradation of ι-carrageenan, a main cell wall polysaccharide of red algae. J Mol Biol 334(3):421–433
40. Mody K, Khambhaty Y, Jha B (2007) Purification and characterization of κ-Carrageenase from a Novel γ-Proteobacterium, Pseudomonas elongate (MTCC 5261) syn. *Microbulbifer elongatus* comb. Nov. Biotechnol. Bioprocess Eng 12(6):668–675

41. Moenne A, Gonzalez A (2021) Chitosan-, alginate- carrageenan-derived oligosaccharides stimulate defense against biotic and abiotic stresses, and growth in plants: a historical perspective. Carbohydr Res 503:108298
42. Mou H, Jiang X, Guan H (2003) A κ-carrageenan derived oligosaccharide prepared by enzymatic degradation containing anti-tumor activity. J Appl Phycol 15:297–303
43. Mou HJ, Jiang XL, Liu ZH, Guan H (2004) Structural analysis of kappa-carrageenan oligosaccharides released by carrageenase from marine *Cytophaga* MCA-2. J Food Biochem 28(3):245–260
44. Ohta Y, Hatada Y (2006) A novel enzyme, λ-carrageenase, isolated from a deep-sea bacterium. J Biochem 140(4):475–481
45. Potin P, Sanseau A, Legall Y, Rochas C, Kloareg B (1991) Purification and characterization of a new κ-carrageenase from a marine *Cytophaga*-like bacterium. Eur J Biochem 201(1):241–247
46. Rebuffet E, Barbeyron T, Jeudy A, Jam M, Czjzek M, Michel G (2010) Identification of catalytic residues and mechanistic analysis of family GH82 ι-carrageenases. Biochemistry 49(35):7590–7599
47. Rinas U, Hoffmann F (2004) Selective leakage of host-cell proteins during high-cell-density cultivation of recombinant and non-recombinant *Escherichia coli*. Biotechnol Prog 20(3):679–687
48. Sarwar G, Matayoshi S, Oda H (1987) Purification of a κ-carrageenase from marine *Cytophaga* species. Microbiol Immunol 31(9):869–877
49. Shen JJ, Chang YG, Dong SJ, Chen F (2017) Cloning, expression and characterization of a ι-carrageenase from marine bacterium *Wenyingzhuangia fucanilytica*: a biocatalyst for producing ι-carrageenan oligosaccharides. J Biotechnol 259:103–109
50. Shen JJ, Chang YG, Chen F, Dong SJ (2018) Expression and characterization of a kappa-carrageenase from marine bacterium *Wenyingzhuangia aestuarii* OF219: A biotechnological tool for the depolymerization of kappa-carrageenan. Int J Biol Macromol 112:93–100
51. Skea GL, Mount Fort DO, Clements KD (2007) Contrasting digestive strategies in four New Zealand herbivorous fishes as reflected by carbohydrase activity profiles. J Comp Biochem Physiol 146:63–70
52. Smith J, Mountfort D, Falshaw R (2005) A zymogram method for detecting carrageenase activity. Anal Biochem 347:336–338
53. Sun F, Ma Y, Wang Y, Liu Q (2010) Purification and characterization of novel κ-carrageenase from marine *Tamlana* sp. HC4. Chin J Oceanol Limnol 28(6):1139–1145
54. Sun HH, Gao L, Xue CH, Mao XZ (2020) Marine-polysaccharide degrading enzymes: status and prospects. Compr Rev Food Sci Food Saf 19(6):2767–2796
55. Wang LN, Li SY, Zhang SL, Li JJ, Yu WG, Gong QH (2015) A new κ-carrageenase CgkS from marine bacterium *Shewanella* sp. Kz7. J Ocean Univ China 14(4):759–763
56. Xiao AF, Xu C, Yan L, Ni H, Cai HN (2016) Preparation and characterization of κ-carrageenase immobilized onto magnetic iron oxide nanoparticles. Electron J Biotechnol 19:1–7
57. Xu XY, Li SY, Yang XM, Yu WG, Han F (2015a) Cloning and characterization of a new κ-carrageenase gene from marine bacterium *Pseudoalteromonas* sp. QY203. J Ocean Univ China 14(6):1082–1086
58. Xu C, Lin Y, Ni H, Zhu Y, Cai H, Xiao A (2015b) Characterization of immobilized carrageenase on superparamagnetic nanoparticles for oligosaccharide preparation. Nanosci Nanotechnol Lett 7(10):855–860
59. Yao Z, Zhang C, Lu F, Bie X, Lu Z (2012) Gene cloning, expression, and characterization of a novel acetaldehyde dehydrogenase from *Issatchenkia terricola* strain XJ-2. Appl Microbiol Biotechnol 93(5):1999–2009
60. Yao Z, Wang F, Gao Z, Jin L, Wu H (2013) Characterization of a κ-carrageenase from marine *Cellulophaga lytica* strain N5-2 and analysis of its degradation products. Int J Mol Sci 14(12):24592–24602
61. Yao Z, Wu H, Zhang S, Du Y (2014) Enzymatic preparation of κ-carrageenan oligosaccharides and their anti-angiogenic activity. Carbohydr Polym 101:359–367

62. Yu Y, Liu ZM, Yang M, Chen M, Wei ZH, Shi LX, Li L, Mou HJ (2017) Characterization of full-length and truncated recombinant κ-carrageenase expressed in *Pichia pastoris*. Front Microbiol 8:1544
63. Yuan HM, Zhang WW, Li XG, Lu XX, Li N, Gao XL, Song JM (2005) Preparation and in vitro antioxidant activity of kappa-carrageenan oligosaccharides and their oversulfated, acetylated and phosphorylated derivatives. Carbohydr Res 340:685–692
64. Zhang YJ, Lang B, Zeng DY, Li ZH, Yang J, Yan RX, Xu XQ, Lin J (2019) Truncation of κ-carrageenase for higher κ-carrageenan oligosaccharides yield with improved enzymatic characteristics. Int J Biol Macromol 130:958–968
65. Zhao YJ, Chi Z, Xu YY, Shi NC, Chi ZM, Liu GL (2018) High-level extracellular expression of kappa-carrageenase in *Brevibacillus choshinensis* for the production of a series of kappa-carrageenan oligosaccharides. Process Biochem 64:83–92
66. Zhao DY, Jiang B, Zhang Y, Sun WH, Pu ZJ, Bao YM (2021) Purification and characterization of a cold-adapted κ-carrageenase from *Pseudoalteromonas* sp. ZDY3. Protein Expr Purif 178:105768
67. Zhou MH, Ma JS, Li J, Ye HR, Huang KX, Zhao XW (2008) A κ-carrageenase from a newly isolated *Pseudoalteromonas*-like bacterium, WZUC10. Biotechnol Bioprocess Eng 13(5):545–551
68. Zhu BW, Ning LM (2016) Purification and characterization of a new κ-carrageenase from the marine bacterium *Vibrio* sp. NJ-2. J Microbiol Biotechnol 26(2):255–262
69. Zhu B, Ni F, Sun Y, Zhu X, Yin H, Yao Z, Du Y (2018a) Insight into carrageenases: major review of sources, category, property, purification method, structure, and applications. Crit Rev Biotechnol 38(8):1261–1276
70. Zhu B, Ni F, Ning L, Yao Z, Du Y (2018b) Cloning and biochemical characterization of a novel kappa-carrageenase from newly isolated marine bacterium *Pedobacter hainanensis* NJ-02. Int J Biol Macromol 108:1331–1338
71. Zhu B, Ni F, Xiong Q, Yao Z (2021) Marine oligosaccharides originated from seaweeds: source, preparation, structure, physiological activity and applications. Crit Rev Food Sci Nutr 61(1):60–74
72. Zhu CH, Mou MJ, Yang LL, Jiang ZD, Zheng MJ, Li ZP, Hong T, Ni H, Li QB, Yang YF, Zhu YB (2023) Enzymatic hydrolysates of κ-carrageenan by κ-carrageenase-CLEA immobilized on amine-modified ZIF-8 confer hypolipidemic activity in HepG2 cells. Int J Biol Macromol 252:126401
73. Ziayoddin M, Lalitha J, Shinde M (2014) Increased production of carrageenase by *Pseudomonas aeruginosa* ZSL-2 using Taguchi experimental design. Int Lett Nat Sci 17:194–207

Fucanase

Yaoguang Chang

Fucanases are hydrolytic enzymes with a wide range of sources. It has been reported that the source strain can be found in seawater, sea mud, surface of algae, and digestive tract of marine organisms, and fucanases mainly exist in marine microorganisms and mollusks [1–3]. Bacteria capable of degrading sulfated fucan were first described in 1959 [4]. The term "fucoidanase" was introduced in 1967 in an article discussing the isolation of the enzyme from the hepatopancreas of Haliotis sp. [5]. Before 2016, fucoidanases were designated as enzymes that catalyze the hydrolysis of the glycoside bonds between sulfated fucose residues in sulfated fucans (also known as fucoidans). Later, some scholars pointed out that enzymes catalyzing the cleavage of any glycoside bonds in sulfated fucans (polysaccharides exclusively composed of fucose residues) should be referred to as fucanases, rather than fucoidanases [6]. Carbohydrate active enzyme (CAZy) database classified glycoside hydrolases based on the similarities of amino acid sequence and secondary structure. According to this classification, fucanases belong to the glycoside hydrolase 107 family (GH107, EC Number 3.2.1.212) and glycoside hydrolase 168 family (GH168, EC Number 3.2.1.211). All members of these families are random endo-acting enzymes. GH107 mainly cleave α-1→4 glycosidic linkage, whereas GH168 cleave α-1→3 glycosidic linkages of sulfated fucans.

Since low-molecular-weight sulfated fucans have lower viscosity, better solubility than sulfated fucans and have been proven to have a variety of biological activities, the preparation of sulfated fucan oligomers is of interest. Enzymes are indispensable tools for structural analysis of sulfated fucans and production of

Y. Chang (✉)
State Key Laboratory of Marine Food Processing and Safety Control, College of Food Science and Engineering, Ocean University of China, Qingdao, China

Laboratory for Marine Drugs and Bioproducts, Pilot National Laboratory for Marine Science and Technology (Qingdao), Qingdao, China
e-mail: changyg@ouc.edu.cn

X. Mao (ed.), *Degrading Enzymes for Marine Polysaccharides and Their Applications*, https://doi.org/10.1007/978-981-96-9517-1_10

oligomers. Enzymatic hydrolysis has advantages of mild conditions, controllable process, green and high efficiency. Due to the high specificity of the action mode of glycoside hydrolase, the products are relatively homogeneous. Therefore, enzymatic degradation of sulfated fucans shows a broad application potential.

1 Source and Classification of Fucanases

Fucanases can be divided into two groups: wild type and recombinant type. Wild-type fucanases have been found in marine organisms: bacteria, invertebrates, and some fungi. *Pseudoalteromonas citrea* KMM 3296 [7], "*Fucophilus fucoidanolyticus*" SI-1234 [8], *Alteromonas* sp. SN-1009 [9], *Marinflexile fucanivorans* SW5^T [1], *Sphingomonas paucimobilis* PIT-1 [10], *Wenyingzhuangia fucanilytica* CZ1127^T [11], *Flavobacterium* sp. F^{-31} [12], *Formosa algae* KMM 3553^T [13], *Bacillus* sp. H-TP2 [14], *Klebsiella* sp. JS3 and JN3 [15], *Flavobacteriacea* sp. RC2-3 [16] are the bacteria strains that can utilize sulfated fucans. While fucanase-producing bacteria *Patinopecten yessoensi* [17], *Lambis* sp. [18] are marine mollusks, *Fusarium* sp. LD8 [19], *Dendryphiella arenaria* TM94 [20], and *Aspergilluswentii* sp. PZ322 belong to fungi.

In 2002, Japanese company Takara Shuzo published two patent sequences Fda1 (GenBank accession number AAO00508.1) and Fda2 (AAO00509.1) with sulfated fucan-degradation function, which were derived from *Alteromonas* sp. SN-1009 [21].

In 2006, French scholars Michel et al. reported recombinant enzyme FcnA (CAI47003.1) [22], which was isolated from marine bacterium *Mariniflexile fucanivorans* SW5^T. Based on the aforementioned research, they established the first fucanase glycoside hydrolase family, GH107. Through the BLASTp search and multiple sequence alignment, it was found that FcnA and patented fucanase Fda1 and Fda2 had high sequence similarity and multiple conserved sites, noted that these three enzymes belonged to the same GH family.

From 2016 to 2017, Russian scholars Kusaykin et al. predicted two fucanase genes *ffa1* (WP_05778421.7) [23] and *ffa2* (WP_05778421.1) [24] by BLASTp compared with FcnA, Fda1 and Fda2. FFA1 and FFA2 of the GH107 family were discovered in the genome of a marine bacterium, the *Formosa algae* KMM 3553^T. In 2020, this research team characterized two recombinant fucanases FWf1 (WP_068825865.1) and FWf2 (WP_068825868.1) [25] from marine bacterium *Wenyingzhuangia fucanilytica* CZ1127^T, which were annotated as GH107 family in CAZy.

In 2018, Japanese scholars Ohshiro et al. purified a fucanase Fct114 (LC378584) from the marine bacterium *Luteolibacter algae* H18 [26]. However, the enzyme displayed no similarity in amino acid sequence to any reported fucanases, and it was not classified in any GH family. In the same year, Danish scholars Stougaard et al. identified three sequences Fp273 (AYC81238.1), Fp277 (AYC81239.1), and Fp279 (AYC81240.1) derived from a marine metagenome. These enzymes were annotated as GH107 family by dbCAN prediction [27]. Subsequently, Canadian scholars Boraston et al. characterized two GH107 family fucanases P5AFcnA (AYF59291.1) and P19DFcnA (AYF59292.1), which were isolated from *Psychromonas* sp. SW5A and *Psychromonas* sp. SW19D, respectively [28].

In 2020, our research group characterized an endo-1,3-fucanase gene *funA* (WP_068826898.1) [29], which was discovered through transcriptomics analysis of the marine bacterium *Wenyingzhuangia fucanilytica* CZ1127^{T}. Since FunA exhibited no similarity to any previously reported glycoside hydrolase domains, a new glycoside hydrolase family GH168 was established. In the same year, Danish scholars Mikkelsen et al. identified a gene coding for a putative fucanase in the genome of the marine bacterium *Formosa haliotis* named Fhf1 (WP_066217780), which shares a high identity with the catalytic domain of GH107 fucanases [30].

2 The Assay Methods of Enzyme Activity

The reported enzyme activity detection methods of fucanases were Nelson-Somogyi method [31], DNS method [32], pHBH method [33], HPLC, and C-PAGE [34] method.

The activities of the wild fucanases secreted by *M. fucanivorans* SW5^{T} and *F. algae* KMM 3553^{T} were detected by using C-PAGE method. The results showed that the enzymes produced by them possess fucanase activities. The enzymes of *W. fucanilytica* CZ1127^{T} and *Flavobacterium* sp. F-31 monitored the decrease of substrate molecular weight after enzymatic hydrolysis by HPLC method, which confirmed the fucanase activity. DNS method was used to detect the activities of enzymes produced by *Bacillus* sp. H-TP2, *Pseudidiomarina* sp. ZJCN121, *Klebsiella* sp. JS3, *Klebsiella* sp. JN3, *Fusarium* sp. LD8, *D. arenaria* TM94, and *Aspergilluswentii* sp. PZ322. The activities of fucanases produced by *Flavobacteriacea* sp. RC2-3, *L. kurila,* and *Lambis sp* were determined by Nelson-Somogyi method. The results showed that the enzymes produced by these bacteria were provided with fucanase activity.

The enzymatic activity of FcnA was assessed through C-PAGE analysis and spectrophotometry at a wavelength of 232 nm. The oligosaccharides of sulfated fucans were detected in C-PAGE. There was no observed increase in absorbance at 232 nm over time, suggesting that the degradation products were not unsaturated oligosaccharides. The result supports the conclusion that FcnA functions as a glycoside hydrolase rather than a polysaccharide lyase. FFA1, FFA2, FWf1, and FWf2 were verified to have the activity on sulfated fucan from *Sargassum horneri* by using C-PAGE. FFA2 and Fhf1 exhibited fucanase activity on sulfated fucan form *Fucus evanescen* by C-PAGE assay method. The activities of Fp273, Fp277, and Fp279 were also verified by C-PAGE. The results showed that these three proteins could degrade sulfated fucans from brown algae Fucales. The results of HPSEC showed that Fct114 could degrade sulfated fucans from *Cladosiphon okamuranus*, and FunA could degrade sulfated fucans from *Isostichopus badionotus* to produce oligosaccharides. In addition, the enzyme activity verification of FunA was also verified using the pHBH method to detect reducing sugar. In 2021, Chinese scholars

Zhu et al. conducted in-depth research on the hydrolysis pattern and enzymatic properties of the enzyme Fda1 through DNS, $K_3Fe(CN)_6$ [35], pHBH, and MBTH [36] methods. The results of these four methods could confirm that Fda1 can degrade sulfated fucans from brown alga *Kjellmaniella crassifolia*.

3 Biochemical Properties

The optimum reaction pH of wild-type fucanase secreted from *P. citrea* KMM 3297 was 6.5–7.0, and an optimum reaction temperature range of 40–50 °C. Wild-type fucanase secreted by *Alteromonas* sp. SN-1009 exhibited its highest activity at 30–35 °C, pH 6.5–8.0. Ca^{2+} was its activator, while Zn^{2+} and Cu^{2+} decreased its activity. The optimal reaction pH of fucanase from *M. fucanivorans* $SW5^T$ was 7.5, and the optimal reaction temperature was 20–25 °C. Ca^{2+} was also its activator. Fucanase from *F. algae* KMM 3553^T displayed the highest enzyme activity at a range of pH 6.5–9.1. Zn^{2+} and Cu^{2+} could decrease the activity, whereas Mg^{2+}, Ca^{2+}, and Ba^{2+} exhibited a significant enhancing effect on its activity. The optimal reaction conditions of wild type fucanase from *Bacillus* sp. H-TP2 were 50 °C, pH 6.5. The optimum reaction pH and temperature of fucanase from *Pseudidiomarina* sp. ZJCN121 were pH 7.5 and 50 °C, respectively. The apparent V_{max} and K_m values were 6.11 $g{\cdot}L^{-1}{\cdot}min^{-1}$ and 5.87 $mg{\cdot}mL^{-1}$, respectively. Ca^{2+} and Ba^{2+} could enhance the activity. The presence of Cu^{2+} and Fe^{2+} resulted in complete inactivation of the enzyme. Mn^{2+}, K^+, and Zn^{2+} decreased the enzyme activity. Mg^{2+} did not affect the activity of the enzyme. The optimum reaction temperatures of fucanase secreted from *Klebsiella* sp. JS3 and *Klebsiella* sp. JN3 were 45 °C and 40 °C respectively, and their optimum reaction pH was 5.0.

Wild fucanase from marine invertebrate *P. yessoensi* showed the highest reaction pH and temperature at 5.0 and 45 °C respectively. Hg^{2+}, Zn^{2+}, and Cu^{2+}decreased the enzyme activity, whereas Mg^{2+}, Ba^{2+}, and Ca^{2+} could enhance the enzyme activity.

The optimal reaction temperature of fucanase from fungi *Fusarium* sp. LD8 was 60 °C, and the optimum reaction pH was 6.0. The apparent K_m and V_{max} values were 8.9 $mg{\cdot}L^{-1}$ and 2.02 $mg{\cdot}min^{-1}{\cdot}mL^{-1}$ respectively. The optimal reaction conditions of wild-type fucanase from *D. arenaria* TM94 were 50 °C, pH 6.0. It was stable when stored at pH 5.0–7.0.

Fda1 showed the highest reaction pH at 7.0, and the highest temperature at 35 °C. The kinetic constants were as follows: K_{cat} 0.82 ± 0.17 min^{-1}; K_m 3.88 ± 0.81 mg/mL; K_{cat}/K_m 0.21 ± 0.09 $mL{\cdot}mg^{-1}{\cdot}min^{-1}$. Cu^{2+} and Co^{2+} decreased the enzyme activity, while Fe^{3+}, Mn^{2+}, and 1.0 M NaCl could enhance the enzyme activity.

FFA1 was most active at 6.5–8.0 and 25–37 °C, respectively. It could be significantly reduced by EDTA, Al^{3+}, Cu^{2+}, Sn^{2+}, and Fe^{3+}, whereas Ca^{2+}, Ba^{2+}, Mg^{2+}, and Mn^{2+} could enhance the activity of the enzyme.

After being incubated at 40 °C for 30 min, the enzyme Fct114 retained 90% of its initial activity, whereas Fct114 was inactive when incubated up to 50 °C. It was stable when stored at pH 5.5–8.5. Ba^{2+}, Fe^{2+}, Ca^{2+}, and Al^{3+} could increase the enzyme activity.

The optimal reaction pH of FWf1 was 6.4–7.2, and the optimal reaction temperature was 24–35 °C. FWf2 exhibited the highest activity at 6.0–6.8, 24–40 °C. Its activity was significantly promoted by 0.1–0.3 M NaCl, Ca^{2+}, Mg^{2+}, and Ba^{2+}, while >0.5 M NaCl, Al^{3+}, Co^{2+}, Cu^{2+}, Fe^{3+}, Mn^{2+}, Ni^{2+}, Pb^{2+}, and Sn^{2+} could reduce the enzyme activities.

The optimum reaction conditions of Fhf1 were pH 8.0 and 37–40 °C. It was not appreciably impacted by NaCl. Ca^{2+}, Zn^{2+}, Mn^{2+}, Ni^{2+}, and Fe^{2+} could enhance the enzyme activity. EDTA could inhibit enzyme activity.

FunA exhibited optimal activity at a temperature of 40 °C and a pH of 8.0. It displayed stability when stored at 4 °C and 25 °C for 24 h, retaining over 80% of its original activity after exposure to pH 6.5–10.0 for 1 h. Moreover, FunA retained 94.12% of its activity after 90 days storage at 4 °C without any protective additives, indicating its remarkable stability. The kinetic parameters of FunA were determined as follows: K_m of 3.26 ± 0.30 μM (1.05 ± 0.10 mg/mL), V_{max} of 25.45 ± 0.97 U/mg, K_{cat} of 20.46 s^{-1}, and K_{cat}/K_m of 6.28 $mM^{-1}\ s^{-1}$.

4 Hydrolysis Pattern

The fucanase from *P. citrea* KMM 3297 could degrade the α-1→3 linkages of sulfated fucans from *Fucus evanescens* and *Saccharina cichorioides*, which were consisted of alternating 1,3- and 1,4-linked α-L-fucopyranose residues.

The intracellular fucanase secreted from "*Fucophilus fucoidanolyticus*" SI-1234 cleaved the α-1 → 3 linkage of sulfated fucans from *Cladosiphon okamuranus*, which composed of repeating units of [3)-α-L-Fucp-($4OSO_3^-$)-1→3-α-L-Fucp-($4OSO_3^-$)-(1→]$_4$; Branches: –β-D-GlcUA-1→2.

In 2004, Japanese scholars reported that the fucanase from *Alteromonas* sp. SN-1009 could degrade the α-1 → 3 linkage of sulfated fucans from *Kjellmaniella crassifolia* in an endo-acting manner and produce the oligosaccharides of sulfated fucan.

The extracellular enzyme of *M. fucanivorans* SW5^T could degrade the sulfated fucans from brown alga *Pelvetia canaliculata, Fucus spiralis,* and *Ascophyllum nodosum*, which were all composed of alternating 1,3- and 1,4-linked α-L-fucopyranose residues.

In 2010, our research group reported the extracellular enzyme of marine bacterium *W. fucanilytica* CZ1127^T, which could degrade the α-1→3 linkage between sulfated fucans from *Acaudina malpadioides* in an endo-acting manner. It reduced the molecular weight of sulfated fucans from 792.6 kDa to 3.7 kDa and produced α-1→3 linked oligosaccharides of sulfated fucan.

According to literature reports, sulfated fucans from *Fucus evanescens* could be degrade by fucanase from *F. algae* KMM 3553^T. *The enzyme from P. yessoensi* could cleave the sulfated fucans backbone from *Nemacystus decipieus* in an endo-acting manner. *The enzyme* secreted by *L. kurila* could cleave the α-1→3 linkages of sulfated fucans from *Fucus distichus.* Fucanase from *Lambis sp* could degrade the α-1→4 linkage of sulfated fucans from brown alga *F. evanescens,* which was composed of alternating 1,3- and 1,4-linked α-L-fucopyranose residues.

In 2011, Chinese scholars reported that the enzymes secreted by *Fusarium* sp. LD8 and *D. arenaria* TM94 were all endo type hydrolase and could cleave the sulfated fucans from *Fucus vesiculosus*.

Besides Fda1, the glycoside hydrolases of the GH107 family could cleave α-1→4 linkages between sulfated fucans consisted of alternating 1,3- and 1,4-linked α-L-fucopyranose residues. Among them, Fp273, Fp277, and Fp279 could also degrade the sulfated fucans from Fucales, which were composed of homogeneous 1,3-α-L-fucopyranose residues. Fda1 specifically cleaved sulfated fucan at the α-1→3 linkages.

The GH168 family enzyme also acted on sulfated fucans composed of homogeneous 1,3-α-L-fucopyranose residues.

5 Preparation Method

At present, the isolation and purification of wild-type fucanases reported various methods: ion-exchange chromatography, affinity chromatography, gel chromatography, and hydrophobic chromatography. Similar to other enzymes, recombinant fucanases can be obtained in two methods: (i) gene sequenced by Edman degradation and PCR after purified from wild-type fucanase and overexpression; (ii) mined from database or metagenome and performing heterologous overexpression.

Fucanase from *P. Citrea* KMM 3296 was purified by Sephadex G-25 column. The intracellular fucanase produced by "*Fucoidanolyticus*" SI-1234 was isolated using DEAE-Cellulofine A-800 column. The extracellular enzyme produced by *Fusarium* sp. LD8 was precipitated with 66.7% acetone and separated by Sephadex G-100.

Fucanases of GH107 (except FcnA) and GH168 were obtained by using method (ii), in brief, the gene sequences heterologously overexpressed in *E. coil*, the recombinant protein isolated and purified using column chromatography aforementioned. The discovery of FcnA was utilized by method (ii). The wild-type enzyme of marine bacterium *Mariniflexile fucanivorans* SW5^{T} through ammonium sulfate precipitation, hydrophobic interaction chromatography on Phenyl Sepharose, ion-exchange chromatography on diethylaminoethyl (DEAE) Sepharose, and gel filtration on Superdex 200. In the SDS-PAGE analysis, the latter protein fraction displayed a single band with an apparent molecular weight of approximately 105 kDa. In particular, Fct114 was purified by cell-free extract.

6 Structure and Catalytic Mechanism of Endo-Fucanases

There are two members of the GH107 family with clear X-ray diffraction quality crystals, namely MfFcnA (Founding member of GH107) and P5AFcnA [28].

6.1 Structural Insights into GH107

The crystal structure of P5AFcnA reveals a single-domain organization, consisting of a (β/α)8-barrel domain (Fig. 1) [28]. This domain is commonly found in glycoside hydrolase families, such as GH1, GH18, and GH27.

The C-terminus of MfFcnA (1-794 aa) is highly dynamic, making it challenging to crystallize. However, the truncated form of MfFcnA (1-734 aa) can be purified and forms crystals suitable for X-ray diffraction. The truncated MfFcnA structure consists of four domains: a large N-terminal (β/α)8-barrel domain connected to three consecutive Ig-like domains (Fig. 2).

The Ig-like domains lack solvent-exposed aromatic amino acid side chains and do not possess carbohydrate-binding sites. Structural similarity searches using the DALI server reveal that the Ig-like domains are most closely related to the C-terminal tandem repetitive modules CDHL-1 and CDHL-2 found in the ligand-binding

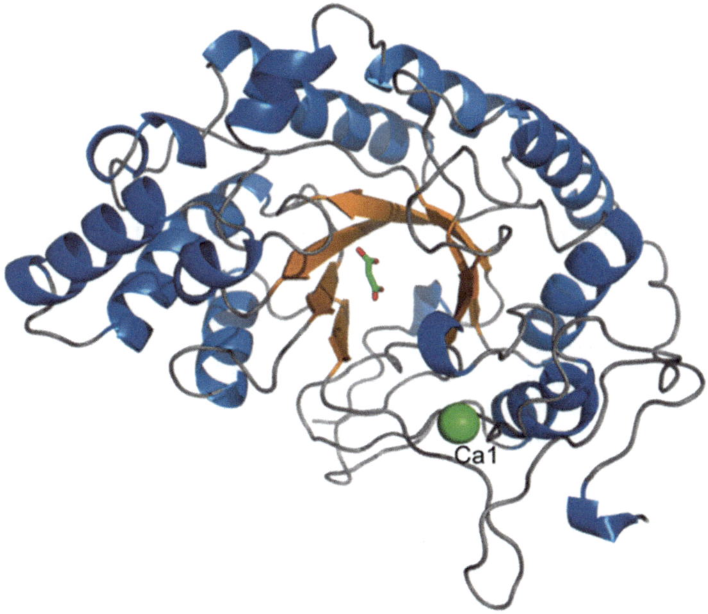

Fig. 1 The structure of P5AFcnA. A bound calcium atom is shown as a *green sphere*, and a bound malonate molecule is shown as *green sticks*. (Reprinted with permission from [28])

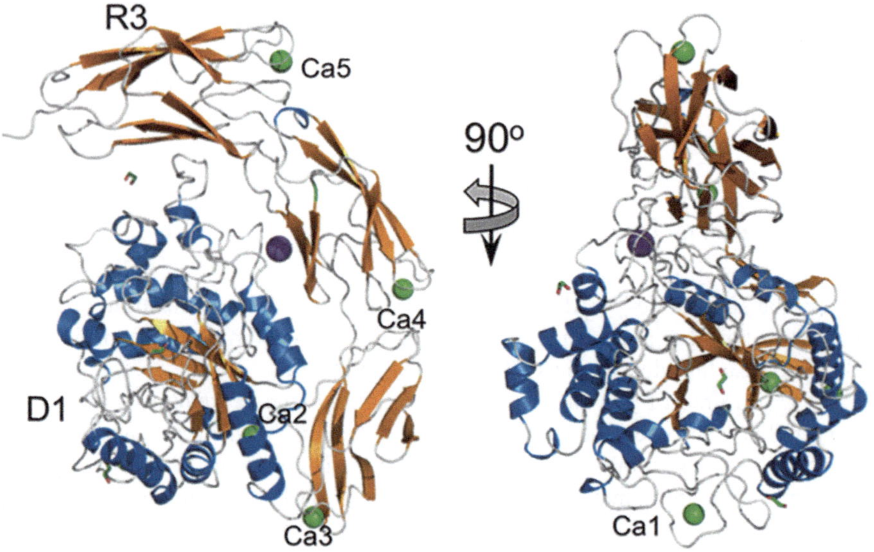

Fig. 2 The structure of MfFcnA. (Reprinted with permission from [28])

region of the Staphylococcus aureus surface-exposed serine-rich repeat glycoprotein (PDB codes 4 M00 and 4 M03). These CDHL modules, similar to calcium-dependent eukaryotic cadherins, function as a rigid stem to extend the N-terminal lectin module of the glycoprotein outward from the S. aureus cell surface. Based on this observation, Vickers proposed that the Ig-like domains in GH107 enzymes may serve as a structural spacer between the N-terminal catalytic domain and the unknown C-terminal domain.

6.2 Catalytic Sites and Substrate Binding of GH107 Enzymes

To date, there is no known structure of GH107 enzymes in complex with substrates or products. Vickers conducted the structural alignment of MfFcnA4 with a complex structure of fucoidanse (BiAfcB) from GH29 family (PDB code 3UET). Asp226 and Asp201 in MfFcnA4 are approximately positioned to attack C1 of fucose in the −1 subsite. Through structural comparison and mutation verification, His294 and His276 were found to be in a suitable position to act as acid/bases (Fig. 3) [28].

Based on NMR results, the −1 subsite in MfFcnA accommodates L-fucopyranose-2-sulfate. However, the unknown structures of P5AFcnA substrates and the distinct surface contouring of the P5AFcnA active site compared to that of MfFcnA make it less clear how sulfate is accommodated in this enzyme. Overall,

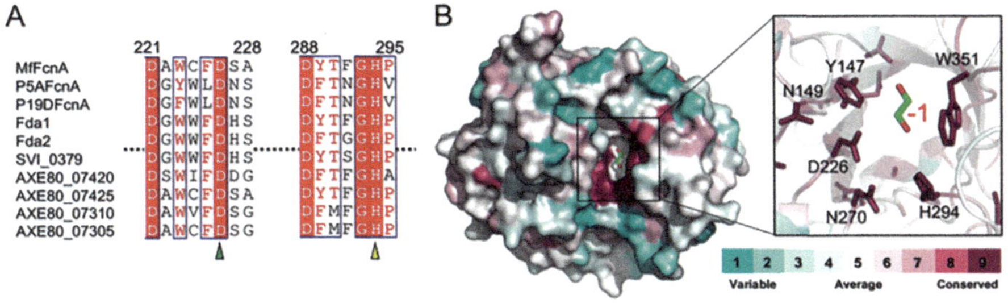

Fig. 3 Conservation of amino acid in GH107 family. (**a**) the conserved putative catalytical residues (indicated by *arrow*) and the tryptophan residue at the −1 subsite. (**b**) the conservation of surface residues of MfFcnA4. The *inset* is the conserved residues in the −1 subsite. Ethylene glycol is shown as *green sticks*. (Reprinted with permission from [28])

GH107 enzymes catalyze the hydrolysis of glycosidic bonds in sulfated fucans using a retaining catalytic mechanism.

7 Applications of Fucanases

7.1 Functional Activity of Degradation Products

The structural heterogeneity, high molecular weight, and viscosity of sulfated fucans limit its application, especially as a medical agent. Bioactive oligosaccharides with low molecular weight are capable of overcoming these problems. Fucanase can specifically degrade sulfated fucans, and the enzymatic reaction is controllable, the reaction conditions are mild, the obtained products are homogeneous, and no side reactions will occur. Therefore, fucanase is the most ideal tool for the preparation of oligosaccharides of sulfated fucan and can be used to study the structure–activity relationship of degradation products.

7.1.1 Prevent Intestinal Mucositis

The polysaccharide sulfated fucans from sea cucumbers have various bioactivities. Studies have shown that different molecular weights of sulfated fucans from *Acaudina molpadioides* (Mw 50–500 kDa) can effectively reverse cyclophosphamide-induced damage in a mouse model of intestinal mucositis. The protective effects of sulfated fucans on cyclophosphamide treatment may be attributed to alterations in the intestinal microbiota structure. The efficiency of absorption may also vary depending on the molecular weight of the polysaccharides, with lower-molecular-weight polysaccharides exhibiting higher absorption efficiency. For instance, 50 kDa sulfated fucans demonstrated superior effects [37, 38].

7.1.2 Prevent the Ethanol-Induced Gastric Ulcer

Sulfated fucans with different molecular weights from *Thelenota ananas* have been shown to effectively prevent ethanol-induced gastric ulcers. The molecular weight of sulfated fucans influences their conformation, with lower-molecular-weight sulfated fucans adopting random coil conformations and higher-molecular-weight sulfated fucans existing as spheres in solution. The mechanism of action involves antioxidation and anti-inflammation, including the restoration of antioxidative enzyme activity levels and the downregulation of pro-inflammatory cytokines and related transcription factors. The gastric protective activity of sulfated fucans is influenced by their molecular weight and chain conformation, with the transition from coil to sphere conformation enhancing their performance [39].

7.1.3 Improve Metabolic Inflammation

The effect of sulfated fucans extracted from *Holothuria tubulosa* (Ht-FUC) with different molecular weights and chain conformations on metabolic inflammation has been investigated. Ht-FUC attenuates the production of inflammatory cytokines, promotes the polarization of co-cultured macrophages toward the M2 phenotype, and enhances lipolysis in co-cultured adipocytes. In vivo, Ht-FUC reduces serum inflammation levels and attenuates inflammatory infiltration in adipose tissue. The performance of Ht-FUC initially improves as the molecular weight decreases and the chain stiffness and molecular size decrease. However, it may decrease due to inadequate chain length [40].

7.1.4 Inhibit the Proliferation and Differentiation of Adipocytes

Sulfated fucans extracted from *Thelenota ananas* (Ta-FUCs) have been found to inhibit the proliferation and differentiation of 3T3-L1 adipocytes. They also down-regulate key transcription factors involved in adipocyte differentiation and lipid accumulation, making them potential candidates for preventing obesity. The anti-adipogenic activities of Ta-FUCs initially increase and then decline with decreasing molecular weight. Among the Ta-FUCs, the 182.4 kDa Ta-FUC exhibits optimal bioactivity [41].

7.1.5 Apply as the Whitening Factor in Cosmetics Development

Sulfated fucans with high sulfate content can be enzymatically degraded to produce favorable tyrosinase inhibitory ability for whitening purposes. The whitening activities of different molecular weight fractions of sulfated fucans were evaluated based on their tyrosinase inhibitory ability, antioxidant activity, and cellular melanogenesis inhibitory ability. Smaller molecular weight sulfated fucans show better whitening activity, with the 5–10 kDa fraction exhibiting the best tyrosinase inhibitory activity, antioxidant activity, and anti-melanogenesis ability in B16 cells. These sulfated fucans can be applied as whitening factors in cosmetics development [42].

7.1.6 Fabricate the Multilayer Emulsion as a Nutrient Delivery System

Sulfated fucans from *Thelenota ananas* can be degraded to different molecular weights using fucanase. The stiffness and dimensions of the sulfated fucans decrease with decreasing molecular weight. Higher-molecular-weight sulfated fucans exhibit a wider pH range of interaction with emulsions and lower concentration required for saturating droplet surfaces. Multilayer emulsions fabricated from higher-molecular-weight sulfated fucans demonstrate better storage stability and salt stability. The digestion rate of these emulsions in vitro gradually reduces as the molecular weight of sulfated fucans decreases. This establishes a structure–function relationship between the molecular weight of polysaccharides and the fabrication of multilayer emulsions with specific functions using polysaccharides [43].

7.2 Key Tools for Structural Analysis of Sulfated Fucans

Fucanases have emerged as valuable tools for investigating the structure of sulfated fucans. Numerous studies have utilized wild-type or heterologously expressed fucanases, in combination with mass spectrometry and nuclear magnetic resonance (NMR) techniques, to elucidate the structure of sulfated fucans in brown algae and sea cucumbers.

In 1999, Daniel [44] obtained enzymatically degraded sulfated fucans from the brown algae *Ascophyllum nodosum* using protein extract from the digestive glands of the marine mollusc *Pecten maximus*. This study provided new insights into the structure of the polysaccharide through the use of 2D ^{1}H NMR data on low-molecular-weight fractions of the algal sulfated fucans. The sulfated fucans from *Ascophyllum nodosum* exhibited a randomly organized structure, composed of α-1→3- and α-1→4-linked unsulfated and 2-sulfated-α-L-fucose residues.

In 2003, sulfated fucans from *Cladosiphon okamuranus* were digested into oligosaccharides using intracellular enzymes from the sulfated fucan-utilizing marine bacterium *Fucophilus sulfated fucanolyticus*, and their structures were determined through NMR analyses [8].

The structure of a variety of sulfated fucans sourced from sea cucumbers, e.g., sulfated fucans from *Aposticous japonicus* [45], *Acaudina molpadioides* [46], *Thelenota ananas* [47], and *Holothuria tubulosa* [48], has been clarified by employing a wild-type endo-1,3-fucanase generated by the marine bacterium *Wenyingzhuangia fucanilytica* CZ1127^{T} by our research group.

In 2016, the structure of sulfated fucans from the brown alga *Turbinaria ornata* was thoroughly investigated using methylation analysis of polysaccharides and mass spectrometry of low-molecular-weight derivatives of sulfated fucan, released by fucanase from the marine mollusk *Patinopecten yessoensis* [49].

However, like most wild-type enzymes, wild-type fucanases have several limitations: (i) they require a considerable amount of purified sulfated fucans from biomass to induce their expression; (ii) the yield of wild-type enzymes is low; (iii) the purification process to remove undesired enzymatic activities from other enzymes, especially sulfatases, is time-consuming and tedious. In contrast, heterologously expressed fucanases have a high-level expression yield and do not require induction with sulfated fucan. The preparation and purification of these fucanases can be completed within a short period. For example, the structure of sulfated fucans from *Sargassum horneri* and its enzymatic transformation products were investigated using the recombinant fucanase FFA1, which was identified, cloned, and produced in *Escherichia coli* [23, 50]. In our research group, we introduced heterologously overexpressed endo-1,3-fucanase (FunA) into the structural identification strategy of sulfated fucans for the first time, integrating glycomics and NMR analysis. As a result, the structural identification of sulfated fucans was achieved by consuming only 10 mg of polysaccharide, which could be prepared from 0.2 g of dried sea cucumber. In contrast, the structural identification of *T. ananas* sulfated fucan required the consumption of 260 mg of sulfated fucans to obtain sufficient enzyme dosage and oligosaccharides preparation [51]. Therefore, the overexpressed fucanase demonstrated advantages in terms of cost and efficiency compared to its wild-type counterpart.

Compared to other polysaccharide degrading enzymes, fucanase research started relatively late, and there are limited related reports. There is a significant shortage of fucanase enzyme preparations in both domestic and foreign markets. Therefore, future research should continue to explore novel fucanases and elucidate their mechanisms of action. With the rapid development of genome sequencing technology, an increasing number of microbial genomes are being analyzed, providing a foundation for the exploration of new fucanases. Furthermore, current research primarily focuses on the basic biochemical properties of fucanases, while the specific catalytic mechanisms and the relationship between structure and function remain unclear, limiting the development and application of sulfated fucans. Investigating the relationship between enzyme structure and function, and uncovering the enzymatic action mechanism, will contribute to the further advancement and application of sulfated fucans and fucanases.

References

1. Descamps V, Colin S, Lahaye M et al (2006) Mar Biotechnol (NY) 8:27. https://doi.org/10.1007/s10126-005-5107-0
2. Bakunina IY, Shevchenko LS, Nedashkovskaya OI et al (2000) Microbiology 69:303. https://doi.org/10.1007/Bf02756738
3. Qianqian W, Shuang M, Hourong X, Min Z, Jingmin C (2011) Evid Based Complement Alternat Med 2011:196190. https://doi.org/10.1155/2011/196190

4. Yaphe W, Morgan K (1959) Nature 183:761. https://doi.org/10.1038/183761b0
5. Thanassi NH (1967) Arch Biochem Biophys 118:172
6. Kusaykin MI, Silchenko AS, Zakharenko AM, Zvyagintseva TN (2016) Glycobiology 26:3. https://doi.org/10.1093/glycob/cwv072
7. Bakunina I, Nedashkovskaia OI, Alekseeva SA et al (2002) Mikrobiologiia 71:49
8. Sakai T, Ishizuka K, Shimanaka K, Ikai K, Kato I (2003) Mar Biotechnol (NY) 5:536. https://doi.org/10.1007/s10126-002-0107-9
9. Sakai T, Kawai T, Kato I (2004) Mar Biotechnol (NY) 6:335. https://doi.org/10.1007/s10126-003-0033-5
10. Woo-Jung KS, Lee Y (2008) J Microbiol Biotechnol 18(4):616–623
11. Chang Y, Xue C, Tang Q, Li D, Wu X, Wang J (2010) Lett Appl Microbiol 50:301. https://doi.org/10.1111/j.1472-765X.2009.02792.x
12. Ohshiro T, Ohmoto Y, Ono Y et al (2010) Biosci Biotechnol Biochem 74:1729. https://doi.org/10.1271/bbb.100327
13. Silchenko AS, Kusaykin MI, Kurilenko VV et al (2013) Mar Drugs 11:2413. https://doi.org/10.3390/md11072413
14. Wang JCP, Qin S, Wu QQ, Wu K, Wang R, Xu D (2004) Food Ferment Ind: 47
15. Liu WXHD, Wang FF, Wang J (2020) J Anhui Agric Sci 48:16
16. Wang Y (2014) J Qingdao Agric Univ (Natural Science) 31:36
17. Kitamura K, Matsuo M, Yasui T (1992) Biosci Biotech Bioch 56:490. https://doi.org/10.1271/bbb.56.490
18. Silchenko AS, Kusaykin MI, Zakharenko AM, et al. (2014) J Mol Catal B Enzym 102:154. https://doi.org/10.1016/j.molcatb.2014.02.007, https://doi.org/10.1007/s10541-005-0264-3
19. KW QQ Wu, J Zhang, JX Tan, RR Pan, JM Cai (2006) Mycosystema 77
20. Wu QQ, Zhang M, Wu K, Liu B, Cai JM, Pan RR (2011) J Appl Phycol 23:197. https://doi.org/10.1007/s10811-010-9588-5
21. Takayama M, Koyama N, Sakai, T, Kato I (2002) Enzymes Capable of Degrading a Sulfated-Fucose-Containing Polysaccharide and Their Encoding Genes. U.S. Patent No US 6,489,155
22. Colin S, Deniaud E, Jam M et al (2006) Glycobiology 16:1021. https://doi.org/10.1093/glycob/cwl029
23. Silchenko AS, Rasin AB, Kusaykin MI et al (2017) Carbohydr Polym 175:654. https://doi.org/10.1016/j.carbpol.2017.08.043
24. Silchenko AS, Ustyuzhanina NE, Kusaykin MI et al (2017) Glycobiology 27:254. https://doi.org/10.1093/glycob/cww138
25. Zueva AO, Silchenko AS, Rasin AB et al (2020) Int J Biol Macromol 164:3025. https://doi.org/10.1016/j.ijbiomac.2020.08.131
26. Nagao T, Arai Y, Yamaoka M et al (2018) J Biosci Bioeng 126:567. https://doi.org/10.1016/j.jbiosc.2018.05.016
27. Schultz-Johansen M, Cueff M, Hardouin K et al (2018) FEBS J 285:4281. https://doi.org/10.1111/febs.14662
28. Vickers C, Liu F, Abe K et al (2018) J Biol Chem 293:18296. https://doi.org/10.1074/jbc.RA118.005134
29. Shen J, Chang Y, Zhang Y, Mei X, Xue C (2020) Front Microbiol 11:1674. https://doi.org/10.3389/fmicb.2020.01674
30. Vuillemin M, Silchenko AS, Cao HTT et al (2020) Mar Drugs 18:562. https://doi.org/10.3390/md18110562
31. Somogyi M (1952) J Biol Chem 195(1):19
32. Miller GL (2002) Anal Chem 31:426. https://doi.org/10.1021/ac60147a030
33. Lever M (1972) Anal Biochem 47:273. https://doi.org/10.1016/0003-2697(72)90301-6
34. Zablackis E, Perez J (1990) Bot Mar 33:273. https://doi.org/10.1515/botm.1990.33.3.273
35. Friedemann TE, Weber CW, Witt NF (1962) Anal Biochem 4:358. https://doi.org/10.1016/0003-2697(62)90137-9

36. Van Wychen S, Long W, Black SK, Laurens LML (2017) Anal Biochem 518:90. https://doi.org/10.1016/j.ab.2016.11.014
37. Shi H, Chang Y, Gao Y et al (2017) Food Funct 8:3383. https://doi.org/10.1039/c7fo00932a
38. Zuo T, Li X, Chang Y et al (2015) Food Funct 6:415. https://doi.org/10.1039/c4fo00567h
39. Xu X, Chang Y, Xue C, Wang J, Shen J (2018) J Agric Food Chem 66:8615. https://doi.org/10.1021/acs.jafc.8b01497
40. Zhu Y, Tian Y, Wang N, Chang Y, Xue C, Wang J (2021) J Food Biochem 45:e13500. https://doi.org/10.1111/jfbc.13500
41. Liu Y, Wang N, Tian Y, Chang Y, Wang J (2021) J Ocean Univ China 20:921. https://doi.org/10.1007/s11802-021-4681-8
42. Chen QR, Kou LY, Wang FW, Wang Y (2019) Carbohydr Polym:225. https://doi.org/10.1016/j.carbpol.2019.115211
43. Shi FF, Tian XJ, McClements DJ, Chang YG, Shen JJ, Xue CH (2021) Food Hydrocoll:113. https://doi.org/10.1016/j.foodhyd.2020.106418
44. Daniel R, Berteau O, Jozefonvicz J, Goasdoue N (1999) Carbohydr Res 322:291. https://doi.org/10.1016/S0008-6215(99)00223-2
45. Yu L, Xue C, Chang Y et al (2015) Food Chem 180:71. https://doi.org/10.1016/j.foodchem.2015.02.034
46. Yu L, Ge L, Xue C et al (2014) Food Chem 142:197. https://doi.org/10.1016/j.foodchem.2013.06.079
47. Yu L, Xue C, Chang Y et al (2014) Food Chem 146:113. https://doi.org/10.1016/j.foodchem.2013.09.033
48. Chang Y, Hu Y, Yu L et al (2016) Carbohydr Polym 136:1091. https://doi.org/10.1016/j.carbpol.2015.10.016
49. Ermakova SP, Menshova RV, Anastyuk SD et al (2016) J Appl Phycol 28:2495. https://doi.org/10.1007/s10811-015-0742-y
50. Rasin AB, Silchenko AS, Kusaykin MI et al (2020) Carbohydr Polym 246:116635. https://doi.org/10.1016/j.carbpol.2020.116635
51. Chen G, Yu L, Zhang Y et al (2021) Carbohydr Polym 272:118480. https://doi.org/10.1016/j.carbpol.2021.118480

Research Progress of Porphyranase

Chunhui Liu, Wenting Zhou, Xiangzhao Mao, and Hong Jiang

The rich nutritional value of porphyra has been widely recognized in recent years, leading to a gradual rise in global consumer demand for its production, with global production of porphyra reaching up to six million tones (by wet weight) in 2020 alone [1]. As one of the major seaweeds consumed by humans, porphyra is rich in functional nutrients, including saccharides, proteins, polyunsaturated fatty acids, vitamins, and minerals [2]. Among them, porphyran is one of the major polysaccharides in porphyra, and its dry weight ratio could be as high as 47.5%. Porphyran plays an important role as functional algal polysaccharides in functional food, cosmetic, and pharmaceutical applications, and its biological activity has been widely studied [3]. Moreover, oligo-porphyran (OP), a degradation product of porphyran, has shown more promising biological activities due to lower molecular weight (MW) and higher degree of sulphation, including activation of plant defenses [4], antioxidant activity, immunomodulation [5], protection against acute renal failure, neuroprotective activity, and antitumor activity [6].

The current method of the preparation of algal oligosaccharides could be divided into two production strategies: polysaccharide degradation and monosaccharide synthesis. For example, Codee et al. successfully constructed a mannosidic acid pentamer using a polymeric orthogonal glycosylation strategy [7]. However, for oligo-porphyran containing complex structural units, this production strategy by organic synthesis is not economically efficient. Therefore, degradation of porphyran is more often used to produce oligo-porphyran, mainly by ultrasonic hydrolysis, acid hydrolysis, oxidative degradation, and enzymatic hydrolysis. Compared with other degradation methods, enzymatic hydrolysis has the advantages of environmental-friendliness, high reaction efficiency, and mild conditions. Moreover,

C. Liu · W. Zhou · X. Mao · H. Jiang (✉)
State Key Laboratory of Marine Food Processing and Safety Control, College of Food Science and Engineering, Ocean University of China, Qingdao, China
e-mail: jh@ouc.edu.cn

X. Mao (ed.), *Degrading Enzymes for Marine Polysaccharides and Their Applications*, https://doi.org/10.1007/978-981-96-9517-1_11

it could retain the functional activity of oligo-porphyran to the maximum extent, and it could also make the molecular weight of the product be limited to a fixed range, which is more conducive to the subsequent purification steps. In this section, we focus on the development of hydrolytic enzymes for the degradation of porphyran in recent years, especially porphyranase, with a view to better application.

1 The Development of Hydrolytic Enzymes for the Degradation of Porphyran

Porphyran and agar, both polysaccharides extract from red algae, are very similar in their natural structures, as shown in Fig. 1. Thus, β-agarase, which recognizes and cleaves the β-1,4 glycosidic bond, could be used to degrade porphyran at the earliest opportunity. For example, Yukihiro et al. in 1996 succeeded in degrading porphyran to neoagarotetraose (NOSP) fraction, sulfated oligosaccharide (DP $\geq$ 6) (OSP3) fraction, sulfated tetrasaccharide (OSP2) fraction, and sulfated disaccharide (OSP1) fraction, using the β-agarase isolated from *Arthrobacter* sp. S-22. The effects of these oligosaccharides, NOSP, OSP3, OSP2, and OSP1, on in vitro digestion, utilization by various gut bacteria, and serum lipid levels in mice were also investigated. However, the low efficiency of porphyran degradation by β-agarase, the inability to obtain sufficient oligo-porphyran products, and the fact that these degradation products are unrepresentative of the entire polysaccharide structure have limited its further exploitation [8, 9]. Subsequently, porphyranases with efficient porphyran

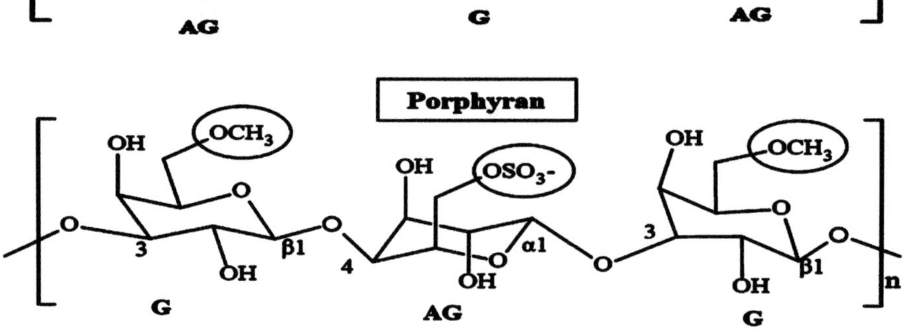

Fig. 1 The structure of agar and porphyran

hydrolase activity were gradually discovered and characterized and were used exclusively for the degradation of porphyran and replaced β-agarase.

Porphyranase (EC 3.2.1.178) is a glycoside hydrolase considered a promising tool for the degradation of porphyran, the major polysaccharides in porphyra, whose structural heterogeneity is mainly due to natural methylation at the O-6 position of the G-unit [3]. Since 2010, only six porphyranases have been discovered, which have hydrolase activity of porphyran to prepare oligo-porphyran. These porphyranases are classified into two glycoside hydrolase families, GH16 and GH86, based on their degree of sequence similarity in the CAZy database. Among them, five belong to the GH16 family, namely PorA, PorB, BpGH16B, Por16A_Wf, and Por16C_Wf. Only one belongs to the GH86 family, BpGH86A.

2 Preparation and Activity Determination of Porphyranase

There are two ways to prepare porphyranase: in vitro recombination and separation with purification. The recombinant enzyme obtained by in vitro recombination is the main way to prepare porphyranase. At present, documented six porphyranases with hydrolase activity of porphyran were prepared by in vitro recombination method.

Detection of enzyme activity was quantified by the para-hydroxybenzoic acid hydrazide method for the released reducing sugars. One unit of porphyranase activity is defined as the amount of enzyme required to release 1.0 μmol of reducing sugar per minute. This is done by incubating the enzyme with porphyran in a buffer solution (pH 6.0) at a certain temperature for a certain period of time to determine the reducing sugar content, where the buffer solution, enzyme dosage, substrate concentration, reaction temperature, and time could be adjusted according to the specific experimental design [10].

3 Porphyranases from Marine Bacteria

3.1 Zobellia Galactanivorans

Hehemann et al. analyzed the genomic information of *Zobellia galactanivorans*, a member of the marine Bacteroidetes, and succeeded in unearthing the first class of porphyranases active on sulfated polysaccharide of marine red algae, which were named PorA and PorB [11]. *Zobellia galactanivorans* is a marine bacterial strain isolated from the surface of the red alga *Delesseria sanguinea*, which is able to grow in minimal media containing agar and could also grow and reproduce using κ- or ι-carrageenan as the sole carbon source [12].

They analyzed the genome of *Zobellia galactanivorans* and found the presence of five genes encoding protein sequences (*Zg1017*, *Zg2600*, *Zg3376*, *Zg3628,* and *Zg3640*) annotated to be classified in the GH16 family. Curiously, they did not belong to β-agarase and κ-carrageenase, as they had very low sequence similarity (both less than 30%) compared to the GH16 family β-agarase AgaA and κ-carrageenase CgkA, and seemed to represent a new type of hydrolase. They then cloned the five genes and expressed them using *E. coli* hosts to determine the substrate specificity of these gene products. Unfortunately, only the catalytic modules encoded by the genes *Zg1017* and *Zg2600* were successfully expressed as soluble proteins, named PorB and PorA, respectively. As expected, they showed no degradation activity on commercial agarose. The two enzymes were further screened for degradation activity on natural polysaccharides extracted from various marine algae and were found to have hydrolytic activity only on extracts of the red algae *Gelidium*, *Gracilaria*, and *Porphyra*. Isolation and purification of the degradation products and product characterization by ^{1}H-NMR spectroscopy showed that both hydrolases cleaved by acting on the β-1,4-glycosidic bond, with the major end product being the disaccharide L6S-G. In a word, PorB and PorA are the earliest porphyranases reported to exhibit significant degradation activity toward porphyran.

In order to determine the substrate recognition mode of these two porphyranases, PorA and PorB were lethally mutated using the targeted mutagenesis technique, and co-crystallization experiments were performed with purified porphyran tetrasaccharide, respectively. These co-crystal structures revealed the key residues for the recognition of the substrate porphyran by these two porphyranases and their mode of action. The crystal structure suggests that subsite-2 in the active pocket of β-porphyranase, which binds to the L6S unit, is the site of action that determines substrate recognition. This is different from the agarase substrate recognition site, which binds to the LA unit. For PorA subsite-2, the L6S unit could be surrounded by W56, R133, and R59 in its active pocket, and for PorB, W67, R133, and R70 in its pocket. Together, these amino acid residues form pockets that are positively charged and hydrophobic, allowing them to conveniently bind the bulky sulfate moiety of the L6S unit. In β-agarase, however, the same position would be occupied by a tyrosine residue, which limits the space for the pocket to bind the sulfate group.

In addition, structural characterization of the oligo-porphyran product showed that PorA strictly recognizes two consecutive neoporphyranobiose units (L6S-G-L6S-G) for degradation and fails to bind the LA-G portion efficiently. This suggested that PorA had a strict recognition mechanism for the sulfate group at the C6 position in its binding subsites. Interestingly, the binding subsite-2 of PorB could tolerate the presence of LA. Thus, PorA has only the disaccharide L6S-G as a hydrolysis product, whereas PorB has both L6S-G and L6S-G-LA-G products. The fact that natural product porphyran could be modified by methyl groups at the C6 position of the G residue explains the high degree of heterogeneity of natural porphyran. Unfortunately, neither PorA nor PorB recognized methylated G (GMe) units [3, 13].

3.2 *Wenyingzhuangia Fucanilytica*

Benefiting from previously isolated, characterized and sequenced genome-wide information on marine bacteria *Wenyingzhuangia fucanilytica* (GenBank: CP014224), Zhang et al. [14, 15] found that the gene information of *Wenyingzhuangia fucanilytica* CZ1127T contained a putative protein annotated as a member of glycoside hydrolase family 16 (GenBank: WP_068825731.1), which was presumed to be a candidate porphyranase. Subsequent studies cloned the encoding gene and expressed it in *E. coli*, and the biochemical properties of the corresponding recombinant protein, Por16A_Wf, were characterized in detail to determine its hydrolysis pattern and specificity. This is the first detailed report to date of information on the biochemical properties of porphyranases [16].

Analysis of the sequence information revealed that Por16A_Wf possesses a total of 513 amino acid residues, including an N-terminal signal peptide of 21 amino acid residues and a catalytic structural domain of the GH16 family located at amino acid residue positions 142-388. Further by the multiple sequence comparison procedure, Por16A_Wf was found to have up to 44% sequence similarity with PorB from *Zobellia galactanivorans*, where the highest coverage was only 69%. Biochemical experiments showed that recombinant Por16A_Wf could hydrolyze porphyran but was inactive against agarose. Por16A_Wf has the highest activity at 40 °C and pH 6.0, with a specific enzyme activity of about 14.08 ± 0.8 U/mg. In addition, the enzyme has strong acid–base tolerance and could maintain stable activity over a wide pH range from 3.5 to 11.0. Mg^{2+}, Cu^{2+}, and Ca^{2+} significantly increased the activity of Por16A_Wf, while Hg^{2+} and SDS decreased its activity. The addition of NaCl or KCl during the reaction also significantly increased the enzyme activity, indicating a degree of salt dependence. This property is prevalent in many enzymes of marine origin [17, 18].

Por16A_Wf is an endo-acting hydrolase, so that a range of porphyran degradation products with varying degrees of polymerization could be prepared by controlling the amount of enzyme and incubation time. Por16A_Wf could completely convert the continuous L6S-G portion of porphyran into disaccharides due to the fact that the presence of porphyran tetrasaccharides $(L6S\text{-}G)_2$ and porphyran hexasaccharides $(L6S\text{-}G)_3$ is not observed in its final product, and it also does not contain LA-G impurities or modified structural parts. This complete hydrolytic nature could facilitate the large-scale targeted preparation of porphyran disaccharides. The total ion chromatogram (TIC) and extracted ion chromatogram of the hydrolyzed final product of Por16A_Wf, determined by HPSEC-MS, showed that three tetrasaccharides and three hexasaccharides were detected in the final products, in addition to the disaccharide L6S-G. The presence of these oligosaccharides was attributed to the fact that their agarose portion was modified by being methylated to form L6S-G/GMe-L6S/LA-G, resulting in their inability to be further degraded by Por16A_Wf [16].

In addition, the authors further found by bioinformatics analysis that the marine bacteria *Wenyingzhuangia fucanilytica* CZ1127T contains a porphyran-utilizing polysaccharide locus (gene locus from AXE80_RS06895 to AXE80_RS07005), where the open reading frame in the polysaccharide-utilizing locus AXE80_RS06955 was annotated as a putative porphyranase and named as Por16C_Wf. Por16C-Wf is located in a separate branch belonging to GH16 subfamily 14 (GH16_14), has the highest sequence homology with the identified porphyranase PorB (only 31%), and it has no available biochemical features or substrate specificity [19]. The enzyme had its highest activity at 35 °C and pH 7.0 with a specific enzyme activity of 56.23 ± 0.5 U/mg. The authors used 75 U of Por16C_Wf to degrade the porphyran and obtained a final product consisting of two disaccharides and three hexasaccharides. Of these, the porphyran disaccharide (L6S-G) was the predominant component, suggesting that the L6S-G portion of the porphyran substrate could be completely converted to porphyran disaccharides by it. The presence of L6S-G_1Me_1 was likewise detected in the product. Interestingly, L6S-G_2Me_2 was monitored to be completely degraded, suggesting that Por16C_Wf is able to degrade $(\text{L6S-G})_n\text{Me}_n$ to $(\text{L6S-G})_1\text{Me}_1$. This novel property of Por16C_Wf suggests that this enzyme could be a key tool to facilitate the complete degradation of porphyran and could be used to study the structural heterogeneity of porphyran. This result highlights the unprecedented β-methylporphyranase activity that accommodates methylated porphyran building blocks [20, 21]. In addition, molecular docking simulations further showed that the glycine and asparagine residue sites in the active cleft of Por16C_Wf form a larger pocket that could accommodate GMe units.

4 Porphyranases from Human Gut Bacterium

4.1 *Bacteroides Plebeius*

Hehemann et al. isolated and characterized two other porphyranases, BpGH16B and BpGH86A, from the human gut bacterium *Bacteroides plebeius*. In the Bacteroidetes, genes encoding CAZymes targeting specific carbohydrates or groups of related carbohydrates are typically found in clusters of genes called polysaccharide utilization loci. The *Bacteroides plebeius* gene information contained a putative porphyran-degrading polysaccharide utilization locus covering up to 40 ORFs from BACPLE_01667 to BACPLE_01706, including 12 putative glycoside hydrolases. These genes encoding possible CAZymes appear to have been acquired from marine microorganisms by horizontal gene transfer, which might target carbohydrates from red seaweeds and therefore might trigger porphyran degradation [11].

Four of these genes were predicted to have endo acting to polysaccharides, two of which were identified as belonging to the GH16 family: BpGH16A (BACPLE_01670) and BpGH16B (BACPLE_01689), with putative agar-hydrolyzing and porphyran-hydrolyzing activities, respectively. The other two named BpGH86A (BACPLE_01693) and BpGH86B (BACPLE_01694) belong to the GH86 family but have unknown substrate specificity. The gene fragments encoding the catalytic structural domains of these four hydrolases were further cloned by PCR reaction and expressed using *E. coli* hosts. Reducing sugar assay was used to detect their activities. The results of the assay showed that BpGH16B and BpGH86A displayed porphyran hydrolysis activity. The crystal structures of these two hydrolases have been successfully resolved.

Further the authors explored the degradation pattern of the substrates of both enzymes by analyzing the structure of the generated products. BpGH16B showed a product generation profile similar to that of the β-porphyranase PorA from *Zobellia galactanivorans*, with L6S-G as its major end product. In contrast, BpGH86A predominantly produces larger oligosaccharides, the smallest of which was the tetra-oligosaccharide L6S-G-L6S-G. The mechanism of this difference has been attributed to the unique structure of BpGH86A.The primary structure of BpGH86A consists of an N-terminal signal peptide, the central portion of which contains the catalytic TIM barrel structural domain of the GH86 family and a C-terminal structural domain of unknown function. Among the structural features of the BpGH86A catalytic TIM barrel were a relatively extended substrate-binding cleft and two accessory β-sandwich structural domains. The crystal structure of the complex with porphyran/agarose hexasaccharide supports the overall retaining mechanism of hydrolysis by this enzyme [22].

References

1. FAO (2023) Fishery and aquaculture statistics – yearbook 2020. FAO Yearbook of Fishery and Aquaculture Statistics, Rome. https://doi.org/10.4060/cc7493en
2. MacArtain P, Gill CI, Brooks M, Campbell R, Rowland IR (2007) Nutritional value of edible seaweeds. Nutr Rev 65(12 Pt 1):535–543. https://doi.org/10.1301/nr.2007.dec.535-543
3. Qiu Y, Jiang H, Fu L, Ci F, Mao X (2021) Porphyran and oligo-porphyran originating from red algae Porphyra: preparation, biological activities, and potential applications. Food Chem 349:129209
4. Hou Y, Wang J, Simerly T, Jin W, Zhang H, Zhang Q (2015) Hydrogen peroxide released from Pyropia yezoensis induced by oligo-porphyrans: mechanisms and effect. J Appl Phycol 27(4):1639–1649
5. Li Y-T, Huo Y-F, Wang F, Wang C, Zhu Q, Wang Y-B, Fu L-L, Zhou T (2020) Improved antioxidant and immunomodulatory activities of enzymatically degraded Porphyra haitanensis polysaccharides. J Food Biochem 44(5). https://doi.org/10.1111/jfbc.v44.510.1111/jfbc.13189
6. He D, Wu S, Yan L, Zuo J, Cheng Y, Wang H, Liu J, Zhang XU, Wu M, Choi JI, Tong H (2019) Antitumor bioactivity of porphyran extracted from Pyropia yezoensis Chonsoo2 on human cancer cell lines. J Sci Food Agric 99(15):6722–6730
7. Codee JDC, van den Bos LJ, de Jong AR, Dinkelaar J, Lodder G, Overkleeft HS, van der Marel GA (2009) The stereodirecting effect of the glycosyl C5-carboxylate ester: Stereoselective synthesis of beta-mannuronic acid alginates. J Org Chem 74:38–47
8. Osumi Y, Kawai M, Amano H, Noda H (1998) Nippon Suisan Gakkaishi 64:98
9. Lahaye M (2001) J Appl Phycol 13:173. https://doi.org/10.1023/a:1011142124213
10. Lever M (1972) Anal Biochem 47:273. https://doi.org/10.1016/0003-2697(72)90301-6
11. Hehemann JH, Correc G, Barbeyron T, Helbert W, Czjzek M, Michel G (2010) Nature 464:908. https://doi.org/10.1038/nature08937

12. Barbeyron T, L'Haridon S, Corre E, Kloareg B, Potin P (2001) Int J Syst Evol Microbiol 51:985. https://doi.org/10.1099/00207713-51-3-985
13. Hehemann JH, Correc G, Thomas F et al (2012) J Biol Chem 287:30571. https://doi.org/10.1074/jbc.M112.377184
14. Chang Y, Xue C, Tang Q, Li D, Wu X, Wang J (2010) Lett Appl Microbiol 50:301. https://doi.org/10.1111/j.1472-765X.2009.02792.x
15. Chen F, Chang Y, Dong S, Xue C (2016) Int J Syst Evol Microbiol 66:3270. https://doi.org/10.1099/ijsem.0.001184
16. Zhang Y, Chang Y, Shen J, Xue C (2019) J Agric Food Chem 67:9307. https://doi.org/10.1021/acs.jafc.9b02941
17. Cui FY, Dong SJ, Shi XC, Zhao X, Zhang XH (2014) Mar Drugs 12:2731. https://doi.org/10.3390/md12052731
18. Mei JF, Tang ZX, Yi Y, Wang H, Wang Q, Ying GQ (2014) Food Sci Biotechnol 23:1605. https://doi.org/10.1007/s10068-014-0218-x
19. Viborg AH, Terrapon N, Lombard V et al (2019) J Biol Chem 294:15973. https://doi.org/10.1074/jbc.RA119.010619
20. Zhang YY, Chang YG, Shen JJ, Mei XW, Xue CH (2020) J Agric Food Chem 68:7032. https://doi.org/10.1021/acs.jafc.0c02404
21. Przybylski C, Correc G, Fer M, Gonnet F, Helbert W, Daniel R (2015) RSC Adv 5:80793. https://doi.org/10.1039/c5ra14449c
22. Hehemann JH, Kelly AG, Pudlo NA, Martens EC, Boraston AB (2012) Proc Natl Acad Sci USA 109:19786. https://doi.org/10.1073/pnas.1211002109

Ulvan Lyase

Peng Wang, Yongzhou Chi, Chunying Du, and Huining Li

1 Introduction

Ulvan is a major component of the cell wall in green algae belonging to the genus *Ulva*. It is a complex water-soluble sulfated polysaccharide composed mainly of rhamnose (Rha), glucuronic acid (GlcA), iduronic acid (IdoA), and xylose (Xyl) [1]. Structurally, it mainly comprises four types of disaccharide repeating units connected by β-(1→4) bonds [2]. The presence of uronic acid and sulfated Rha distinguishes ulvan from other seaweed polysaccharides and imparts similarities to mammalian glycosaminoglycans [2]. These unique structural properties make ulvan an attractive candidate for application in health products, biopharmaceuticals, and drug delivery systems [3–5]. However, the application of high-molecular-weight polysaccharides is limited owing to their poor solubility and low bioavailability [6]. Consequently, technologies for the efficient preparation of suitable oligosaccharides have been receiving increasing attention.

Generally, the original structure of polysaccharides is destroyed during chemical degradation owing to strict reaction conditions, especially for polysaccharides containing sulfate groups, such as ulvan [7]. Conversely, enzymatic degradation has the advantages of mild reaction conditions, high substrate specificity, and high efficiency and is considered a promising method for oligosaccharide preparation [8]. Polysaccharides containing uronic acid can be degraded via a β-elimination mechanism by a group of enzymes called polysaccharide lyases (PLs). To date, various ulvan lyases from different sources have been identified and characterized [9–12]. Based on protein sequence similarity, ulvan lyases have been classified into five PL families in the Carbohydrate-Active Enzymes (CAZy) database, namely PL24,

P. Wang (✉) · Y. Chi · C. Du · H. Li
State Key Laboratory of Marine Food Processing and Safety Control, College of Food Science and Engineering, Ocean University of China, Qingdao, China
e-mail: pengwang@ouc.edu.cn

X. Mao (ed.), *Degrading Enzymes for Marine Polysaccharides and Their Applications*, https://doi.org/10.1007/978-981-96-9517-1_12

PL25, PL28, PL37, and PL40. These enzymes have distinct structures and catalytic modes and exhibit different enzymatic properties. In this chapter, we review and summarize the sources, classification, preparation methods, biochemical characteristics, structure features, and action patterns of ulvan lyases.

2 Sources and Categories of Ulvan Lyase

2.1 Sources of Ulvan Lyase

In 1997, Lahaye et al. isolated a bacterial strain from mud containing decomposing *Ulva* sp. that produced an enzyme to degrade ulvan into oligosaccharides via a β-elimination reaction, and this enzyme was subsequently named ulvan lyase [6]. In the last decade, reports on ulvan lyase have gradually increased. Ulvan lyases are primarily isolated from bacteria found in seawater, algal surfaces, and marine animal intestines. The microbial sources of partial ulvan lyases are listed in Table 1, with *Alteromonas* accounting for a relatively large proportion. For instance, *Alteromonas* sp. LOR was isolated from the feces of *Aplysia punctate* [9, 13–15], while *Alteromonas* sp. 76-1 was isolated from surface seawater [16]. Additionally, *Thalassomonas* sp. LD5 [17], *Pseudoalteromonas* sp. PLSV [3, 11, 15], *Nonlabens*

Table 1 Sources of partial ulvan lyases from different polysaccharide lyase families

PL family	Source microorganism	Protein name	Reference
PL24	*Alteromonas* sp. LOR	LOR_61	[9, 15]
		LOR_107	
	Alteromonas sp. 76-1	ALT761_04143	[16]
		ALT761_04144	
		ALT761_04188	
	Alteromonas sp. KUL17	UlIA	[20]
	Catenovulum maritimum	Uly1	[12]
	Pseudoalteromonas sp. PLSV	PLSV_3875	[15]
		PLSV_3925	
PL25	*Alteromonas* sp. LOR	LOR_29	[13]
	Alteromonas sp. 76-1	ALT761_04173	[16]
	Alteromonas sp. A321	ALT3695	[4]
	Pseudoalteromonas sp. PLSV	PLSV_3936	[3, 11]
	Nonlabens ulvanivorans	IL45_03835	[21]
	Thalassomonas sp. LD5	TsUly25B	[17]
PL28	*Formosa agariphila* KMM 3901	FaUL	[5]
	Nonlabens ulvanivorans	NLR42	[10]
		NLR48	
PL37	*Formosa agariphila* KMM 3901	Cdf79930	[18]
PL40	*Formosa agariphila* KMM 3901	BN863_21990	[19]

ulvanivorans [10, 14], and *Formosa agariphila* KMM 3901 [5, 18, 19] are important sources of the currently characterized ulvan lyases.

With the development of techniques for studying microbial genomics, the genetic resources of marine microorganisms have been thoroughly analyzed and mined, and many functional genes have been discovered, including those encoding polysaccharide-degrading enzymes. In the CAZy database (www.cazy.org), there are more than 200 entries for ulvan lyases, 95% of which have not been characterized. Except for three ulvan lyases identified in the archaea of the genus *Halosimplex*, the rest were identified in bacteria, including *Bacteroides*, *Mucilaginibacter*, *Phocaeicola*, and *Vibrio*.

2.2 Categories of Ulvan Lyase

Ulvan lyases identified so far are all endo-type lyases that cleave the (1→4) glycosidic linkage at the non-reducing end of the uronic acid, forming 4-deoxy-L-*threo*-hex-4-enopyranosiduronic acid (ΔUA) at the non-reducing end of the oligosaccharide products. Ulvan lyases can be divided into two types on the basis of substrate specificity. As shown in Fig. 1a, the first type specifically cleaves the (1→4) glycosidic linkage between Rha3S and GlcA, such as LOR_61 and LOR_107 obtained from *Alteromonas* sp. LOR, as well as PLSV_3875 and PLSV_3925 obtained from *Pseudoalteromonas* sp. PLSV [15]. Their main degradation end products are ΔUA-Rha3S, ΔUA-Rha3S-IdoA-Rha3S, and ΔUA-Rha3S-Xyl-Rha3S. The second type of ulvan lyase cleaves the (1→4) glycosidic linkage between Rha3S and either GlcA or IdoA, such as LOR_29 obtained from *Alteromonas* sp. LOR [13], PLSV_3936 obtained from *Pseudoalteromonas* sp. PLSV [11], and Cdf79930 obtained from *Formosa agariphila* KMM 3901 [18]. The primary degradation end products are ΔUA-Rha3S and ΔUA-Rha3S-Xyl-Rha3S, with no ΔUA-Rha3S-IdoA-Rha3S detected (Fig. 1b).

Additionally, ulvan lyases are categorized according to the similarity of protein sequence. In the CAZy database, they are classified into five PL families, namely PL24, PL25, PL28, PL37, and PL40. Ulvan lyases of the PL24 family have been the most extensively studied by researchers. Five enzymes from this family have been characterized, including LOR_61 and LOR_107 obtained from *Alteromonas* sp. LOR [9, 15], PLSV_3875 and PLSV_3925 obtained from *Pseudoalteromonas* sp. PLSV [15], and Uly1 obtained from *Catenovulum maritimum* [12]. Notably, the crystal structures of LOR_107 and Uly1 have been determined. In the PL25 family, only PLSV_3936 obtained from *Pseudoalteromonas* sp. PLSV has been characterized, and its crystal structure was resolved [11]. All the ulvan lyases in the PL28, PL37, and PL40 families that have been characterized are listed in Table 1. Functionally, ulvan lyases belonging to the PL24 family specifically cleave the (1→4) glycosidic linkage between Rha3S and GlcA, whereas those belonging to the PL25, PL28, and PL37 families target the (1→4) glycosidic linkage between Rha3S and either GlcA or IdoA. The PL40 family is the most recently discovered ulvan

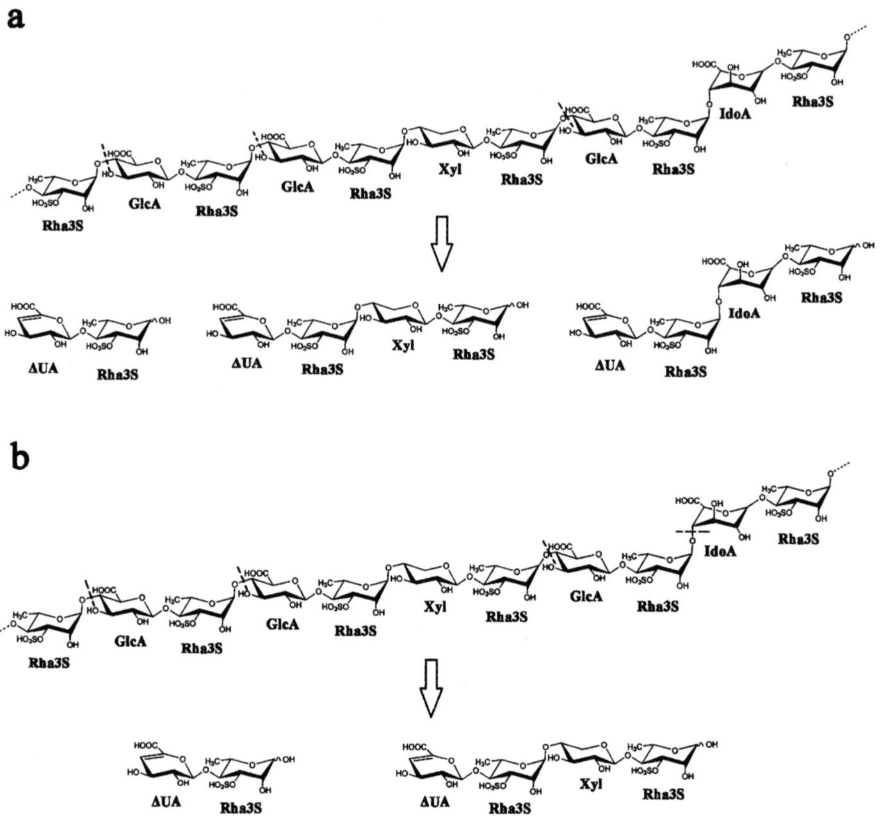

Fig. 1 The action mode of two types of ulvan lyases

lyase family, with 138 genes encoding ulvan lyases that have been mined and included to date. This database provides an important resource support for the identification of ulvan lyases with excellent properties.

2.3 Sequence Analysis of Ulvan Lyase

Protein sequence similarities between ulvan lyases belonging to the same family are relatively high. For ulvan lyases in the PL24 family, the sequence similarity of LOR_107 with Uly1, PLSV_3875, ALT761_04144, and UIIA is 43.6%, 86.0%, 86.6%, and 65.1%, respectively. The multiple sequence alignments of these enzymes are shown in Fig. 2. As representative members of the PL24 family, the structures and catalytic mechanisms of LOR_107 and Uly1 have been characterized. The key amino acid residues involved in catalysis such as His146, His167, Tyr243, and Arg259 in LOR_107 [9] and His128, His149, Tyr223, and Arg239 in Uly1 [12] are

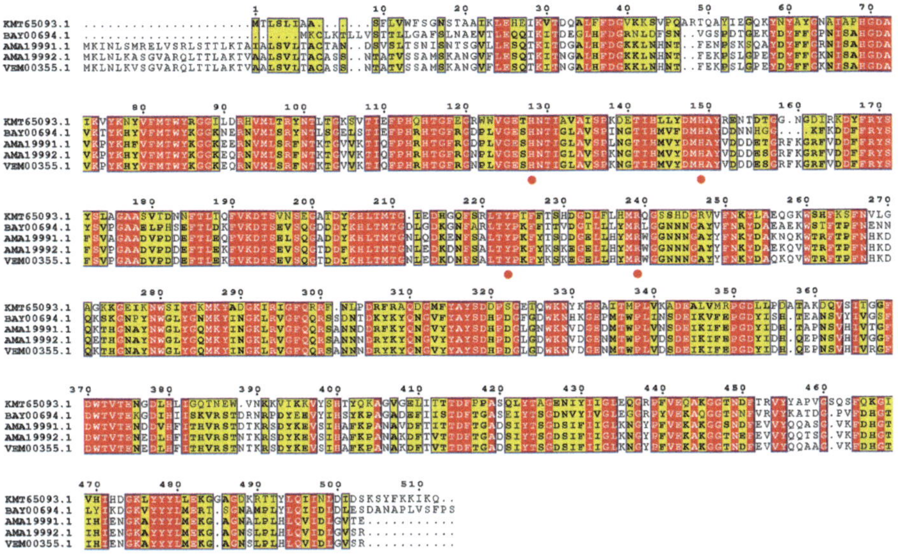

Fig. 2 Multiple sequence alignments of ulvan lyases from PL24 family, including PLSV_3875 (AMA19992.1), ALT761_04144 (VEM00355.1), LOR_107 (AMA19991.1), UIIA (BAY00694.1), and Uly1 (KMT65093.1). Red dots indicate the key residues involved in catalysis

highly conserved. Multiple sequence alignments of ulvan lyases from the PL25 and PL28 families are shown in Figs. 3 and 4, and the key highly conserved amino acid residues involved in catalysis are marked with red dots. The proposed catalytic residues of PLSV_3936 from the PL25 family are His123, His143, Tyr188, Arg204, and Tyr246 [11], which are highly similar to those of PL24 family ulvan lyases. This similarity suggests that ulvan lyases from the PL24 and PL25 families may exhibit a similar catalytic mechanism. The highly conserved catalytic residues of ulvan lyases from the PL28 family are Arg117, Gln160, Lys162, and Tyr281 [10]. However, protein sequence similarity between ulvan lyases from different families is generally low, at less than 15%.

3 Preparation of Ulvan Lyase

3.1 Screening and Fermentation of Ulvan Lyase-Producing Bacteria

Screening for ulvan lyase-producing bacteria is a prerequisite for the preparation of ulvan lyases. Researchers generally select the feces and intestines of marine animals fed with *Ulva* [20, 22, 23], decayed *Ulva* [24, 25], or seawater and mud containing decayed *Ulva* [6, 26] as sources. The screening media used are listed in Table 2, which are mainly composed of inorganic salts and ulvan, with the occasional

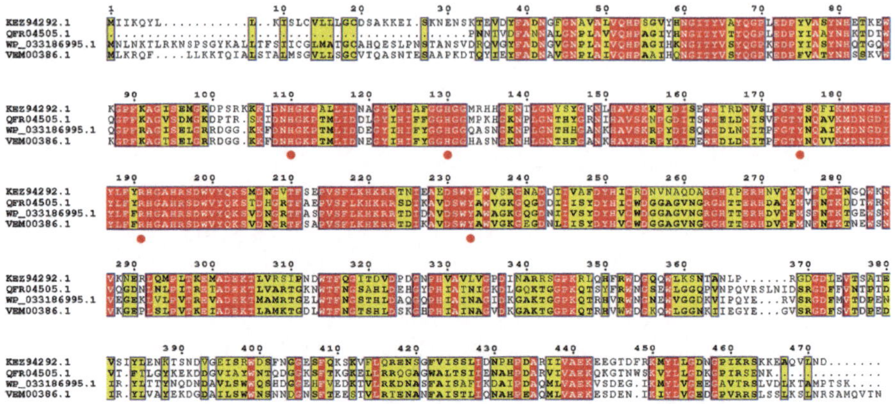

Fig. 3 Multiple sequence alignments of ulvan lyases from PL25 family, including PLSV_3936 (WP_033186995.1), ALT761_04173 (VEM00386.1), ALT3695 (QFR04505.1), and IL45_03835 (KEZ94292.1). Red dots indicate the residues involved in catalysis

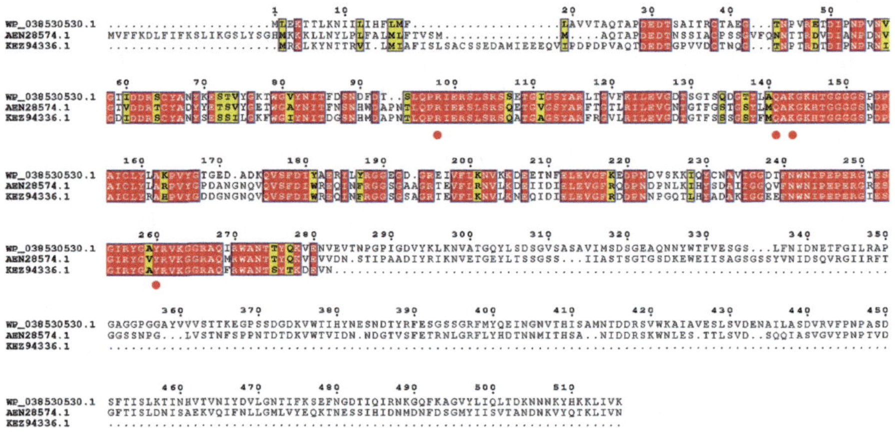

Fig. 4 Multiple sequence alignments of ulvan lyases from PL28 family, including NLR42 (AEN28574.1), NLR48 (KEZ94336.1), and FA2219 (WP_038530530.1). Red dots indicate the residues involved in catalysis

addition of organic nutrients like peptone and yeast extract. He et al. [20] prepared a screening medium based on the 1/2 Marine Broth 2216 medium by adding 0.98% NaCl, 0.2% ulvan and 1.5% agar [20]. Coste et al. [22] prepared a screening medium directly from seawater containing 1% ulvan, without adding agar, relying on the gel-forming properties of ulvan itself [22].

Several methods can be used to determine whether a strain produces ulvan lyase during plate screening. The simplest method is to observe whether a clear zone forms around the colony. A clear zone indicates that the bacteria are degrading ulvan into oligosaccharides, with a larger clear zone corresponding to higher degradation activity. If a clear zone is difficult to observe, chemical methods can be used.

Table 2 Screening and fermentation of ulvan lyase-producing bacteria

Microorganism	Source	Screening medium	Fermentation medium	References
Unknown gram-negative bacterium	Muds containing decomposing *Ulva* sp.	22 g NaCl, 3.7 g Na_2SO_4, 0.6 g KCl, 0.1 g KBr, 10 g $MgCl_2{\cdot}6H_2O$, 2.94 g $CaCl_2{\cdot}2H_2O$, 0.16 g $NaHCO_3$, 5 mg $Na_2HPO_4{\cdot}12H_2O$, 25 mg $NaNO_3$, 2.5 g Casamino-acids, 20 g ulvan and 15 g agar in 1 L deionized water	22 g NaCl, 3.7 g Na_2SO_4, 0.6 g KCl, 0.1 g KBr, 10 g $MgCl_2{\cdot}6H_2O$, 2.94 g $CaCl_2{\cdot}2H_2O$, 0.16 g $NaHCO_3$, 5 mg $Na_2HPO_4{\cdot}12H_2O$, 25 mg $NaNO_3$, 2.5 g Casamino-acids, 20 g ulvan and 2 g *Ulva* sp. flakes in 1 L deionized water	[6]
Alteromonas sp. KUL17	Feces of mollusks and shrimps fed with *Ulva ohnoi*	18.7 g Difco Marine Broth 2216, 9.8 g NaCl, 2 g ulvan and 15 g agar in 1 L deionized water	8.7 g Difco Marine Broth 2216, 9.8 g NaCl and 0.5 g ulvan in 1 L deionized water	[20]
Alteromonas sp. GIIUL2	Gut of *Gammarus insensibilis*	1% ulvan in 35 ppm sterilized natural seawater	1% ulvan in 35 ppm sterilized natural seawater	[22]
Persicivirga ulvanivorans	Feces of the gastropod *Aplysia punctata* fed with *Ulva* sp.	–	ZoBell medium containing 0.4% (w/v) of ulvan	[23]
Alteromonas sp. A321	Surfaces of decayed *Ulva prolifera* and sediments	Mineral salts medium containing 0.5% ulvan and 1% agar	0.15% K_2HPO_4, 0.9% ulvan, 0.4% $NaNO_3$, 1.0% NaCl and 0.05% $MgSO_4$	[24]
Catenovulum sp. LP	Seawater containing decayed *Ulva prolifera*	10 g ulvan, 1 g yeast extract, 2 g K_2HPO_4, 15 g NaCl, 0.1 g $CaCl_2$, 0.5 g $MgSO_4$ and 10 g agar in 1 L deionized water	10 g ulvan, 1 g yeast extract, 2 g K_2HPO_4, 15 g NaCl, 0.1 g $CaCl_2$ and 0.5 g $MgSO_4$ in 1 L deionized water	[26]
Diverse microorganisms	Fresh and rotting *Ulva* spp.	Difco™ Marine Broth 2216 medium containing 100 mM Tris-HCl (pH 8.5), 200 mM NaCl, 0.1% ulvan and 1% agar	–	[25]

For example, cetylpyridinium chloride reacts with acidic polysaccharides to form insoluble precipitates. He et al. [20] used this approach by flooding plates with 5% cetylpyridinium chloride to determine whether ulvan was degraded [20]. Another chemical method involves ruthenium red, a colorant that binds to acidic polysaccharides via electrostatic interactions. Rodrigues et al. [25] used 0.05% ruthenium red to develop plates for 10–15 min, resulting in a pink background, and the area around the ulvan lyase-producing bacteria remained colorless [25]. Additionally, Coste et al. [22] utilized an ulvan gel medium, in which gel liquefaction indicated the production of ulvan lyase [22].

After screening the ulvan lyase-producing strains, the enzyme was produced by liquid fermentation. The fermentation medium composition was similar to that of the screening medium. In some cases, researchers optimized the composition to increase enzyme production. For example, Li et al. [24] optimized the fermentation medium composition for *Alteromonas* sp. A321, resulting in an optimum medium consisting of 0.15% K_2HPO_4, 0.9% ulvan, 0.4% $NaNO_3$, 1.0% NaCl, and 0.05% $MgSO_4$ [24]. This optimization led to a 90% increase in enzyme activity. Currently, the preparation of ulvan lyase by fermentation is limited to shake flasks, and a few studies have used fermenters. Qiao et al. [27] systematically investigated the fermentation process for the ulvan lyase-producing bacterium *Catenovulum* sp. LP in a 5-L fermenter using a series of strategies [27]. The authors used *U. prolifera* powder instead of ulvan as the carbon source to reduce the cost of fermentation. Temperature is an important factor affecting the growth and metabolite synthesis of microorganisms. During fermentation of *Catenovulum* sp. LP, a higher temperature (32 °C) during the early stage stimulated the accumulation of cell biomass. However, prolonged fermentation at this temperature resulted in decreased cell biomass and enzyme activity in the fermentation broth. The later stage of fermentation was better suited to a temperature of 28 °C. Mixing by agitation is another key factor in achieving maximum productivity during microbial fermentation. In the early stage of fermentation, the focus was on maintaining the growth of the seed liquid, with a low oxygen requirement and an optimal agitation speed of 200 rpm. Then, the oxygen demand gradually increased with the growth of microorganisms. As microbial growth progressed, the oxygen demand increased. An agitation speed of 400 rpm maintained the dissolved oxygen levels at 32%, which was conducive to microorganism growth and metabolite biosynthesis. The pH of the medium can affect the metabolic product distribution, strain, and ulvan lyase performance. The optimal pH for *Catenovulum* sp. LP fermentation was 7.0, and losses of 32.9%, 19.6%, and 71.7% in ulvan lyase production were observed at pH 6.0, 8.0, and 9.0, respectively. Therefore, the optimal process was two-stage controlled fermentation, with an initial pH of 7.0 and aeration rate of 1.5 vvm for 0–12 h at 32 °C with a speed of 200 rpm and for 12–24 h at 28 °C with a speed of 400 rpm. After optimization, the fermentation time was reduced to 24 h. The kinetic models, including cell growth, product synthesis, and substrate consumption under optimal conditions, were as follows:

$$\frac{dX}{dt} = 0.27\left(1 - \frac{x}{1.14}\right)X$$

$$\frac{dP}{dt} = 1.35\frac{dx}{dt} + 0.024x$$

$$-\frac{dS}{dt} = 7.68\frac{dx}{dt} + 2.37\frac{dp}{dt}$$

Based on the established fermentation kinetic models, ulvan lyase fermentation was performed in 30-L fermenter to investigate the possibility of industrial-scale production. The results showed that the activity of ulvan lyase reached a maximum within 20 h, and the enzyme activity increased by 10.09% compared with that at the same time of fermentation in a 5-L fermenter. These positive results indicate the potential for industrial-scale production of ulvan lyase.

3.2 *Heterologous Expression of Ulvan Lyase*

Most ulvan lyases isolated and identified from wild bacteria are obtained by overexpression. Target genes obtained after gene annotation are usually genetically analyzed by researchers using online prediction tools such as NCBI, and the genes are divided into different structural domains on the basis of functional differences. The choice of structural domains can greatly affect enzyme expression and activity. In the laboratory, to ensure the correct folding of the protein, the target gene is heterologously expressed after the prediction and removal of the signal peptide using the SignalP server, which is one of the most common tools used to study ulvan lyase [9, 10, 15, 18, 19]. Additionally, other functional structural domains of the target gene can be predicted using the NCBI database, and gene domains composed of different structural domains can be formed by performing a polymerase chain reaction [20, 28]. For example, He et al. [20] cloned four domains, namely full-length ($UllA_{KUL17}$), full-length without signal peptide ($UllA1_{KUL17}$), length beyond the catalytic domain ($UllA2_{KUL17}$), and truncated version of UllA ($UllA3_{KUL17}$), among which $UllA3_{KUL17}$ exhibited higher activity than other ulvan lyase structural domains [20].

Heterologous expression techniques are commonly used to study gene function and their interactions. In addition to the optimization of the structural domain, the selection of an expression system for heterologous expression techniques is particularly important for enhancing the enzyme activity. Because most genes encoding ulvan lyases are obtained from wild bacteria, they are more suitable for expression in bacteria. The prokaryotic expression system represented by the *E. coli* expression system has the advantages of a clear genetic background, low cost, high expression capacity, and relatively simple isolation and purification of expression products, making it one of the most widely used expression systems at present. A study reported that ulvan lyase was most highly expressed in *E. coli* BL21 (DE3), and a few ulvan lyases were cloned in *E. coli* Rosetta2 (DE3) [29]. Recombinant plasmids were constructed by restriction cloning or seamless cloning of genes and were transformed into host receptor cells for ulvan lyase overexpression. Isopropyl β-d-1-thiogalactopyranoside (IPTG) or lactose was used to induce the lac operator element in pET, pProEXHTa, and the other series of plasmids for the normal translation of the enzyme. However, the optimal expression of ulvan lyase may require different concentrations of IPTG, with final concentrations generally ranging from 0.01 to 1 mM.

The culture conditions for enzymes vary because of the different research directions. A high concentration of the purified enzyme broth is required for protein crystallization in order to study the crystal structure of ulvan lyase. Therefore, it was obtained by fermenting and concentrating a large amount of the enzyme broth. Ulaganathan et al. [9] performed protein crystallization experiments artificially, using 1 L of the medium for inducing fermentation and purification [9]. The enzymatic properties of ulvan lyase can be studied by fermenting small volumes of the fermentation broth. For instance, Gao et al. [4] studied enzymatic properties such as optimum temperature and temperature stability after the overexpression of ulvan lyase using 50 mL of Luria-Bertani medium [4].

For ulvan lyase produced by wild bacteria, obtaining highly active enzymes by shake flask fermentation is impossible because of low enzyme activity. Qiao et al. [27] performed batch fermentation using the wild bacterium *Catenovulum* sp. LP in 5-L fermenter and obtained the enzyme with an activity of 1.2 U/mL [27]. Research on ulvan lyases has mainly focused on the study of enzyme structures and the acquisition of individual enzymes. Large-scale fermentation of ulvan lyases is not a popular research topic. However, improving the enzyme activity by adopting the fermentation approach is crucial (Table 3).

3.3 *Purification of Ulvan Lyases*

The vast majority of ulvan lyases produced by the fermentation of wild bacteria are extracellular enzymes, and the first step in their purification is to separate the bacteria from the fermentation broth, which is usually performed in the laboratory by high-speed centrifugation. Cell lysis is a necessary step for the subsequent purification of intracellular enzymes produced by recombinant bacterial fermentation. Ultrasonic disruption is the most widely used method under laboratory conditions. The bacteria were collected by centrifugation, resuspended in a lysis buffer, disrupted using an ultrasonicator, and centrifuged to remove cell debris. The lysis buffer and sonication conditions are critical for the protection of the enzyme structure and activity. The most common lysis buffer used by researchers in the disruption of ulvan lyase producing bacteria is 20 mM Tris-HCl (pH 7.5–8.5), sometimes with the addition of certain concentrations of NaCl (200–500 mM), dithiothreitol (1 mM), imidazole (5–20 mM), or glycerol (5%) [3, 5, 9, 10, 15, 30]. The ultrasonic disruption was performed in an ice bath with alternating outputs and pauses. For example, Qin et al. [30] performed recombinant expression of ulvan lyase AsPL, which was resuspended in 20 mM Tris-HCl (pH 8.0), containing 20 mM imidazole, 500 mM NaCl, and 1 mM dithiothreitol, and then disrupted using an ultrasonicator set at 3 s output and 2.5 s pause for 10 min [30].

Table 3 Heterologous expression information of ulvan lyases

Genbank	Structure domain	Plasmid/strain	Insert site	Growth conditions	IPTG concentration	Induction conditions	References
BAY00694.1	$UlIA_{KUL17}$	pET-21a/BL21 (DE3)	Seamless cloning	–	1 mM	4 h at 20 °C	[20]
	$UlIA1_{KUL17}$						
	$UlIA2_{KUL17}$						
	$UlIA3_{KUL17}$						
BAY00693.1	UlIA	pET-21a	Seamless cloning	LB medium	1 mM	20 °C	[20, 28]
	UlIB						
	UlIC						
BAY00695.1	Full-length	pET-21a/BL21 (DE3)	Seamless cloning	–	1 mM	4 h at 20 °C	[20]
AMA19991.1	Full-length (LOR_107)	pQE80L/BL21 (DE3)	Restriction cloning (BamH I and Hind III)	120 rpm at 37 °C in LB medium until OD_{600} reached 0.8	0.5 mM	18 h at 16 °C	[15, 30]
	Without signal peptide (LOR_107d)	pET-22/pET-28	Restriction cloning (Nco I/ Xho I)	37 °C for 2.5 h in LB medium until OD_{600} reached 0.6–0.8	0.01 mM	18 h at 16 °C	[9, 15]
AMA19992.1	Full-length	pET-28a/BL21 (DE3)	Restriction cloning (Nde I/ EcoR I)	220 rpm at 37 °C in 200 mL LB medium until OD_{600} reached 0.6–0.8	0.2 mM	16–18 h at 16 °C	[3]
WP 032096165.1	Full-length	BL21 (DE3)	Restriction cloning	150–200 rpm at 37 °C for 2.5 h until OD_{595} reached 0.6–0.8.	0.3 mM	18 h at 16 °C	[13]
WP 033186995.1	Full-length						
WP 052010178.1	Full-length						
WP 036580476.1	Full-length						

(continued)

Table 3 (continued)

Genbank	Structure domain	Plasmid/strain	Insert site	Growth conditions	IPTG concentration	Induction conditions	References
QFR04505.1	Full-length	pProEXHTa/BL21 (DE3)	Seamless cloning	200 rpm at 37 °C in 50 mL LB medium until OD_{600} reached 0.7	1 mM	18 h at 28 °C	[4]
WP_038530530.1	Full-length	pET-28/BL21 (DE3)	Restriction cloning (Not I/ Nde I)	180 rpm at 30 °C in 50 mL TB medium until OD_{600} reached 0.6	1 mM	24 h at 20 °C	[19]
	Without T9SS signal peptide						
AEN28574.1	UL	pET-22b/Rosetta 2 (DE3) (pLysSRARE2)	–	180 rpm at 37 °C for 4 h	Autoinduc-tion solutions	18 h at 20 °C	[23, 29, 31]
	BD						
	CD						
KEZ94336.1	Without signal peptide	pET-28/BL21 (DE3)	Restriction cloning	6 L terrific Broth medium until OD_{600} reached 1.5	0.5 mM	18 h at 16 °C	[10]
		pGEX/BL21 (DE3)	Ligation-independent cloning				
CDF79930.1	Without signal peptide	pET-28a/BL21 (DE3)	Seamless cloning	37 °C in 1 L LB medium until OD_{600} reached 0.8	0.2 mM	16 °C	[18]

Column chromatography is a common technique used for enzyme purification and includes affinity, gel filtration, and ion-exchange chromatography. For recombinant ulvan lyases, a His-tag is usually placed at the C-terminus [9, 11, 12] or a GST-tag at the N-terminus [10], which facilitates their purification by affinity chromatography. Ulvan lyases with a C-terminal His-tag can chelate various transition metal ions (Cu^{2+}, Ni^{2+}, Co^{2+}, and Zn^{2+}), and researchers prefer to purify them using immobilized metal-chelating affinity chromatography. Commonly used chromatography columns or resins are the Histrap-Excel column, Ni-NTA resin, and TALON metal affinity resin (Table 4). After the sample is loaded onto the column, unbound or weakly bound proteins are first washed with a wash buffer. The wash buffer is usually a lysis buffer containing low concentrations of imidazole (30–60 mM). After washing, the protein is eluted with an elution buffer containing a high concentration of imidazole. If the target protein is being purified for the first time, a linear gradient elution with imidazole is necessary to determine the appropriate imidazole concentration. Compared with traditional column chromatography purification methods, magnetic beads purify His-tagged proteins by magnetic

Table 4 Chromatographic purification techniques for recombinant ulvan lyases

Chromatography Technology	Chromatography columns/resins	Elution buffer	References
Affinity chromatography	Histrap-excel column	20 mM Tris (pH 8.0), 500 mM NaCl, 500 mM imidazole	[18]
		20 mM Tris (pH 8.5), 500 mM NaCl, 500 mM imidazole	[5]
	Ni-NTA Superflow resin	20 mM Tris-HCl (pH 8.0), 300 mM NaCl, 1 mM DTT, 500 mM imidazole	[3, 30]
		Sodium phosphate buffer (pH 8.0), 300 mM NaCl, 250 mM imidazole	[28]
	Ni-NTA resin	–	[12]
	Ni-Sepharose column	5–500 mM imidazole	[15]
	TALON metal affinity resin	40 mM imidazole	[9, 11]
	GST resin	His-TEV protease	[10]
Gel filtration chromatography	Sephacryl S-100 HR column	20 mM Tris buffer (pH 8.0) and 150 mM NaCl	[18]
		20 mM Tris-HCl (pH 8.5) and 150 mM NaCl	[5]
	Superdex 200 increase column	20 mM Tris-HCl (pH 8.0), 150 mM NaCl, 1 mM DTT	[30]
	Superdex 200 column	10 mM Tris-HCl (pH 8.0) and 100 mM NaCl	[12]
	Sephacryl S-200 column	–	[9, 11]
Ion exchange chromatography	MonoQ ion exchange resin	0 to 1 M NaCl in running buffer (20 mM Tris-HCl (pH 8.0), 1 mM DTT)	[6]

separation without the need for expensive column chromatography equipment, thus providing easy access to high-purity target proteins. Gao et al. [4] used this method to purify recombinant ulvan lyase ALT3695 [4].

To improve the protein purity, the fraction eluted by affinity chromatography was further purified by gel filtration chromatography. The gel chromatography columns were selected based on the molecular weight of the target protein. The molecular weights of the recombinant ulvan lyases are generally around 60 kDa, and the commonly used columns are listed in Table 4, such as the Sephacryl S-100 HR column and the Superdex 200 increase column. The most frequently used elution buffer is 20 mM Tris-HCl (pH 8.0) and 150 mM NaCl. In addition to affinity and gel filtration chromatography, ion-exchange chromatography is occasionally used to purify recombinant ulvan lyases. For example, Qin et al. [30] used the Ni-NTA Superflow resin as the first step in the purification of the recombinant ulvan lyase AsPL, which was then loaded onto the MonoQ ion exchange resin and eluted with a linear gradient of NaCl from 0 M to 1 M in a running buffer (20 mM Tris-HCl pH 8.0, 1 mM DTT). Subsequently, the collected enzyme was purified the third time using the Superdex 200 increase column [30]. The fractions with enzymatic activity were collected after each purification step, and their purity was assessed by SDS-PAGE.

Ulvan lyases produced by the fermentation of wild bacteria are generally purified without affinity chromatography but through a combination of organic solvent precipitation, ion-exchange chromatography, gel filtration chromatography, or hydrophobic chromatography. For ulvan lyase produced by *Alteromonas* sp. KUL17, researchers first added ammonium sulfate to the fermentation supernatant up to 70% saturation to separate the protein from other impurities [20]. The protein precipitate was dissolved in a small volume of 20 mM Tris-HCl (pH 8.0) and dialyzed in the same buffer to remove excess ammonium sulfate. The second purification step was performed by ion-exchange chromatography. The enzyme solution was loaded onto the Q Sepharose Fast Flow column and eluted with a linear gradient of NaCl from 0 M to 0.5 M in 20 mM Tris-HCl buffer (pH 8.0). The third step involved hydrophobic chromatography. Ammonium sulfate was added to the sample at a final concentration of 2 M, loaded onto the Phenyl Sepharose 6 Fast Flow column and eluted with a linear gradient of ammonium sulfate from 2 M to 0 M in 20 mM Tris-HCl buffer (pH 8.0).

4 Characterization of Ulvan Lyase

4.1 Ulvan Lyase Activity Assay

Ulvan lyase activity was examined using two methods. The depolymerization of ulvan produces unsaturated uronic acid via a β-elimination mechanism. The first activity measured the change in unsaturated double bonds by determining absorbance at 235 nm, according to the method described by Collén et al. [23]. One unit of ulvan lyase activity is defined as the amount of enzyme required to generate 1 μmol of unsaturated glucuronide residues per minute. The formation of reducing sugars is another feature of ulvan degradation, and the reducing sugars are quantified by the Miller method using 3,5-dinitrosalicylic acid (DNS) [32]. One unit of ulvan lyase activity is defined as the amount of the degradase enzyme needed to produce 1 μmol of reducing sugars (glucose equivalent) in 1 min under a certain condition. A study has shown that ulvan lyase activity measured by the formation of double bonds is consistent with that observed using the DNS method.

4.2 *Effect of Temperature and pH on Enzyme Activity*

All ulvan lyases identified to date have been reported in the CAZy database. Ulvan lyases, obtained from marine bacteria by isolation and purification, include UlIA from *Alteromonas* sp. KUL17 (55 kDa), PLSV_3875 from *Pseudoalteromonas* sp. PLSV (59 kDa), LOR_29 from *Alteromonadales* sp. LOR (52 kDa), NLR_492 from *Nonlabens ulvanivorans* (55 kDa), ALT3695 from *Alteromonas* sp. A321 (53 kDa), FaUL from *Formosa agariphila* KMM 3901 (55 kDa), NLR42 from *Nonlabens ulvanivorans* (46 kDa), and NLR48 from *Nonlabens ulvanivorans* (30 kDa) [3, 4, 10, 13, 20, 33]. As shown in Table 5, the exact molecular weights of these ulvan lyases are distributed in the range of 30–50 kDa, all of which are intermediate molecular weights. The theoretical molecular weights of ulvan lyase are shown in parentheses. The theoretical molecular weights of some ulvan lyases are higher than their exact molecular weights. This may indicate the molecular weight of the entire polysaccharide utilization site in the gene. For example, the highest ulvan lyase activity was obtained by knocking out certain amino acid sequences in KUL17, and its molecular weight was consistent with the exact molecular weight of 55 kDa [20].

Statistics on the optimum reaction temperature of ulvan lyase showed that the enzyme was concentrated when the temperature ranged from approximately 30 °C to 50 °C. Among them, NLR42 obtained from *Nonlabens ulvanivorans* and ALT3695 obtained from *Alteromonas* sp. A321 required a higher optimum temperature of 50 °C [4, 23], whereas *Fa*PL28 obtained from *Formosa agariphila* KMM 3901^T required a lower optimum temperature of 29.5 °C [34]. Other PLs in the ocean have been rationally designed to increase their optimum temperature. No change was observed in ulvan lyases when the optimum temperature increased. The optimum pH for most ulvan lyases is 8.0. LOR_29 from *Alteromonadales* sp. LOR required the lowest optimum pH of 7.5, whereas NLR42 obtained from *Nonlabens ulvanivorans* required the highest optimum pH of 9.0 [13, 23].

Table 5 Properties and characteristics of ulvan lyases

PL family	Source	Genbank	Mw (kDa)	Optimal temperature (°C)	Optimal pH	Products	References
PL24	*Alteromonas* sp. KUL17	BAY00694.1	55 (110.85)	–	–	DP2,4,6	[20]
	Glaciecola sp. KUL10	BAY00693.1	(112.74)	–	–	–	[20]
	Alteromonas sp. KUL42	BAY00695.1	(111.23)	–	–	–	[20]
	Alteromonadales sp. LOR_107	AMA19991.1	(59.64)	40	8	DP2, 4	[9]
	Alteromonadales sp. LOR_61	WP_032096165.1	(110.9)	–	–	DP2, 4	[13]
	Pseudoalteromonas sp. PLSV_3875	AMA19992.1	59 (59.6)	35	8	DP2, 4	[3, 15]
	Pseudoalteromonas sp. PLSV_3925	WP_033186955.1	(111.4)	–	–	DP2, 4	[20]
PL25	*Pseudoalteromonas* sp. PLSV_3936	WP_033186995.1	(54.28)	–	–	DP2, 4	[11]
	Alteromonadales sp. LOR_29	WP_052010178.1	52 (53.72)	45	7.5	DP2, 4	[13]
	Nonlabens ulvanivorans NLR_492	WP_036580476.1	55 (53.61)	–	–	DP2, 4	[13]
	Alteromonas sp. A321	QFR04505.1	53	50	8	DP2, 4	[4]
PL28	*Formosa agariphila* KMM 3901	CDF79931.1	55 (54.73)	45	8.5	DP2–6	[33]
	Formosa agariphila KMM 3901^T	WP_038530530.1	–	29.5	8.5	DP2, 4	[34]
	Nonlabens ulvanivorans PLR	AEN28574.1	46 (58.96)	50	9	DP2, 4	[23]
	Nonlabens ulvanivorans PLR	KEZ94336.1	30 (33.43)	40	8	DP2, 4	[10]
PL37	*Formosa agariphila* KMM 3901	CDF79930.1	(69.31)	40	8	DP2, 4	[18]
PL40	*Formosa agariphila* KMM 3901	CDF79911.1	(92.86)	–	–	–	[19]

4.3 Effect of Different Metal Ions on Enzyme Activity

The catalytic activity of ulvan lyase is closely related to the metal ions and reagents. This needs to be measured by the addition of metal ions to the optimum buffer to ensure that the structure of the protein is in the optimum state. The effects of different metal ions on the enzyme activity are different. Table 6 shows that Ca^{2+}, Mg^{2+}, K^{+}, and Na^{+} can activate enzyme activity, whereas Zn^{2+}, Cu^{2+}, Co^{2+}, and Mn^{2+} can inhibit it. The crystal structures of ulvan lyases show that certain divalent cations can coordinate with the enzyme and stabilize the structure. DTT and EDTA also inhibited this enzyme. The reduction in enzyme activity with DTT was due to the presence of cysteine in the enzyme structure, which is involved in the formation of disulfide bonds and plays an important role in enzyme activity and stability. The decrease in activity after the addition of EDTA was due to the elimination of divalent cations bound by the enzyme, indicating that ulvan lyase is a metal-dependent enzyme. For example, Reisky et al. [34] selected the mixture of 50 mM Tris-HCl buffer (pH 8.5) and 83 mM NaCl as the best buffer and added 1 mM divalent metal cations, DTT, and EDTA to study their effect on enzyme activity. The addition of the ions had no significant effect on the enzyme activity. However, the addition of DTT and EDTA reduced the enzyme activity by 36% and 63%, respectively. After the removal of divalent cations using the Chelex 100 cation exchange resin, the addition of Ca^{2+} and Mg^{2+} significantly activated the enzyme. Ca^{2+} may be a natural cofactor in the enzyme structure. This is the first validation that the PL28 family of ulvan lyases exists and that divalent metal cations can stabilize the protein [34].

There was some variation in the effects of the same metal ions on enzyme activity. For example, different families of ulvan lyases react differently to Co^{2+}, Fe^{2+}, Fe^{3+}, Mn^{2+}, and Cu^{2+}. Co^{2+} increases the enzyme activity of PsPL, whereas the same ion decreases the activity of FaPL28 and ALT3695 [3, 4, 34]. Cu^{2+} activates Uly1 while inhibiting ALT3695 [4, 12]. This may be due to significant differences in enzyme structures between the families.

Table 6 Effect of different metal ions on the activity of ulvan lyases

Source	Activators	Inhibitors	References
Alteromonadales sp. LOR_107	Ca^{2+}, Mg^{2+}, Ni^{2+}, Fe^{2+}	Cu^{2+}, Co^{2+}, Mn^{2+}, Zn^{2+}, DTT, EDTA	[30]
Catenovulum maritimum	Mg^{2+}, Li^{+}, K^{+}, Ca^{2+}, Cu^{2+}	Mn^{2+}, Co^{2+}	[12]
Pseudoalteromonas sp. PLSV_3875	Mg^{2+}, Ca^{2+}, Co^{2+}, Fe^{3+}, Ni^{2+}	Cu^{2+}, Mn^{2+}, Zn^{2+}, EDTA	[3]
Alteromonas sp. A321	K^{+}, Ca^{2+}, Mg^{2+}, Ba^{2+}, Tween-80,Triton X-100, 1,10-phenanthroline	Zn^{2+}, Cu^{2+}, Fe^{2+}, Co^{2+}, Cd^{2+}, Hg^{2+}, Fe^{3+}	[4]
Formosa agariphila KMM 3901^{T}	Ca^{2+}, Co^{2+}, Cu^{2+}, Fe^{2+}, Mg^{2+}, Mn^{2+}, Ni^{2+}, Na^{+}	Zn^{2+}, DTT, EDTA	[34]

Several studies have investigated the effects of heavy metal ions and reagents on enzyme activity. Heavy metal ions exert significant inhibitory effects on enzyme activity. Some reagents can activate enzymes. For example, Gao et al. [4] found that Hg^{2+} inactivated the enzyme, whereas Tween-80, Triton X-100, and 1,10-phenanthroline activated ALT3695 at low concentrations [4].

4.4 Kinetic Measurements

The kinetic parameters such as K_m, k_{cat}, and V_{max} depended on the substrate species. For example, Konasani et al. [5] studied the K_m values of FaUL for different substrates. When incubated with ulvan isolated from *U. armoricana*, *U. intestinalis*, and *U. lactuca*, the K_m values of FaUL were 0.60 ± 0.03 mg·mL^{-1}, 2.6 ± 0.1 mg·mL^{-1}, and 0.70 ± 0.02 mg·mL^{-1}, respectively [5]. The K_m value is closely related to the substrate, and the size and molecular weight of the substrate affect the catalytic efficiency of the enzyme, which may be caused by hindering the electron transfer between the substrate and the active site of the enzyme. Mondal et al. [28] investigated the enzyme kinetic parameters of KUL_10 using disaccharide and tetrasaccharide substrates. The K_m was 5.1 μM under disaccharide substrate conditions and 24 μM under tetrasaccharide conditions. These results indicated that the affinity of KUL_10 for disaccharides was higher than that for tetrasaccharides, and the specific activity for tetrasaccharides was higher than that for disaccharides [28]. However, the kinetic parameter values also depend on the enzyme type. While the members of PL25 and PL28 interact with ulvan blocks with GlcA and IdoA (substrates), the representative of PL24 is specific to GlcA-containing blocks. The K_m values of KUL_10 and KUL_17 were 24 μM and 0.72% (m/v), respectively, using *U. ohnoi* as a substrate [20, 28].

The k_{cat} value of ulvan lyase varied with the protein state in different buffers. For example, Reisky et al. [34] monitored the apparent kinetics of the ulvan lyase *Fa*PL28 by adding 100 or 600 mM NaCl in 50 mM Tris-HCl buffer (pH 8.5). The results of fitting the Michaelis-Menten kinetic model showed a K_m of 0.26 ± 0.06 g·L^{-1} and a k_{cat} of 24.8 ± 1.7 s^{-1} at 100 mM NaCl and a K_m of 0.75 ± 0.19 g·L^{-1} and a k_{cat} of 20.1 ± 2.2 s^{-1} at 600 mM NaCl concentration. This indicated that a high concentration of NaCl reduced the affinity of FaPL28 for ulvan and had a detrimental effect on its apparent kinetics [34] (Table 7).

Table 7 Kinetic parameters of ulvan lyases

Source	K_m (mg·mL^{-1})	V_{max}	k_{cat} (S^{-1})	K_{cat}/K_m (mL·mg^{-1}·s^{-1})	References
Pseudoalteromonas sp. PLSV_3875	2.1 ± 0.31	12.74 ± 0.86 U·mg^{-1}	7.08 ± 0.12	3.37	[3]
Glaciecola sp. KUL10	24 μM	119 unit	–	–	[28]
Alteromonas sp. A321	0.43	0.11 μmol·min^{-1}·mL^{-1}	–	–	[4]
Formosa agariphila KMM 3901	0.77	540 μg·min^{-1}·mL^{-1}	–	–	[5, 18]
Formosa agariphila KMM 3901^T (100 mM NaCl)	0.26 ± 0.06	–	24.8 ± 1.7	–	[34]
Formosa agariphila KMM 3901^T (600 mM NaCl)	0.75 ± 0.19	–	20.1 ± 2.2	–	
Alteromonas sp. KUL17	0.72	185 μmol·min^{-1}·mg protein^{-1}	–	–	[20]
Alteromonadales sp. LOR_107	1.29 ± 0.08	–	45.93 ± 1.23	35.6	[30]
Nonlabens ulvanivorans PLR	0.0051 ± 0.0002	142 ± 11 mM·cut bonds min^{-1}·mg^{-1}	–	–	[23]

5 Structure and Catalytic Mechanism of Ulvan Lyases

5.1 Structure of Ulvan Lyases

The three-dimensional structures of ulvan lyases are shown in Fig. 5. The overall structure of LOR_107 is a seven-bladed β-propeller [9]. Each blade is composed of four antiparallel β-strands, where blades 1–6 are formed by consecutive residues, and blade 7 is formed by three C-terminal β-strands and one N-terminal β-strand. The loops within blades 1, 2, 4, 5, and 7 are particularly long, whereas the loops within blades 3 and 6 are shorter. These long loops created a deep canyon at the top of the propeller. Highly conserved residues are exposed on the surface of the canyon, which is defined as the substrate-binding site of LOR_107. Moreover, a Ca^{2+} is

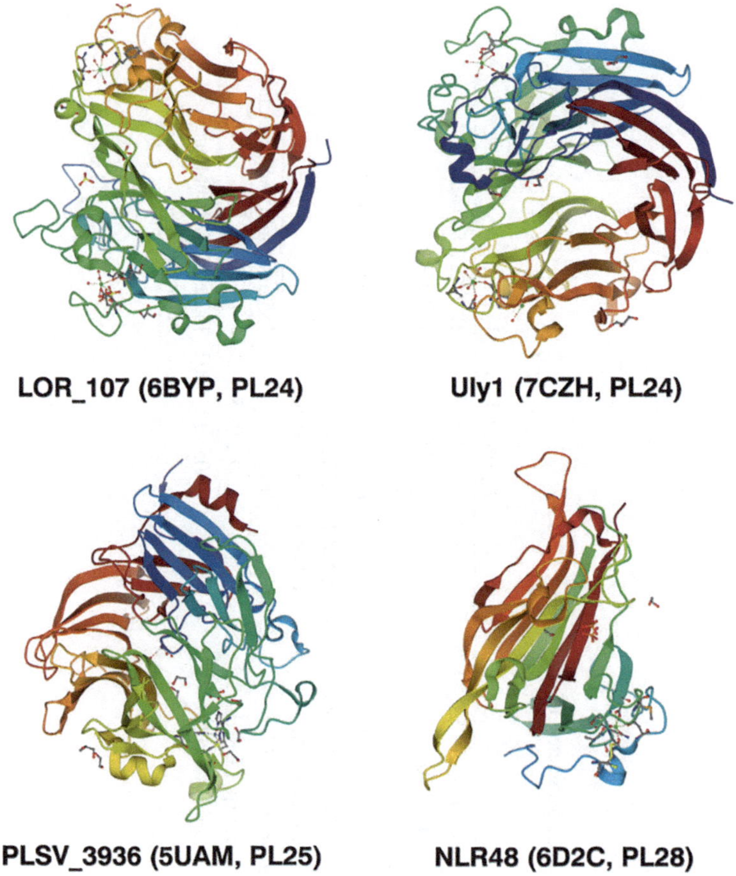

Fig. 5 The ulvan lyases' overall structure from RCSB Protein Data Bank (RCSB.org) of LOR_107 (PL24, PDB ID: 6BYP) [9], Uly1 (PL24, PDB ID: 7CZH) [12], PLSV_3936 (PL25, PDB ID: 5UAM) [11], and NLR48 (PL28, PDB ID: 6D2C) [10]

located at the periphery of the propeller, away from the conserved residues in the canyon, which may play a strictly structural role. Uly1 is another ulvan lyase in the PL24 family with a crystalline structure [12]. The root mean square deviation value between Uly1 and LOR_107 is 1.32 Å, implying a high structural homology between them. Uly1 also shows a seven-bladed β-propeller fold, and each propeller consists of four antiparallel β-strands. The catalytic center is located in the canyon at the top of the propeller. In contrast to LOR_107, each Uly1 monomer contains only approximately 0.1 Ca^{2+}, which are most likely derived from $Ca(CH_3COO)_2$ in the crystallization solution.

The protein sequence similarity between PLSV_3936 and LOR_107 or Uly1 was only 11%; however, its overall structure was similar to those of LOR_107 and Uly1. As resolved by Ulaganathan et al. [11], the overall structure of the PLSV_3936 monomer is a seven-bladed β-propeller, and each blade is composed of four antiparallel β-strands [11]. The long connecting loops at the top of the propeller formed a deep and elongated canyon. A Zn^{2+} was located between the second and third blades of the propeller and formed a tetrahedral coordination with the side chains of His208, His264, His278, and Cys266. The Zn^{2+} plays a structural role, and its removal destabilizes the substrate-binding sites, thereby inactivating the enzyme.

The PL28 family ulvan lyase NLR48 adopts the fold of a β-sandwich jelly roll with two concave β-sheets stacking on each other [10]. Each β-sheet consists of seven antiparallel and strongly bent β-strands. The inner β-sheet (strands β1–β4–β13–β6–β7–β8–β9) forms a deep crevice with mainly basic and polar side chains exposed on the surface, which may be the substrate binding site for NLR48. The N-terminal amino acids 21–50 contain no secondary structure and are fixed to the loop connecting strands β1 and β2 by a disulfide bond between Cys32 and Cys59. Furthermore, the Ca^{2+} near the disulfide bond provides additional stability to the N-terminal region.

5.2 *Catalytic Mechanism of Ulvan Lyase*

Ulvan lyase catalyzes the degradation of ulvan via a β-elimination mechanism, a process in which the 4-O-glycosidic bond is cleaved while generating oligosaccharides containing ΔUA at the non-reducing end. Metal-assisted catalysis and the His/Tyr-dependent catalysis [35, 36] are the two classical mechanisms of β-elimination. The metal ions in ulvan lyase are distant from the active center and do not participate in the catalytic action. The reported ulvan lyases of the PL24, PL25, and PL28 families all adopt the His/Tyr-dependent catalytic mechanism; however, the amino acid residues involved in catalysis are different.

The catalytic process of PLs involves the following three typical steps: (1) neutralization of the negative charge on the carboxyl group, (2) abstraction of the proton on C5 by a base, and (3) donation of the proton to the leaving group by an acid. In the PL24 family ulvan lyase LOR_107, Arg259 is the neutralizer, and His167 is critical for the correct orientation of the carboxyl group of GlcA in its interaction with Arg259 (Fig. 6a). The Arg259 side chain promotes the movement of the substrate to the depth of the canyon, and its absence causes the substrate to remain at the initial binding site [9]. His146 most likely acts both as an acid and a base by abstracting the proton from C5 of GlcA and donating it to bridging O4. In contrast to the classical His/Tyr-dependent catalytic mechanism, Tyr243 of LOR_107 does not directly participate in proton transfer, but plays a supporting role in properly orienting His146.

PLSV_3936 cleaves the glycosidic linkage between Rha3S and either GlcA or IdoA. Ulaganathan et al. (2017) investigated the degradation of GlcA-containing substrates using PLSV_3936 [11]. Arg204 is the neutralizer that neutralizes the

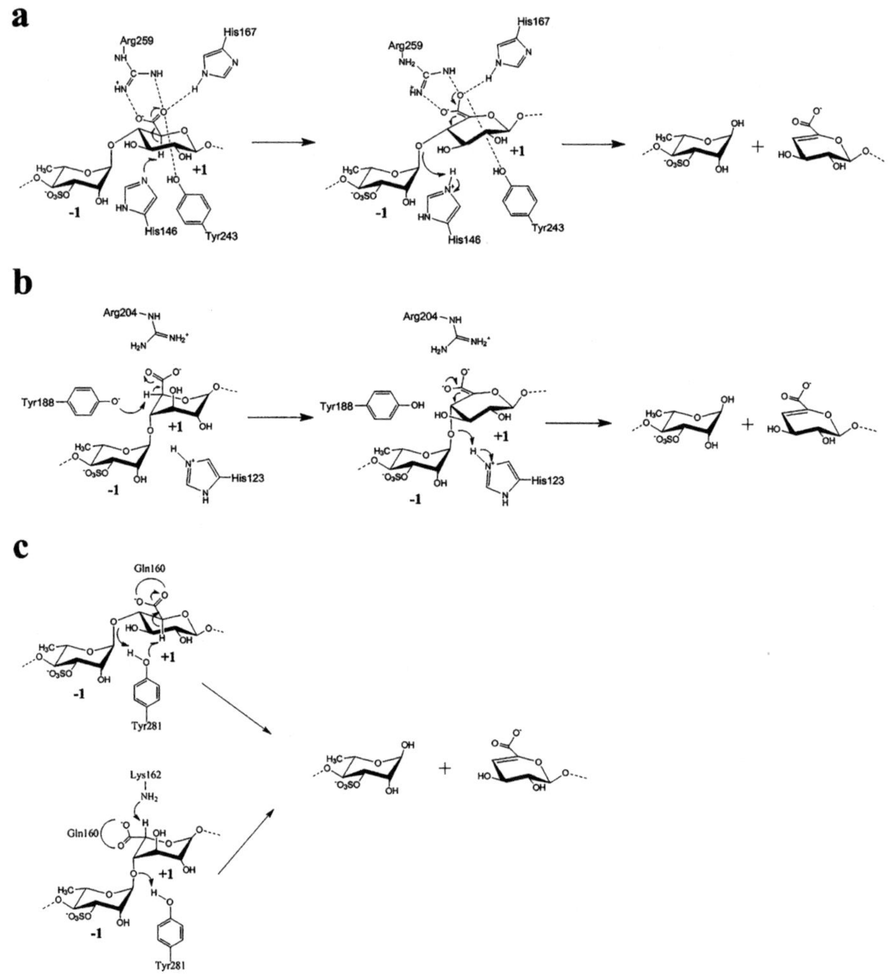

Fig. 6 Proposed catalytic mechanisms of ulvan lyases LOR_107 from PL24 family (**a**), PLSV_3936 from PL25 family (**b**), and NLR48 from PL28 family (**c**)

acidic group of GlcA in the +1 position with the help of His143 and Tyr246 (Fig. 6b). Tyr188 is the most likely candidate base, whereas His123 is the most likely general acid, transferring one proton to the bridging oxygen. Subsequently, the C4–O4 bond is broken, and the C4=C5 double bond is formed. The orientation and conformation of IdoA at the +1 position are different from those of GlcA, indicating that amino acid residues involved in catalysis are distinct. His208 and Tyr246 in the vicinity of +1 sugar may act as acids and bases in degrading GlcA-containing substrates.

As shown in Fig. 6c, the key catalytic residues of NLR48 are Gln160, Tyr281, and Lys162 [10]. Gln160 neutralizes the acidic group of +1 GlcA/IdoA and promotes protonation by forming two hydrogen bonds with the substrate. Tyr281 can perform proton abstraction and donate it to the leaving group when the +1 position is GlcA. With IdoA in the +1 position, Lys162 serves as a base and Tyr281 serves as an acid for the leaving group, with oxygen bridging the -1 and +1 sugars.

6 Application of Ulvan Lyases

Ulvan lyase can be used to prepare ulvan oligosaccharides, which play an important role in biotechnological applications. PLs degrade polysaccharides into oligosaccharides, fully exposing their active functional groups and improving their bioactivities [37]. Ulvan possesses various bioactivities owing to its unique monosaccharide composition and abundant active functional groups. However, the high molecular weight and viscosity of ulvan are not conducive to its bioavailability and bioactivity [6]. Ulvan oligosaccharides obtained via specific enzymatic degradation are superior in terms of bioavailability and bioactivity. In addition, ulvan lyase catalyzes the degradation of ulvan via a β-elimination mechanism that produces unsaturated double bonds (C=C) at the non-reducing end of products, which has important implications for the bioactivities of the oligosaccharides. In contrast, oligosaccharides obtained by degradation using physical or chemical methods do not exhibit these properties.

Ulvan lyase is a key tool in the preparation of bioethanol from *Ulva* spp. green algae. As the population continues to grow, the global energy demand has increased dramatically. Fossil fuel resources are finite and will certainly be replaced by bioenergy. Seaweed resources are enormous, and their cell walls are rich in carbohydrates. The ethanol yield of seaweed is much higher than that of crops such as corn and sugarcane, making it a major raw material for the production of third-generation bioethanol [38, 39]. The conversion of biomass into bioethanol typically involves three steps, namely raw material pretreatment, hydrolysis, and bioethanol fermentation. Among them, the efficiency of hydrolysis directly affects the efficiency of the entire conversion process and ethanol yield [38]. The ulvan content in green algae can exceed 30% of their dry weight [2]. Thus, converting ulvan into oligosaccharides using ulvan lyase and obtaining natural microorganisms or engineered strains that can ferment oligosaccharides will further increase ethanol production by *Ulva* spp. green algae.

Ulvan lyase can be used to prepare protoplasts of *Ulva* spp. green algae. Protoplasts have the advantages of a simple structure, synchronized development, and easy access to exogenous genetic material, which are important materials for basic fields such as plant breeding, trait improvement, and genetic research [40, 41]. During the preparation of green algal protoplasts using cellulases, the highly viscous ulvan present in the cell wall and intercellular matrix of green algae is not conducive to separating protoplasts from impurities. The application of ulvan lyase combined with cellulase can rapidly reduce the viscosity of the solution and improve the efficiency of the green algal protoplast preparation.

Ulvan lyase can also be used for the structural elucidation of ulvan. The basis of the biological activity of polysaccharides is their chemical structure; therefore, the study of their physiological functions presupposes the resolution of their chemical structure. The combination of mass spectrometry and NMR spectroscopy is a powerful technique for the detailed structural analysis of polysaccharides, which requires their hydrolysis [42]. For heteropolysaccharides like ulvan, acid hydrolysis can alter substituents and branches, complicating accurate structural analysis [43]. In contrast, enzymatic hydrolysis under mild conditions with homogeneous hydrolysates offers a more effective method for structural analysis, particularly for examining repeating units [37, 44]. For example, Chi et al. [44] analyzed the structure of ulvan extracted from *U. clathrata* with the assistance of an ulvan lyase belonging to the PL25 family [44]. A series of disaccharides to octasaccharides was obtained via enzymatic degradation and purification. Based on mass spectrometry and NMR spectroscopy analyses, it was determined that the backbone of this ulvan consisted mainly of →4)-β-D-GlcA-(1→4)-α-L-Rha3S-(1→ and →4)-β-D-Xyl-(1→4)-α-L-Rha3S-(1→ in a ratio of approximately 4:1.

Seaweed extract, rich in polysaccharides, betaine, phenolic compounds, and phytohormones, can be used as a plant biostimulant [45]. The presence of a large amount of viscous polysaccharides in the algal cell wall limits the extraction rate of active substances such as phenolic compounds and betaine. For *Ulva* spp. green algae, ulvan lyase-assisted extraction can break down the cell walls and effectively release bioactive compounds from the cells. Furthermore, the regulatory effect of oligosaccharides on plant growth is generally superior to that of polysaccharides [46]. Li et al. [47] prepared *U. prolifera* extract through ulvan lyase degradation, which enhanced the antioxidant capacity of the plant, reduced lipid peroxidation, and improved wheat seedling growth under drought conditions [47].

Compared with the degrading enzymes of brown and red algae polysaccharides, research on green algae polysaccharide-degrading enzymes is still in the initial stage. The in-depth exploration of ulvan lyases enriches the enzyme database for seaweed polysaccharide degradation and contributes to the high-value utilization of seaweed resources.

References

1. Lahaye M, Robic A (2007) Biomacromolecules 8:1765
2. Kidgell JT, Magnusson M, de Nys R, Glasson CR (2019) Algal Res 39:101422
3. Qin H-M, Xu P, Guo Q et al (2018) RSC Adv 8:2610
4. Gao J, Du C, Chi Y, Zuo S, Ye H, Wang P (2019) Mar Drugs 17:568
5. Konasani VR, Jin C, Karlsson NG, Albers E (2018) Algal Res 36:106
6. Lahaye M, Brunel M, Bonnin E (1997) Carbohydr Res 304:325
7. Yaich H, Garna H, Besbes S et al (2014) Food Hydrocoll 40:53
8. Hardouin K, Bedoux G, Burlot A-S et al (2016) Algal Res 16:233
9. Ulaganathan T, Helbert W, Kopel M, Banin E, Cygler M (2018) J Biol Chem 293:4026
10. Ulaganathan T, Banin E, Helbert W, Cygler M (2018) J Biol Chem 293:11564
11. Ulaganathan T, Boniecki MT, Foran E et al (2017) ACS Chem Biol 12:1269
12. Xu F, Dong F, Sun X-H et al (2021) Appl Environ Microbiol 87:e00412
13. Foran E, Buravenkov V, Kopel M et al (2017) Algal Res 25:39
14. Kopel M, Helbert W, Henrissat B, Doniger T, Banin E (2014) Genome Announc 2:e01081
15. Kopel M, Helbert W, Belnik Y, Buravenkov V, Herman A, Banin E (2016) J Biol Chem 291:5871
16. Koch H, Freese HM, Hahnke RL, Simon M, Wietz M (2019) Front Microbiol 10:504
17. Wang D, Li Y, Han L, Yin C, Fu Y, Zhang Q, Zhao X, Li G, Han F, Yu W (2022) Mar Drugs 20(3):168
18. Konasani VR, Jin C, Karlsson NG, Albers E (2018) Sci Rep 8:1
19. Mann AJ, Hahnke RL, Huang S et al (2013) Appl Environ Microbiol 79:6813
20. He C, Muramatsu H, Kato S-I, Ohnishi K (2017) Biosci Biotechnol Biochem 81:2145
21. Kopel M, Helbert W, Henrissat B, Doniger T, Banin E (2014) Genome Announc 2:e00793
22. Coste O, Malta E-J, López JC, Fernández-Díaz C (2015) Algal Res 10:224
23. Collén PN, Sassi J-F, Rogniaux H, Marfaing H, Helbert W (2011) J Biol Chem 286:42063
24. Li Y, Wang J, Yu Y et al (2013) Carbohydr Polym 98:988
25. Rodrigues V, Onime L, Huws S, Odaneth A, Lali A (2017) J Appl Biotechnol Bioeng 2:136
26. Li Y, Huang Z, Qiao L et al (2015) Process Biochem 50:799
27. Qiao L, Yang X, Xie R et al (2020) Algal Res 46:101812
28. Mondal R, Ohnishi K (2020) Biochem Biophys Res Commun 523:441
29. Melcher RL, Neumann M, Werner JPF, Gröhn F, Moerschbacher BM (2017) Sci Rep 7:1
30. Qin H-M, Gao D, Zhu M et al (2020) Int J Biol Macromol 147:1309
31. Barbeyron T, Lerat Y, Sassi J-F, Le Panse S, Helbert W, Collen PN (2011) Int J Syst Evol Microbiol 61:1899
32. Miller GL (1959) Anal Chem 31:426
33. Qi H, Zhao T, Zhang Q, Li Z, Zhao Z, Xing R (2005) J Appl Phycol 17:527
34. Reisky L, Stanetty C, Mihovilovic MD, Schweder T, Hehemann J-H, Bornscheuer UT (2018) Appl Microbiol Biotechnol 102:6987
35. Garron M-L, Cygler M (2014) Curr Opin Struct Biol 28:87
36. Garron M-L, Cygler M (2010) Glycobiology 20:1547
37. Chi Y, Zhang M, Wang X, Fu X, Guan H, Wang P (2020) Int J Biol Macromol 157:75
38. Kostas ET, White DA, Cook DJ (2020) Bioenergy Res 13:271
39. Osman ME, Abo-Shady AM, Elshobary ME, Abd El-Ghafar MO, Abomohra AE-F (2020) Environ Sci Pollut Res 27:32481
40. Avila-Peltroche J, Won BY, Cho TO (2020) J Appl Phycol 32:2233
41. Avila-Peltroche J, Won BY, Cho TO (2021) J Genet Eng Biotechnol 19:1
42. Yu L, Xu X, Xue C et al (2013) Food Chem 139:702
43. Heyraud A, Gey C, Leonard C, Rochas C, Girond S, Kloareg B (1996) Carbohydr Res 289:11
44. Chi Y, Li H, Wang P et al (2020) Carbohydr Polym 229:115497
45. Battacharyya D, Babgohari MZ, Rathor P, Prithiviraj B (2015) Sci Hortic 196:39
46. Nagasawa N, Ha VTT, Hien NQ, Nakanishi TM (2009) Radiat Phys Chem 78:796
47. Li F, Zuo S, Chi Y et al (2020) J Renewable Mater 8:1519

Batch number: 10725823

Printed by Printforce, the Netherlands